T0227330

The Finite Element Method
A Practical Course

Second Edition

G. R. Liu
School of Aerospace Systems
University of Cincinnati, USA

S. S. Quek
Institute of High Performance Computing
Singapore

ELSEVIER

AMSTERDAM • BOSTON • HEIDELBERG • LONDON • NEW YORK
OXFORD • PARIS • SAN DIEGO • SAN FRANCISCO
SINGAPORE • SYDNEY • TOKYO

Butterworth-Heinemann is an imprint of Elsevier

Butterworth-Heinemann is an imprint of Elsevier
The Boulevard, Langford Lane, Kidlington, Oxford, OX51GB, UK
225 Wyman Street, Waltham, MA 02451, USA

First edition 2013

Notice
No responsibility is assumed by the publisher for any injury and/or damage to persons or property as a matter of products liability, negligence or otherwise, or from any use or operation of any methods, products, instructions or ideas contained in the material herein. Because of rapid advances in the medical sciences, in particular, independent verification of diagnoses and drug dosages should be made.

British Library Cataloguing in Publication Data
A catalog record for this book is available from the British Library.

Library of Congress Cataloging-in-Publication Data
A catalog record for this book is available from the Library of Congress.

ISBN: 978-0-08-098356-1

For information on all Butterworth-Heinemann publications
visit our website at books.elsevier.com

Printed and bound by CPI Group (UK) Ltd, Croydon, CR0 4YY

Working together
to grow libraries in
developing countries

ELSEVIER   Book Aid International

www.elsevier.com • www.bookaid.org

*To Zuona, Yun, Kun, Run, and my extended family
for their love and support.*

*To all my fellow students
for their company in studying this subject*

G. R. Liu

*To my wife, Lingzhi, and my children, Matthew and Michelle
for their love and support.*

*To my mentor, Dr. Liu
for his guidance*

S. S. Quek

Biography

G.R. Liu

Dr Liu received his PhD from Tohoku University, Japan, in 1991. He was a Postdoctoral Fellow at Northwestern University, USA. He was a Professor at the Department of Mechanical Engineering, National University of Singapore. He founded the Centre for Advanced Computations in Engineering Science (ACES), National University of Singapore, and served as the Director of the ACES from 1998 to 2010. He founded the Association for Computational Mechanics (Singapore) (SACM) and served as the President of the SACM from 2002 to 2010. He is an executive council member of the International Association for Computational Mechanics, and the President of the Asia-  Pacific Association for Computational Mechanics (APACM). He is currently a Professor, School Faculty Chair and Ohio Eminent Scholar at the School of Aerospace Systems, University of Cincinnati, USA. He has provided consultation services to many national and international organizations. He has authored more than 500 technical publications including more than 380 international journal papers and eight books, including two bestsellers: "Meshfree Methods: Moving Beyond the Finite Element Method" and "Smoothed Particle Hydrodynamics: A Meshfree Particle Method." He is the Editor-in-Chief of the *International Journal of Computational Methods*, Associate Editor of *Inverse Problems in Science and Engineering*, and has served as an editorial member of five other journals including the IJNME. He is the recipient of the Outstanding University Researchers Awards, the Defence Technology Prize (National award), Silver Award at the CrayQuest competition, the Excellent Teachers Awards, the Engineering Educator Awards, the APACM Award for Computational Mechanics, the JSME Award for Computational Mechanics, and the ASME Ted Belytschko Award of Applied Mechanics. His research interests include aerospace systems, bio-mechanical systems, computational mechanics, mesh free methods, nano-scale computation, micro-biosystem computation, vibration and wave propagation in composites, mechanics of composites and smart materials, inverse problems and numerical analysis.

S.S. Quek

Dr Quek received his PhD from the National University of Singapore. He joined the Institute of High Performance Computing (IHPC) under the Agency for Science, Technology, and Research (A*STAR), Singapore, and participated in the prestigious Visiting Investigator Program led by Professor David J. Srolovitz (currently at the University of Pennsylvania, USA). Through this program, Dr Quek established himself as a competent researcher in the field of computational mechanics and, in particular, computational modeling of dislocation plasticity. Dr Quek is currently a scientist at IHPC and also an Adjunct Assistant Professor at the Department of Mechanical Engineering,

National University of Singapore. He has authored several technical publications including a book chapter and prestigious international journal papers. His other research interests include microstructure effects on the mechanics of materials, thin film mechanics, dislocation dynamics, nano-scale mechanics, and microstructure evolution.

Contents

Preface to the First Edition

For the past half a century, the Finite Element Method (FEM) has been developed into an indispensable technology in the modeling and simulation of engineering systems. In the development of an advanced engineering system, engineers have to go through a very rigorous process of modeling, simulation, visualization, analysis, designing, prototyping, testing, and finally, fabrication/construction. As such, techniques related to modeling and simulation play an increasingly important role in building advanced engineering systems quickly and effectively, and therefore the applications of the FEM have multiplied rapidly.

This book provides unified and detailed course material on the FEM for engineers and university students to solve linear problems in mechanical and civil engineering, with the main foci on structural mechanics and heat transfer. The aim of the book is to provide the necessary concepts, theories, and techniques of the FEM for readers, such that, at the end of this course, they are able to understand and use a commercial FEM package comfortably. Important fundamental and classical theories are introduced in a straightforward and easy to understand way. Modern, state-of-the-art treatment of engineering problems in designing and analysing structural and thermal systems, including microstructural systems, are also discussed. Useful key techniques in FEMs are described in depth, and case studies are provided to demonstrate the theory, methodology, techniques, and practical applications of the FEM (through the introduction of data input files of a typical FEM software package). Equipped with the concepts, theories, and modeling techniques described in this book, readers should be able to use a commercial FEM software package for solving engineering structural problems in a professional manner.

The general philosophy governing the book is to make all the topics insightful but simple, informative but concise, theoretical but applicable.

The book unifies topics on mechanics for solids and structures, energy principles, weighted residual approach, the FEM, and techniques of modeling and computation, as well as the use of commercial software packages. The FEM was originally invented for solving mechanics problems in solids and structures. It is thus appropriate to learn the FEM via problems involving the mechanics of solids. Mechanics for solid structures comprises a vast subject by itself, which needs volumes of books to describe it thoroughly. This book will devote one chapter to briefly cover the mechanics of solids and structures by presenting the important basic principles. It focuses on the derivation of key governing equations for 3D solids. Drawings are used to illustrate all the field variables in solids, as well as the relationships between them. Equations for various types of solids and structure components, such as 2D solids, trusses, beams, and plates, are then deduced from the general equations for 3D solids. It has been found from our teaching practices that this method of delivering the basics of the mechanics of solid structures is very effective. The introduction of the general 3D equations before examining the other structural components actually gives students a firm fundamental background, from which the other equations can be easily derived and naturally understood. Understanding is then enforced by studying the examples and case studies that are solved using the FEM in other chapters. Our practice of teaching in the past few years has shown that most students managed to understand the fundamental basics of mechanics without too much difficulty, and many of them do not even possess an engineering background.

We have also observed that, over the past few years of handling industrial projects, many engineers are asked to use commercial FEM software packages to simulate engineering systems. Many do not have proper knowledge of the FEM, and are willing to learn via self-study. They thus need a book that describes the FEM in their language, and not with the use of overly technical symbols and terminology.

Without such a book, many would end up using the software packages blindly like a "black box." This book therefore aims to throw light into the "black box" so that users can see clearly what is going on inside by relating things that are done in the software with the theoretical concepts in the FEM. Detailed description and references are provided in case studies to show how the FEM's formulation and techniques are implemented in the software package.

We believe that being informative need not necessarily mean being exhaustive. There are a large number of techniques developed during the last five decades in the area of FEM, and it is not practically possible to be exhaustive. Our past experience has found that very few of the vast number of FEM techniques are still frequently used. This book does not want to be an encyclopedia (as that is already available), but intends to be informative enough for the really useful modeling techniques that are and will be alive for years to come. Useful techniques are also often very interesting, and by describing the key features of these lively techniques, this book is written to instill an appreciation of them for solving practical problems. It is with this appreciation that we hope readers will be enticed even more to the FEM.

Theories can be well accepted and appreciated if their applications can be demonstrated explicitly. The case studies used in the book also serve the purpose of demonstrating the finite element theories. They include a number of applications of the FEM for the modeling and simulation of microstructures and microsystems. Most of the case studies are idealized practical problems to clearly bring forward the concepts of the FEM, and will be presented in a manner that makes it easier for readers to follow. Following through these case studies, ideally in front of a workstation, helps the reader to easily understand the important concepts, procedures, and theories.

A picture tells a thousand words. Numerous drawings and charts are used to describe important concepts and theories. This is very important and will definitely be welcomed by readers, especially those from non-engineering backgrounds.

The book provides practical techniques for using commercial software packages, (e.g., ABAQUS and ANSYS). The case studies and examples calculated using one software package could easily be repeated using any other commercial software package. Commonly encountered problems in modeling and simulation using commercial software packages are discussed, and rules-of-thumb and guidelines are also provided to solve these problems effectively in professional ways.

Note that the focus of this book is on developing a good understanding of the fundamentals and principles of linear FE analysis. We have chosen ABAQUS for most of the examples as it can easily handle linear analyses. However, with further reading, readers could also extend the use of ABAQUS for projects involving non-linear FE analyses too. A chapter-by-chapter description of the book is given below.

Chapter 1: Highlights the role and importance of the FEM in computational modeling and simulation required in the design process for engineering systems. The general aspects of computational modeling and simulation of physical problems in engineering are discussed. Procedures for the establishment of mathematical and computational models of physical problems are outlined. Issues related to geometrical simplification, domain discretization, numerical computation, and visualization that are required in using the FEM are discussed.

Chapter 2: Describes the basics of mechanics for solids and structures. Important field variables of solid mechanics are introduced, and the key dynamic equations of these variables are derived. Mechanics for 2D and 3D solids, trusses, beams, frames, and plates are covered in a concise and easy to understand manner. Readers with a mechanics background may skip this chapter.

Chapter 3: Introduces the general finite element procedure. Concepts of strong and weak forms of a system equations and the construction of shape functions for interpolation of field variables are described. The properties of the shape functions are also discussed with an emphasis on the sufficient requirement of shape functions for establishing finite element equations. Hamilton's principle is introduced and applied to establish the general forms of the finite element equations. Methods to solve the finite element equation are discussed for static, eigenvalue analysis, as well as transient analysis.

Chapter 4: Details the procedure used to obtain finite element matrices for truss structures. The procedures to obtain shape functions, the strain matrix, local and global coordinate systems, and the assembly of global finite element system equations are described. Very straightforward examples are used to demonstrate a complete and detailed finite element procedure to compute displacements and stresses in truss structures. The reproduction of features and the convergence of the FEM as a reliable numerical tool are revealed through these examples.

Chapter 5: Deals with finite element matrices for beam structures. The procedures carried out to obtain shape functions and the strain matrix are described. Elements for thin beam elements are developed. Examples are presented to demonstrate application of the finite element procedure in a beam microstructure, using both ABAQUS and ANSYS.

Chapter 6: Shows the procedure for formulating the finite element matrices for frame structures, by combining the matrices for truss and beam elements. Details on obtaining the transformation matrix and the transformation of matrices between the local and global coordinate systems are described. An example is given to demonstrate the use of frame elements to solve practical engineering problems.

Chapter 7: Formulates the finite element matrices for 2D solids. Matrices for linear triangular elements, bilinear rectangular and quadrilateral elements are derived in detail. Area and natural coordinates are also introduced in the process. Iso-parametric formulation and higher order elements are also described. An example of analysing a micro device is used to study the accuracy and convergence of triangular and quadrilateral elements.

Chapter 8: Deals with finite element matrices for plates and shells. Matrices for rectangular plate elements based on the more practical Reissner–Mindlin plate theory are derived in detail. Shell elements are formulated simply by combining the plate elements and 2D solid plane stress elements. Examples of analyzing a micro device using ABAQUS are presented.

Chapter 9: Finite element matrices for 3D solids are developed. Tetrahedron elements and hexahedron elements are formulated in detail. Volume coordinates are introduced in the process. Formulation of higher order elements is also outlined. An example of using 3D elements for modeling a nano-scaled heterostructure system is presented.

Chapter 10: Special purpose elements and recent advanced methods are introduced and briefly discussed. Crack tip elements for use in many fracture mechanics problems are derived. Infinite elements formulated by mapping and a technique of using structure damping to simulate an infinite domain are both introduced. The finite strip method and the strip element method are also discussed. In addition, the meshfree methods and the smoothed finite element methods are discussed.

Chapter 11: Modeling techniques for the stress analyses of solids and structures are discussed. Use of symmetry, multipoint constraints, mesh compatibility, the modeling of offsets, supports, joints, and the imposition of multipoint constraints are all covered. Examples are included to demonstrate use of the modeling techniques.

Chapter 12: The FEM procedure for solving partial differential equations based on the weighted residual methods is presented. In particular, heat transfer problems in 1D and 2D are formulated. Issues in solving heat transfer problems are discussed. Examples are presented to demonstrate the use of ABAQUS for solving heat transfer problems.

Chapter 13: The basics of using commercial packages are discussed. Both ABAQUS and ANSYS are outlined so as to help a beginner to get a head start on using an FEM software package. We use

ABAQUS to introduce a typical FEM data input file, and use ANSYS for the introduction of creating an FEM model using a graphical user interface (GUI). An example is presented to provide a step-by-step outline of the procedure of creating an ABAQUS input file. Important information required by most FEM software packages is highlighted. A similar model is also created, executed, and post-processed using ANSYS with GUI. This chapter is meant for readers doing self-study on using an FEM package and ABAQUS and ANSYS are used to provide a basic idea of how to begin. For university courses, it is most effective to conduct hands-on sessions in computer rooms, together with students, when teaching about commercial software packages.

Most of the materials in the book are selected from lecture notes prepared for classes conducted by the first author since 1995 for both under- and post-graduate students. Those lecture notes were written using materials from many excellent, existing books on the FEM (as the number of references used are too many to be cited explicitly, they are listed in the Reference section of the book, while regrettably noting that some may have been unintentionally left out), and evolved over years of lecturing at the National University of Singapore, the Institute of Mechanics at the Chinese Academy of Science, Hunan University (China), Taiyuan University of Technology (China) and University of Cincinnati (USA). The materials have indeed been presented to readers/audiences of various backgrounds. The authors wish to express their sincere appreciation to the authors of all the existing FEM books, which some of these materials may have been referenced from (they are listed in the Reference section without explicitly citing them). FEM has been well developed and documented in detail in various existing books. In view of this, the authors have tried their best to limit the information in this book to the necessary minimum required to make it useful for those applying FEM in practice. Readers seeking more advanced theoretical material are advised to refer to books such as those by Zienkiewicz and Taylor (2000). The authors would also like to thank the students (both past and current) for their company in the study of the subject of FEM over the years, which has helped to develop this course into what it is today.

Preparing lectures for FEM courses is a very time-consuming task, as many drawings and pictures are required to explain all these theories, concepts, and techniques clearly. A set of colorful PowerPoint slides for the materials in the book has therefore been produced by the authors for lecturers to use. These slides can be found at the following website: www.textbooks.elsevier.com. It is aimed at reducing the amount of time taken in preparing lectures using this textbook. All the slides are grouped according to the chapters. The lecturer has the full freedom to cut and add slides according to the level of the class and the hours available for teaching the subject, or to simply use them as provided. Our suggestions on the use of material for undergraduate and graduate semester courses (typically $\sim$40 lecture hours) are as follows.

Undergraduate Courses	Graduate Courses (FEM entry level)
Chapter 1 (With detailed coverage)	Chapter 1 (Skimming through)
Chapter 2 (With detailed coverage, may skip Section 2.6)	Chapter 2 (Skimming throughdetailed discussion on Section 2.6)
Chapter 3 (Detail with slower pace, may skip the proofs in Section 3.4.4)	Chapter 3 (May skip the proofs in Section 3.4.4)
Chapters 4–7 (Detail with slower pace)	Chapters 4–12
Chapter 9 (Informative)	Chapter 13 (Preferably done in a computer lab with hand-on sessions)
Chapter 11 (Selective)	
Chapter 13 (Preferably done in a computer lab with hand-on sessions)	

It is advisable that an undergraduate FEM course may focus more on the basics for both mechanics and FEM procedure, so that the students can have a solid knowledge for understanding and using the FEM. This is important, especially for undergraduate courses during the lower years. A graduate FEM course may focus on the FEM theory and formulation, so that the students can have a solid foundation for advanced use of FEM and even develop their new techniques for their graduate projects.

Finally, it is the authors' belief that people can rarely become an expert by reading a book. They may become an expert by hard work and persistent creative practices. In addition, the broad subject of the FEM is evolving continuously. This book aims to bring the reader into a 'workshop' with clear introductions to the basic tools there. It is up to the reader to perfect these tools and use them to innovate, in terms of using the FEM, or even to develop novel computational techniques.

G.R. Liu and S.S. Quek

Preface for the Second Edition

Since the first edition of this book was published in 2003, it has been used by many universities all over the world as a textbook and/or reference book. It has also been translated into other languages. We have subsequently received much encouragement, important comments, and suggestions. We would like to take this opportunity to thank all the users and readers for the use of this book and the constructive comments and suggestions that they have provided to us. Based on this feedback, the second edition contains the following major changes:

1. Editorial revision has been thoroughly carried out for the entire book.
2. Chapter 2 has been enriched for better and easier review on mechanics of solids and structures.
3. The number of exercises (review questions) has been significantly increased (more than double).
4. Coverage of axial symmetric problems has been added to Chapter 7.
5. Applications of ABAQUS have been updated.
6. Brief descriptions on the use of ANSYS have been added.
7. Briefings on advanced meshfree methods and the smoothed finite element methods have been added.

For instructors using this book for a course, ancillary materials, such as PowerPoint lecture slides and a solutions manual (for instructors only), are available by registering at www.textbooks.elsevier.com.

G.R. Liu and S.S. Quek

Computational Modeling

CHAPTER OUTLINE HEAD

1.1 Introduction

The Finite Element Method (FEM) has developed into a key indispensable technology in the modeling and simulation of advanced engineering systems in various fields like housing, transportation, communications, and so on. In building an advanced engineering system, engineers and designers go through a sophisticated process of modeling, simulation, visualization, analysis, design, prototyping, testing, and lastly, fabrication. More often than not, much work is involved before the fabrication of the final product or system. This is to ensure the workability of the finished product, as well as for cost effectiveness in the manufacturing process. This process is illustrated as a flowchart in Figure 1.1. It is often iterative in nature, meaning that some of the procedures are repeated based on the results obtained at a current stage, so as to achieve an optimal performance at the lowest cost for the system to be built. Therefore, techniques related to modeling and simulation in a rapid and effective way play an increasingly important role, and the FEM becomes very much a standard tool in any major development of a system or product.

This book deals with topics related mainly to modeling and simulation, which are underscored in Figure 1.1. The focus will be on the techniques of physical, mathematical, and computational modeling, and other aspects of computational simulation. A good understanding of these techniques plays an important role in building an advanced engineering system in a rapid and cost-effective way.

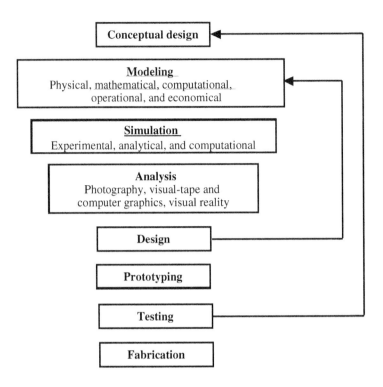

FIGURE 1.1

Processes leading to the fabrication of advanced engineering systems.

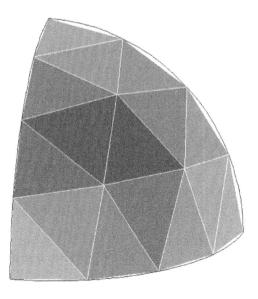

FIGURE 1.2

Hemispherical section discretized into several shell elements.

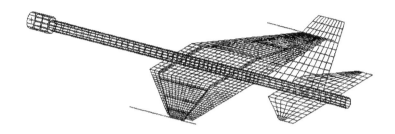

FIGURE 1.3

Mesh for the design of a scaled model of an aircraft for dynamic testing in the laboratory (Quek 1997–98).

So what is the FEM? The FEM was first used to solve problems of solid and structural analysis, and has since been applied to many problems like thermal analysis, fluid flow analysis, piezoelectric analysis, and many others. Basically, the analyst seeks to determine the distribution of some field variable, for example, the displacement in structural mechanics analysis, the temperature or heat flux in thermal analysis, the electrical charge in electrical analysis, and so on. The FEM is a numerical method that seeks an approximated solution of the distribution of field variables in the problem domain that is often difficult to obtain analytically. It is done by first dividing the problem domain into a number of small pieces called elements, often of simple geometry, as shown in Figures 1.2 and 1.3. Physical principles/laws are then applied to each small element. Figure 1.4 shows a schematic illustration of a distribution of a field, $F(x)$, in one dimension that is approximated using the FEM. In this case, $F(x)$ is a continuous function that is approximated using piecewise linear functions in an element. In this one-dimensional case, the ends of each element are termed *nodes*. The unknown variables in the FEM are simply the discrete values of the field variable at the nodes. Physical and mathematical principles are then followed to establish governing equations for each element, after which the elements are "tied" to one another to describe the distribution of the field in the entire geometry. This process leads to a set

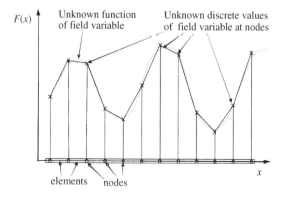

FIGURE 1.4

Finite element approximation for a one-dimensional case. A continuous function is approximated using piece-wise linear functions in each sub-domain/element.

of linear algebraic simultaneous equations for the entire system that can be solved easily to yield the required field variable.

This book aims to bring across the various concepts, methods, and principles used in the formulation of FE equations in a manner that is easy to understand. Worked examples and case studies using the well-known commercial software package ABAQUS and ANSYS will be discussed, and effective techniques and procedures will be highlighted.

1.2 Physical problems in engineering

There are numerous physical problems in an engineering system. As mentioned earlier, although the FEM was initially used for solid and structural analysis, many other physical problems can be solved using the FEM. Mathematical models of the FEM have been formulated for the numerous physical phenomena that occur in engineering systems. Common physical problems solved using the standard FEM include:

- Mechanics for solids and structures
- Heat transfer
- Acoustics
- Fluid mechanics
- Combinations of the above
- Others

This book first focuses on the formulation of finite element equations for the mechanics of solids and structures, since that is what the FEM was initially designed for. FEM formulations for heat transfer problems are then described. Nevertheless, the conceptual understanding of the methodology of the FEM is the most important and will be emphasized in both solid mechanics and heat transfer problems. The application of the FEM to all other physical problems utilizes similar concepts.

Computer modeling using the FEM consists of the major steps discussed in the next section.

1.3 Computational modeling using FEM

The behavior of a physical phenomenon in a system depends upon the *geometry* or *domain* of the system, the *property* of the *material* or *medium*, and the *boundary*, *initial*, and *loading conditions*. For an engineering system, the geometry or domain can be very complex. Furthermore, the boundary and initial conditions can also be very complicated. It is therefore, in general, very difficult to solve the governing differential equation via *analytical* means. In practice, most of the problems are solved using numerical methods. Among these, the methods of *domain discretization* championed by the FEM are the most popular due to their reliability, practicality, versatility, and robustness.

The procedure of computational modeling using the FEM broadly consists of four steps:

- Modeling of the geometry
- Meshing (discretization)
- Specification of material property
- Specification of boundary, initial, and loading conditions

1.3.1 **Modeling of the geometry**

Most physical structures, components or domains are in general very complicated, and usually made up of multiple components. One can imagine the complexity that goes into making an automobile, an aircraft, an ocean-going vessel, and so on. It is therefore common, and often a good practice, to simplify parts of the geometry so that modeling is more manageable. The geometry and boundary of a structure can be made up of curved surfaces/lines, but as we perform FEM based modeling, it is important to bear in mind that the geometry is eventually represented by a collection of elements, and the curved lines/surfaces may be approximated by piecewise straight lines or flat surfaces, if these elements are assumed to be flat/straight pieces/segments (that is, linearity is assumed). Figure 1.2 shows an example of a curved boundary represented by the straight lines of the edges of triangular elements. The accuracy of the representation of the curved part, as in Figure 1.2, is controlled by the number of elements used. It is obvious that with more elements, the representation of the curved parts by straight edges would be smoother and more accurate. On the other hand, with more elements, longer computational time is required. Unfortunately, modelers often face constraints in terms of available computational resources and it is often necessary to limit the number of elements used in the model. As such, compromises are usually made in order to decide on an optimum number of elements used. These compromises usually result in the omission of fine details of the geometry unless very accurate results are required for those regions. The analysts will then have to interpret the results of the simulation with these geometric approximations in mind.

Depending on the software used, there are many ways to create a proper geometry in the computer for the FE mesh. Points can be created simply by keying in the coordinates. Lines and curves can be created by connecting the points or nodes. Surfaces can be created by connecting, rotating or translating the existing lines or curves. Solids can be created by various operations of connecting, rotating or translating the existing surfaces. Points, lines and curves, surfaces and solids can be translated, rotated or reflected to form new ones.

More often than not, the use of a graphic interface helps in the creation and manipulation of the geometrical objects on computers. There are numerous Computer Aided Design (CAD) software packages used for engineering design. These CAD packages can generate appropriate files containing the geometry of the designed engineering system and these files can then be read by modeling software packages, in which appropriate discretization of the geometry into elements can be carried out. However, in many cases, complex objects read directly from a CAD file may need to be modified and simplified before performing discretization. It may be worth mentioning that there are CAD packages which incorporate modeling and simulation packages, and these are useful for the rapid prototyping of new products.

While most modeling software packages try to make modeling a breeze by developing excellent user interfaces, it is equally important to have knowledge, experience, and good engineering judgment. This is what distinguishes a good modeler from one who is just proficient in using the software. For example, finely detailed geometrical features often only play an aesthetic role, and have negligible effects on the functional performance of the engineering system. These features can sometimes be omitted, ignored or simplified, but it takes good engineering judgment to decide if any geometrical assumptions/simplifications will have a negligible effect on the overall simulation results.

Possessing sufficient knowledge and engineering judgment enables one to recognize a physical component, and possibly simplify that component mathematically, so that modeling can be more effective. For example, a plate has three dimensions geometrically. The plate in the plate theory of mechanics is represented mathematically only in two dimensions (the reason for this will be elaborated in Chapter 2). Therefore the geometry of a "mechanics" plate is a two-dimensional flat surface. *Plate elements* will be used in meshing these surfaces. A similar situation can be found in shells. A physical beam also has three dimensions. The beam in the beam theory of mechanics is represented mathematically in only one dimension, therefore the geometry of a "mechanics" beam is a one-dimensional straight line. *Beam elements* have to be used to mesh the lines in models. This is also true for truss structures. In fact, the entire physical world that we visualize in space is in three dimensions, yet there are often components like the plates, shells, beams, and so on that we can approximate with a lower dimension that makes modeling so much easier and still provides acceptable accuracy for engineering design purposes.

1.3.2 Meshing

When we discretize the geometry or domain into small pieces which we have come to call *elements* or *cells*, the process is called meshing. Why do we mesh? The rationale behind this is in fact based on a very logical understanding. We can expect the solution for an engineering problem to be very complex, and to vary in a way that is very unpredictable using functions across the whole domain of the problem. However, if the problem domain can be divided (*meshed*) into small elements or cells using a set of *grids* or *nodes*, the variation of the solution within an element can be approximated very easily using simple functions such as polynomials. The collective variations of the solutions for all of the elements thus form the variation of the solution for the whole problem domain.

Proper theories are needed for *discretizing* the governing equations based on the discretized domains. The theories used are different from problem to problem, and will be covered in detail in this book for various types of problems. But before that, we need to generate a mesh for the problem domain.

Mesh generation is a very important task in *pre-processing*. It can be a very time consuming task to the analyst, and usually an experienced analyst will produce a more credible mesh for a complex problem. The domain has to be meshed properly into elements of specific shapes such as triangles and quadrilaterals in 2D, and tetrahedrons and hexahedrons in 3D. Information such as *element connectivity* must be created during the meshing for use later in the formation of the FEM equations. It is ideal to have a fully automated mesh generator, which by itself is a big research field, but at this moment, it is still not commercially available. A semi-automatic mesh generator is available in most pre-processors as part of commercial FEM software packages. There are also packages designed mainly for meshing. Such packages can generate files of a mesh, which can be read by other modeling and simulation packages.

Triangulation is the most flexible and well-established way to create meshes with triangular elements. It can be made almost fully automated for two-dimensional planes and even three-dimensional spaces. Therefore it is commonly available in most pre-processors. The additional advantage of using triangles is the flexibility in adapting complex geometry and its boundaries. One can easily visualize fitting triangles into an acute corner of geometry, but it will be less obvious how one can do it with, say, quadrilateral elements without severely distorting the elements. However, the disadvantage of using triangle elements is that the accuracy of the simulation results based on triangular elements is often significantly lower[1] than that obtained using quadrilateral elements. Quadrilateral

---

[1] This has been overcome by the Smoothed Finite Element methods (S-FEM). See Liu and Trung, *Smoothed Finite Element Methods*, CRC Press, 2010.

element meshes are in general more difficult to generate in an automated manner. Some examples of meshed solids and structures are given in Figures 1.3–1.7.

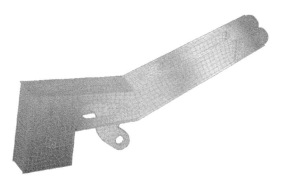

FIGURE 1.5

Mesh for a boom showing the stress distribution. (Picture used by courtesy of EDS PLM Solutions.)

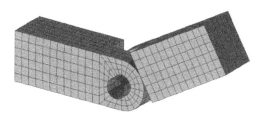

FIGURE 1.6

Mesh of a hinge joint.

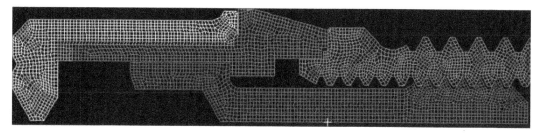

FIGURE 1.7

Axisymmetric mesh of part of a dental implant. (The CeraOne® abutment system, Nobel Biocare.)

1.3.3 **Material or medium properties**

Many engineering systems consist of multiple components and each component can be of a different material. In fact, even within a single component, there can be multiple materials, as in the case of a composite material. Properties of materials can therefore be defined for a group of elements or even for individual elements if needed. The FEM can therefore work very conveniently for systems with multiple materials, which is a significant advantage of the FEM. For different phenomena or physics to be simulated, different sets of material properties are required. For example, Young's modulus and shear modulus are required for the stress analysis of solids and structures, whereas the thermal conductivity coefficient will be required for a thermal analysis. The input of material properties into a pre-processor is usually straightforward; all the analyst needs to do is to key in the relevant inputs (usually with a convenient user interface) for the required material properties and specify to which region of the geometry or which elements the data applies. Nevertheless, obtaining these properties is not always easy. There are commercially available material databases for standardized materials, but experiments are usually required to accurately determine the property of special materials to be used in the system. In some cases, perhaps another simulation (at a smaller length scale, like molecular dynamics simulation and/or quantum mechanics-based first principles calculations) is required to first obtain the material's properties, which are then used as inputs for the FEM model. This, however, is beyond the scope of this book, and throughout, we simply assume that the material property is known.

1.3.4 **Boundary, initial, and loading conditions**

Boundary, initial, and loading conditions play a decisive role in the simulation. Prescribing these conditions is usually done easily using commercial pre-processors, and it is often interfaced with graphics. Users can specify these conditions either to the geometrical identities (points, lines or curves, surfaces, and solids) or to the mesh identities (nodes, elements, element edges, element surfaces). Again, to accurately simulate these conditions for actual engineering systems requires experience, knowledge, and proper engineering judgments. The boundary, initial, and loading conditions are different from problem to problem, and will be covered in detail in subsequent chapters.

1.4 **Solution procedure**
1.4.1 **Discrete system equations**

Based on the mesh generated, a set of discrete simultaneous system equations can be formulated using existing approaches. There are a few types of approach for establishing the simultaneous equations. The first is based on energy principles, such as Hamilton's principle (Chapter 3), the minimum potential energy principle, and so on. The traditional FEM is established based on these principles. The second approach is the weighted residual method, which is also often used for establishing FEM equations for many physical problems, and will be demonstrated for heat transfer problems in Chapter 12. The third approach is based on the Taylor series, which led to the formation of the traditional Finite Difference Method (FDM). The fourth approach is based on the control of conservation laws on each finite volume (elements) in the domain. The Finite Volume Method (FVM) is established using this approach. Another approach is by integral representation used in some mesh-free methods (Liu, 2009), in particular the Smoothed Particle Hydrodynamics (SPH) method (Liu and Liu, 2003). Engineering practice has so far shown that the first two approaches are most often used for solid and structures, and the other two approaches are often used for fluid flow simulation. However, the FEM has also been used to develop commercial packages for fluid flow and heat transfer problems, and FDM can be used for solids and structures. It is interesting to note, without going into detail, that the mathematical foundation of all these approaches can be based on the *residual method*. An appropriate choice of the "test" and "trial" functions in the residual method can lead to the FEM, FDM or FVM formulation.

This book first focuses on the formulation of finite element equations for the mechanics of solids and structures based on energy principles. FEM formulations for heat transfer problems are then described so as to demonstrate how the weighted residual method can be used for deriving FEM equations. This will provide the basic knowledge and key approaches into the FEM for dealing with other physical problems.

1.4.2 **Equation solvers**

After an FEM model has been created, it is then fed into a *solver* to solve the discretized system of equations—simultaneous equations for the field variables at the nodes of the mesh. This is the most computer hardware-demanding process. Different software packages use different algorithms depending upon the physical phenomenon to be simulated. There are two very important considerations when choosing algorithms for solving a system of equations: One is the storage required, and another is the central processing unit (CPU) time needed.

There are two main types of method for solving simultaneous equations: direct methods and iterative methods. Commonly used direct methods include the Gauss elimination method and the LU decomposition method (factorizing a matrix into a product of a Lower triangular matrix and a Upper triangular matrix). Those methods work well for relatively small equation systems. Direct methods operate on fully assembled system equations, and therefore demand far larger storage space. It can also be coded in such a way that the assembling of the equations is done only for those elements involved in the current stage of equation solving. This can reduce the requirements on storage significantly.

Iterative methods include the Gauss-Jacobi method, the Gauss-Seidel method, the successive over-relaxation (SOR) method, generalized conjugate residual methods, the line relaxation method, and so on. These methods work well for relatively larger systems. Iterative methods are often coded in such a way as to avoid full assembly of the system matrices in order to save significantly on the storage. The performance in terms of the rate of convergence of these methods is usually problem dependent. In using iterative methods, pre-conditioning plays a very important role in accelerating the convergence process. For nonlinear problems, another iterative loop is needed. The nonlinear equation has to be properly formulated (linearized) into a set of linear equations through the iterations of the extra iterative loop.

For time-dependent problems, time stepping is necessary. Starting from the given initial state, the time history of the solution is computed by marching forward to the next time steps until the desired time is reached. There are two main approaches to time stepping: the implicit and explicit approaches. Implicit approaches are usually more stable numerically but less efficient computationally than explicit approaches for one step. Moreover, contact algorithms can be developed easily using explicit methods. Details of these issues will be given in Chapter 3.

1.5 **Results visualization**

The result generated after solving the system equation usually comprises a vast volume of digital data. The results have to be visualized in such a way that it is easy to interpolate, analyze, and present. The visualization is performed through a so-called post-processor usually packaged together with the software. Most of these processors allow the user to display 3D objects in many convenient and colorful ways on screen. The object can be displayed in the form of wire-frames, groups of elements, and groups of nodes. The user can rotate, translate, and zoom into and out from the objects. Field variables can be plotted on the object in the form of contours, fringes, wire-frames, and deformations. Usually, there are also tools available for the user to produce iso-surfaces, or vector fields of variables. Tools to enhance visual effects are also available, such as shading, lighting, and shrinking. Animations and movies can also be produced to simulate the dynamic aspects of a problem. Outputs in the form of tables, text files, and x-y plots are also routinely available. Throughout this book, worked examples with various post-processed results are given.

Advanced visualization tools, such as virtual reality and 3D visualization, are available nowadays. These advanced tools allow users to display objects and results in a much more realistic 3D form. The platform can be a goggle, inversion desk or even a room. When the object is immersed in a room, analysts can walk through the object, go to the exact location and view the results. Figures 1.8 and 1.9 show an airflow field in a virtually designed building.

To summarize, this chapter has briefly given an overview of the steps involved in computer modeling and simulation using FEM. With rapidly advancing computer technology, using FEM with

FIGURE 1.8

Immersed, 3D visualization of air flow field in virtually designed building. (Courtesy of the Institute of High Performance Computing.)

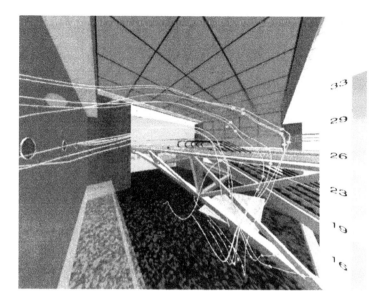

FIGURE 1.9

A 3D rendition of air flow in a virtually designed building system. (Courtesy of the Institute of High Performance Computing.)

the aid of computers is indispensable in any design process of an advanced engineering system. Software developers have been making it easier and more user-friendly for anyone to use simulation software. The subsequent chapters in this book will discuss the "behind-the-scenes" processes that the software is developed to execute when an engineer sits in front of the computer and performs an FEM analysis.

Briefing on Mechanics for Solids and Structures

2

CHAPTER OUTLINE HEAD

2.1 Introduction

The concepts and classical theories of the mechanics of solids and structures are readily available in numerous textbooks (see, for example, Timoshenko, 1940; Fung, 1965; Timoshenko and Goodier, 1970; Xu, 1979; Hu, 1982). This chapter serves to introduce these basic concepts and classical theories in a brief and easy to understand manner, so that readers are well prepared to appreciate the application of the finite element method to solve solid mechanics problems.

Solids and structures are stressed when they are subjected to *loads* or *forces*. The *stresses* are, in general, not uniform, and lead to *strains*, which can be measured using strain gauges, or even observed as either *deformation* or *displacement*. *Solid mechanics* and *structural mechanics* deal with the relationships between stresses and strains, displacements and forces, and stresses (strains) and forces, for a set of properly given boundary conditions for solids and structures. These relationships are vital in modeling, simulating, and designing engineered structural systems.

Forces can be static and/or dynamic. *Statics* deals with the mechanics of solids and structures subjected to static loads such as the dead weight on the floor of buildings and their own weight. Solids and structures will experience vibration under the action of dynamic forces varying with time, such as excitation forces generated by a running machine on the floor. In this case, the stress, strain, and displacement will be functions of time, and the principles and theories of *dynamics* must apply. As statics can be treated as a special case of dynamics, the static equations can be derived by simply dropping out the dynamic terms in the general dynamic equations. This book will adopt this approach of first deriving the dynamic equations, and the static equations are then obtained as a special case of the dynamic equations, by simply omitting the dynamic terms.

Depending on the property of the material, solids can be *elastic*, meaning that the deformation in the solids disappears fully if it is unloaded. There are also solids that are considered *plastic*, meaning that the deformation in the solids cannot be fully recovered when it is unloaded. *Elasticity* deals with solids and structures of elastic materials, and *plasticity* deals with those of plastic materials. The scope of this book deals mainly with solids and structures of elastic materials, so that the focus can be concentrated in presenting the basic theories and procedures for the finite element method. In addition, this book deals only with problems of very small deformation, where the deformation and load has a *linear relationship*, meaning that the deformation grows proportionally with the growth of the external forces or loadings. Therefore, our problems will mostly be *linear elastic*.

Materials can be *anisotropic*, meaning that the material property varies with direction. Deformation in anisotropic material caused by a force applied in a particular direction may be different from that caused by the same magnitude of force applied in another direction. Composite materials are often anisotropic. A large number of material constants have to be used to define the material property of anisotropic materials. Many engineering materials are, however, *isotropic*, where the material property is not direction-dependent. Isotropic materials are a special case of anisotropic material. There are only two independent material constants for isotropic materials: usually the Young's modulus and Poisson's ratio. This book deals mostly with isotropic materials. Nevertheless, most of the finite element method (FEM) formulations are also applicable to anisotropic materials.

Boundary conditions are another important consideration in mechanics. There are displacement and force boundary conditions for solids and structures. For heat transfer problems there are temperature, heat flux, and convection boundary conditions. Treatment of the boundary conditions is a very important topic, and will be covered in detail in this chapter and also throughout the rest of the book.

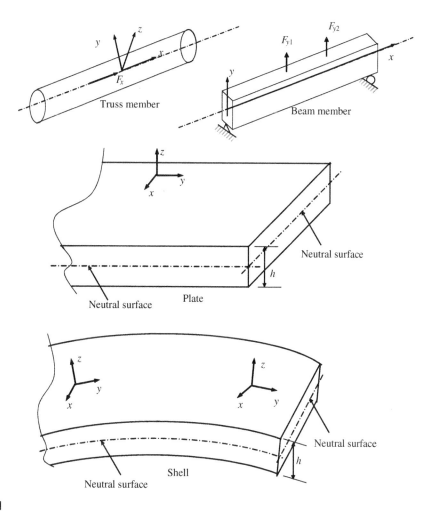

FIGURE 2.1

Four common types of structural components. Their geometrical features are made use of to derive dimension reduced system equations.

Structures are made of structural components that are in turn made of solids. There are generally four most commonly used structural components: truss, beam, plate, and shell, as shown in Figure 2.1. In physical structures, the main purpose of using these structural components is to effectively utilize the material and reduce the weight and cost of the structure. A practical structure can consist of different types of structural components, including solid blocks. Theoretically, the principles and methodology in solid mechanics can be applied to solve a mechanics problem for all structural components, but this is usually not a very efficient method. Theories and formulations for taking geometrical advantages of the structural components have therefore been developed. Formulations for a truss, a beam, 2D solids, and plate structures will be discussed in this chapter. In engineering practice, plate elements are often used together with two-dimensional solids for modeling shells. Therefore in this book, shell structures will be modeled by combining plate elements and 2D solid elements, without the use of classic theories for (curved) shells.

2.2 Equations for three-dimensional solids

2.2.1 Stress and strain

Let's consider a continuous three-dimensional (3D) elastic solid with a volume V and a surface S, as shown in Figure 2.2. The surface of the solid is further divided into two types: A surface on which the *external* forces are prescribed is denoted S_F; and a surface on which the displacements are prescribed is denoted S_d. The solid can also be loaded by body force f_b and surface force f_s in any distributed fashion in the volume of the solid.

At any point in the solid, the components of stress are indicated on the surface of an "infinitely" small cubic volume, as shown in Figure 2.3. On each surface, there will be the normal stress component, and two components of shearing stress. The sign convention for the subscript is that the first letter represents the surface on which the stress is acting, and the second letter represents the direction of the stress. The directions of the stresses shown in the figure are taken to be the positive directions. By taking moments of forces about the central axes of the cube at the state of equilibrium, it is easy to confirm the *shear equivalence* relations:

$$\sigma_{xy} = \sigma_{yx}; \quad \sigma_{xz} = \sigma_{zx}; \quad \sigma_{yz} = \sigma_{zy} \tag{2.1}$$

Therefore, there are six independent stress components in total at a point in 3D solids. These stresses are often called *stress tensors* (because they obey the rules of coordinate transformation for tensors). They can often be written (especially in the FEM) in a vector form of:

$$\boldsymbol{\sigma}^T = \left\{ \sigma_{xx} \ \sigma_{yy} \ \sigma_{zz} \ \sigma_{yz} \ \sigma_{xz} \ \sigma_{xy} \right\} \tag{2.2}$$

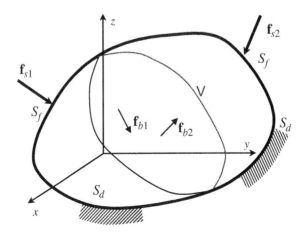

FIGURE 2.2

Solid subjected to forces applied within the solid (body force) and on the surface of the solid (surface force).

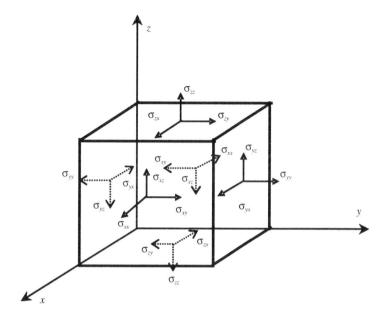

FIGURE 2.3

Six independent stress components at a point in a solid viewed on the surfaces of an infinitely small cubic block.

Corresponding to the six stress tensors, there are six strain components at any point in a solid, which can also be written in a similar vector form of:

$$\boldsymbol{\varepsilon}^T = \left\{ \varepsilon_{xx} \;\; \varepsilon_{yy} \;\; \varepsilon_{zz} \;\; \gamma_{yz} \;\; \gamma_{xz} \;\; \gamma_{xy} \right\} \tag{2.3}$$

Notice that we use the engineering notation of γ_{ij} (engineering shear strain) instead of the tensor notation of ε_{ij} ($=\gamma_{ij}/2$) for shear strain components in the vector form of strains. Strain is essentially the change of displacements per unit length, and therefore the components of strain in a 3D solid can be obtained from the derivatives of the displacements as follows:

$$\varepsilon_{xx} = \frac{\partial u}{\partial x}; \quad \varepsilon_{yy} = \frac{\partial v}{\partial y}; \quad \varepsilon_{zz} = \frac{\partial w}{\partial z};$$

$$\gamma_{xy} = 2\varepsilon_{xy} = \frac{\partial u}{\partial y} + \frac{\partial v}{\partial x};$$

$$\gamma_{xz} = 2\varepsilon_{xz} = \frac{\partial u}{\partial z} + \frac{\partial w}{\partial x}; \tag{2.4}$$

$$\gamma_{yz} = 2\varepsilon_{yz} = \frac{\partial v}{\partial z} + \frac{\partial w}{\partial y}$$

where u, v, and w are the displacement components in the x, y, and z directions, respectively. The six strain–displacement relationships in Eq. (2.4) can be rewritten in the following matrix form:

$$\boldsymbol{\varepsilon} = \mathbf{LU} \tag{2.5}$$

where $\mathbf{U}$ is the displacement vector, and has the form of:

$$\mathbf{U} = \left\{ \begin{array}{c} u \\ v \\ w \end{array} \right\} \tag{2.6}$$

and $\mathbf{L}$ is a matrix of partial differential operators obtained simply by inspection of Eq. (2.4):

$$\mathbf{L} = \begin{bmatrix} \partial/\partial x & 0 & 0 \\ 0 & \partial/\partial y & 0 \\ 0 & 0 & \partial/\partial z \\ 0 & \partial/\partial z & \partial/\partial y \\ \partial/\partial z & 0 & \partial/\partial x \\ \partial/\partial y & \partial/\partial x & 0 \end{bmatrix} \tag{2.7}$$

2.2.2 Constitutive equations

The constitutive equation gives the relationship between the stress and strain in the material of a solid. It is often termed Hooke's law. The original Hooke's law is derived only for a 1D bar. The generalized Hooke's law for 3D anisotropic materials can be given in the following matrix form:

$$\boldsymbol{\sigma} = \mathbf{c}\boldsymbol{\varepsilon} \tag{2.8}$$

where $\mathbf{c}$ is a matrix of material constants, which are normally obtained through experiments. The constitutive equation can be written explicitly as:

$$\left\{ \begin{array}{c} \sigma_{xx} \\ \sigma_{yy} \\ \sigma_{zz} \\ \sigma_{yz} \\ \sigma_{xz} \\ \sigma_{xy} \end{array} \right\} = \begin{bmatrix} c_{11} & c_{12} & c_{13} & c_{14} & c_{15} & c_{16} \\ & c_{22} & c_{23} & c_{24} & c_{25} & c_{26} \\ & & c_{33} & c_{34} & c_{35} & c_{36} \\ & & & c_{44} & c_{45} & c_{46} \\ & sy. & & & c_{55} & c_{56} \\ & & & & & c_{66} \end{bmatrix} \left\{ \begin{array}{c} \varepsilon_{xx} \\ \varepsilon_{yy} \\ \varepsilon_{zz} \\ \gamma_{yz} \\ \gamma_{xz} \\ \gamma_{xy} \end{array} \right\} \tag{2.9}$$

Note that, since $c_{ij} = c_{ji}$, there are altogether 21 independent material constants c_{ij}, which is the case for a fully anisotropic material. For *isotropic* materials, however, $\mathbf{c}$ can be reduced to:

$$\mathbf{c} = \begin{bmatrix} c_{11} & c_{12} & c_{12} & 0 & 0 & 0 \\ & c_{11} & c_{12} & 0 & 0 & 0 \\ & & c_{11} & 0 & 0 & 0 \\ & & & (c_{11} - c_{12})/2 & 0 & 0 \\ & sy. & & & (c_{11} - c_{12})/2 & 0 \\ & & & & & (c_{11} - c_{12})/2 \end{bmatrix} \tag{2.10}$$

where

$$c_{11} = \frac{E(1-v)}{(1-2v)(1+v)}; \quad c_{12} = \frac{Ev}{(1-2v)(1+v)}; \quad \frac{c_{11} - c_{12}}{2} = G \tag{2.11}$$

in which E, v, and G are Young's modulus, Poisson's ratio, and the shear modulus of the material, respectively. There are only two independent constants among these three constants. The relationship between these three constants is:

$$G = \frac{E}{2(1+v)} \tag{2.12}$$

That is to say, for any isotropic material, given any two of the three constants, the other one can be calculated using the above equation.

2.2.3 Dynamic equilibrium equations

To formulate the dynamic equilibrium equations, let's consider an infinitesimal block of solid, shown in Figure 2.4. As in forming all equilibrium equations, equilibrium of forces is required in all directions. Note that, since this is a general, dynamic system, we also have to consider the inertial forces of the block. The equilibrium of forces in the x direction gives.

$$(\sigma_{xx} + d\sigma_{xx})dy\,dz - \sigma_{xx}dy\,dz + (\sigma_{yx} + d\sigma_{yx})dx\,dz - \sigma_{yx}dx\,dz$$
$$+ (\sigma_{zx} + d\sigma_{zx})dx\,dy - \sigma_{zx}dx\,dy + \underbrace{f_x dx\,dy\,dz}_{\text{external force}} = \underbrace{\rho \ddot{u}dx\,dy\,dz}_{\text{inertial force}} \tag{2.13}$$

where the term on the right-hand side of the equation is the inertial force term with ρ being the density of the material, and f_x is the *distributed* external body force (density) evaluated at the center of the small block. Note that:

$$d\sigma_{xx} = \frac{\partial \sigma_{xx}}{\partial x}dx, \quad d\sigma_{yx} = \frac{\partial \sigma_{yx}}{\partial y}dy, \quad d\sigma_{zx} = \frac{\partial \sigma_{zx}}{\partial z}dz \tag{2.14}$$

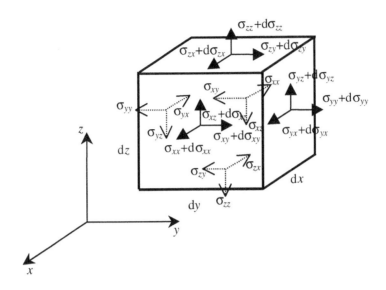

FIGURE 2.4

Stresses on an infinitesimal block. Equilibrium equations are derived based on this state of stresses.

Hence, Eq. (2.13) becomes one of the equilibrium equations, written as:

$$\frac{\partial \sigma_{xx}}{\partial x} + \frac{\partial \sigma_{yx}}{\partial y} + \frac{\partial \sigma_{zx}}{\partial z} + f_x = \rho \ddot{u} \tag{2.15}$$

Similarly, the equilibrium of forces in the y and z directions results in two other equilibrium equations:

$$\frac{\partial \sigma_{xy}}{\partial x} + \frac{\partial \sigma_{yy}}{\partial y} + \frac{\partial \sigma_{zy}}{\partial z} + f_y = \rho \ddot{v} \tag{2.16}$$

$$\frac{\partial \sigma_{xz}}{\partial x} + \frac{\partial \sigma_{yz}}{\partial y} + \frac{\partial \sigma_{zz}}{\partial z} + f_z = \rho \ddot{w} \tag{2.17}$$

This set of three equilibrium equations, Eqs. (2.15) to (2.17), can be written in a concise matrix form:

$$\mathbf{L}^T \boldsymbol{\sigma} + \mathbf{f}_b = \rho \ddot{\mathbf{U}} \tag{2.18}$$

where $\mathbf{f}_b$ is the vector of external body forces in the x, y, and z directions:

$$\mathbf{f}_b = \begin{Bmatrix} f_x \\ f_y \\ f_z \end{Bmatrix} \tag{2.19}$$

Using Eqs. (2.5) and (2.8), the equilibrium equation Eq. (2.18) can be further written in terms of displacements:

$$\mathbf{L}^T \mathbf{cLU} + \mathbf{f}_b = \rho \ddot{\mathbf{U}} \qquad (2.20)$$

The above is the general form of the dynamic equilibrium equation expressed as a matrix equation. We note that since $\mathbf{L}$ is a matrix of *first order* differential operators [see Eq. (2.7)]; $\mathbf{L}^T \mathbf{cL}$ will be a matrix of *second order* differential operators. Therefore, Eq. (2.20) is a set of three second order partial differential equations (PDEs), which can be seen more clearly if we expand equation Eq. (2.20) in detail (for isotropic materials):

$$\frac{E}{2(1+\upsilon)}\left[\frac{1}{1-2\upsilon}\left(\frac{\partial^2 u}{\partial x^2}+\frac{\partial^2 v}{\partial x\,\partial y}+\frac{\partial^2 w}{\partial x\,\partial z}\right)+\left(\frac{\partial^2 u}{\partial x^2}+\frac{\partial^2 u}{\partial y^2}+\frac{\partial^2 u}{\partial z^2}\right)\right]+f_x=\rho\frac{\partial^2 u}{\partial t^2}$$

$$\frac{E}{2(1+\upsilon)}\left[\frac{1}{1-2\upsilon}\left(\frac{\partial^2 u}{\partial x\,\partial y}+\frac{\partial^2 v}{\partial y^2}+\frac{\partial^2 w}{\partial y\,\partial z}\right)+\left(\frac{\partial^2 u}{\partial x^2}+\frac{\partial^2 u}{\partial y^2}+\frac{\partial^2 u}{\partial z^2}\right)\right]+f_y=\rho\frac{\partial^2 v}{\partial t^2}$$

$$\frac{E}{2(1+\upsilon)}\left[\frac{1}{1-2\upsilon}\left(\frac{\partial^2 u}{\partial x\,\partial z}+\frac{\partial^2 v}{\partial y\,\partial z}+\frac{\partial^2 w}{\partial z^2}\right)+\left(\frac{\partial^2 u}{\partial x^2}+\frac{\partial^2 u}{\partial y^2}+\frac{\partial^2 u}{\partial z^2}\right)\right]+f_z=\rho\frac{\partial^2 w}{\partial t^2}$$

$$(2.20a)$$

If the loads applied on the solid are static, the only concern is then the static status of the solid. Hence, the static equilibrium equation can be obtained simply by dropping the dynamic term (the inertial force term) in Eq. (2.20):

$$\mathbf{L}^T \mathbf{cLU} + \mathbf{f}_b = 0 \qquad (2.21)$$

2.2.4 Boundary conditions

There are two types of boundary conditions: displacement (*essential*) and force/stress (*natural*) boundary conditions. The displacement boundary condition can be simply written as:

$$u = \bar{u} \quad \text{and/or} \quad v = \bar{v} \quad \text{and/or} \quad w = \bar{w} \qquad (2.22)$$

on displacement boundaries. The bar stands for the prescribed value for the displacement component. For most of the actual simulations, the displacement is used to describe the support or constraints on the solid, and hence the prescribed displacement values are often zero. In such cases, the boundary condition is termed a *homogenous boundary condition*. Otherwise, they are *inhomogenous boundary conditions*.

The force boundary conditions are often written as:

$$\mathbf{n}\boldsymbol{\sigma} = \bar{\mathbf{t}} \qquad (2.23)$$

on force/stress boundaries, where $\mathbf{n}$ is given by:

$$\mathbf{n} = \begin{bmatrix} n_x & 0 & 0 & 0 & n_z & n_y \\ 0 & n_y & 0 & n_z & 0 & n_x \\ 0 & 0 & n_z & n_y & n_x & 0 \end{bmatrix} \qquad (2.24)$$

in which n_i ($i = x, y, z$) are cosines of the outwards normal on the boundary. The bar stands for the pre-scribed value for the force component. A force boundary condition can also be both homogenous and inhomogenous. If the condition is homogenous, it implies that the boundary is a free surface.

The displacement boundary condition and the force boundary condition are usually "complementary:" when the displacements are pre-specified on a part of the boundary, the force there is in general not known, and when the force is pre-specified on a part of the boundary, the displacements there are, in general, not known. In other words, we cannot pre-specify both of them at the same time at the same location on the boundary, but only one of them. Otherwise, they can be "contradictary," and the problem may become an ill-posed one. Special measures are required for dealing with such ill-posed problems (see Liu and Han, 2003).

The reader may naturally ask why the displacement boundary condition is called an essential boundary condition and the force boundary condition is called a natural boundary conditions. This is also related to the complementary nature of these two types of boundary conditions. The terms "essential" and "natural" come from the use of the so-called *weak form* formulation (such as the one used in FEM) for deriving system equations using assumed displacement methods. In such a formulation process, the displacement condition has to be satisfied *first* before derivation starts, or the process will fail. Therefore, the displacement condition is *essential*. As long as the essential (displacement) condition is satisfied, the weak formulation process will lead to the equilibrium equations as well as the force boundary conditions. This means that the force boundary condition is *naturally* derived from the process, and it is therefore called the natural boundary condition. Since the terms essential and natural boundary do not describe the physical meaning of the problem, it is actually a mathematical term, and they are also used for problems other than in mechanics.

With a well defined set of equilibrium equations with a body force, displacement, and force boundary conditions, analytical means can then be applied to solve the problem for displacements and stress, if the setting of the problem is sufficiently simple and problem domain is "regular." Detailed descriptions can be found in any book on elasticity such as the ones by Fung (1965), and Timoshenko and Goodier (1970). This book aims to solve the equations system through the FEM for problems of general complexity.

Equations obtained in this section are applicable to 3D solids. The objective of most analysts is to solve the equilibrium equations and obtain the solution of the field variable, which in this case is the displacement. Theoretically, these equations can be applied to all other types of structures such as trusses, beams, plates, and shells, because physically they are all 3D in nature. However, treating all the structural components as 3D solids makes computation very expensive, and sometimes practically impossible. Therefore, theories for taking geometrical advantage of different types of solids and structural components have been developed. Application of these theories in a proper manner can reduce the analytical and computational effort drastically. A brief description of these theories is given in the following sections.

2.3 Equations for two-dimensional solids

2.3.1 Stress and strain

Three-dimensional problems can be drastically simplified if they can be treated as a two-dimensional (2D) solid. For representation as a 2D solid, we basically remove one coordinate (usually the z axis), and hence assume that all the dependent variables are independent of the z axis, and all the external loads are also independent of the z coordinate, and applied only in the x–y plane. Therefore, we are left with a system with only two coordinates, x and y.

There are primarily two types of 2D solids. One is a *plane stress* solid, and the other is a *plane strain* solid. Plane stress solids are solids whose thickness in the z direction is very small compared with dimensions in the x and y directions. External forces are applied only in the x–y plane, and stresses in the z direction (σ_{zz}, σ_{xz}, σ_{yz}) are all zero, as shown in Figure 2.5. Plane strain solids are those solids whose thickness in the z direction is very large compared with the dimensions in the x and y directions. External forces are applied evenly along the z axis, and the movement in the z direction at any point is constrained. The strain components in the z direction (ε_{zz}, γ_{xz}, γ_{yz}) are, therefore, all zero, as shown in Figure 2.6.

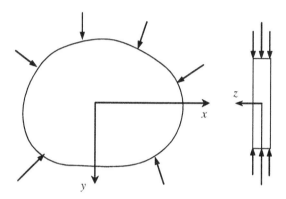

FIGURE 2.5

Plane stress problem. The dimension of the solid in the thickness (z) direction is much smaller than that in the x and y directions. All the forces are applied within the x–y plane and hence the displacements are functions of x and y only.

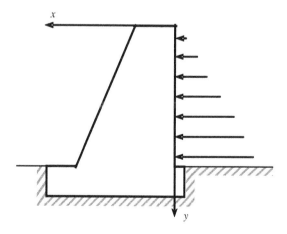

FIGURE 2.6

Plane strain problem. The dimension of the solid in the thickness (z) direction is much larger than that in the x and y directions, and the cross-section and the external forces do not vary in the z direction. A cross-section can then be taken as a representative cell, and hence the displacements are functions of x and y only.

Note that for the plane stress problems, the strains γ_{xz} and γ_{yz} are zero, but ε_{zz} will not be zero. It can be recovered easily using Eq. (2.9) after the in-plane stresses are obtained. Similarly, for the plane strain problems, the stresses σ_{xz} and σ_{yz} are zero, but σ_{zz} will not be zero. It can be recovered easily using Eq. (2.9) after the in-plane strains are obtained.

The system equations for 2D solids can be obtained immediately by omitting terms related to the z direction in the system equations for 3D solids. The stress components are:

$$\boldsymbol{\sigma} = \begin{Bmatrix} \sigma_{xx} \\ \sigma_{yy} \\ \sigma_{xy} \end{Bmatrix} \tag{2.25}$$

There are three corresponding strain components at any point in 2D solids, which can also be written in a similar vector form:

$$\boldsymbol{\varepsilon} = \begin{Bmatrix} \varepsilon_{xx} \\ \varepsilon_{yy} \\ \gamma_{xy} \end{Bmatrix} \tag{2.26}$$

The strain–displacement relationships are:

$$\varepsilon_{xx} = \frac{\partial u}{\partial x}; \quad \varepsilon_{yy} = \frac{\partial v}{\partial y}; \quad \gamma_{xy} = \frac{\partial u}{\partial y} + \frac{\partial v}{\partial x} \tag{2.27}$$

where u and v are the displacement components in the x and y directions, respectively. The strain–displacement relation can also be written in the following matrix form:

$$\boldsymbol{\varepsilon} = \mathbf{L}\mathbf{U} \tag{2.27a}$$

where the displacement vector has the form of:

$$\mathbf{U} = \begin{Bmatrix} u \\ v \end{Bmatrix} \tag{2.28}$$

and the differential operator matrix is obtained simply by inspection of Eq. (2.27) as:

$$\mathbf{L} = \begin{bmatrix} \partial/\partial x & 0 \\ 0 & \partial/\partial y \\ \partial/\partial y & \partial/\partial x \end{bmatrix} \tag{2.29}$$

2.3.2 Constitutive equations

Hooke's law for 2D solids has the following matrix form with $\boldsymbol{\sigma}$ and $\boldsymbol{\varepsilon}$ from Eqs. (2.25) and (2.26):

$$\boldsymbol{\sigma} = \mathbf{c}\boldsymbol{\varepsilon} \tag{2.30}$$

where **c** is a matrix of material constants, which have to be obtained through experiments. For plane stress, isotropic materials, we have:

$$\mathbf{c} = \frac{E}{1 - \upsilon^2} \begin{bmatrix} 1 & \upsilon & 0 \\ \upsilon & 1 & 0 \\ 0 & 0 & (1 - \upsilon)/2 \end{bmatrix} \quad \text{(Plane stress)} \tag{2.31}$$

To obtain the plane stress **c** matrix above, the conditions of $\sigma_{zz} = \sigma_{xz} = \sigma_{yz} = 0$ are imposed on the generalized Hooke's law for isotropic materials. For plane strain problems, $\varepsilon_{zz} = \gamma_{xz} = \gamma_{yz} = 0$ are imposed, or alternatively, replace E and v in Eq. (2.31), respectively, with $E/(1 - \upsilon^2)$ and $\upsilon/(1 - \upsilon^2)$, which leads to:

$$\mathbf{c} = \frac{E(1 - \upsilon)}{(1 - \upsilon)(1 - 2\upsilon)} \begin{bmatrix} 1 & \frac{\upsilon}{1-\upsilon} & 0 \\ \frac{\upsilon}{1-\upsilon} & 1 & 0 \\ 0 & 0 & \frac{1-2\upsilon}{2(1-\upsilon)} \end{bmatrix} \quad \text{(Plane strain)} \tag{2.32}$$

The displacement and force boundary conditions can be written in a similar way as for the 3D solids.

2.3.3 Dynamic equilibrium equations

The dynamic equilibrium equations for 2D solids can be easily obtained by removing the terms related to the z coordinate from the 3D counterparts of Eqs. (2.15)–(2.17):

$$\frac{\partial \sigma_{xx}}{\partial x} + \frac{\partial \sigma_{yx}}{\partial y} + f_x = \rho \ddot{u} \tag{2.33}$$

$$\frac{\partial \sigma_{xy}}{\partial x} + \frac{\partial \sigma_{yy}}{\partial y} + f_y = \rho \ddot{v} \tag{2.34}$$

These equilibrium equations can be written in a concise matrix form of:

$$\mathbf{L}^T \boldsymbol{\sigma} + \mathbf{f}_b = \rho \ddot{\mathbf{U}} \tag{2.35}$$

where $\mathbf{f}_b$ is the external force vector given by:

$$\mathbf{f}_b = \begin{Bmatrix} f_x \\ f_y \end{Bmatrix} \tag{2.36}$$

The expanded equation of Eq. (2.35) for 2D plane stain and isotropic materials becomes:

$$\frac{E}{2(1+\upsilon)}\left[\frac{1}{1-2\upsilon}\left(\frac{\partial^2 u}{\partial x^2}+\frac{\partial^2 v}{\partial x\,\partial y}\right)+\left(\frac{\partial^2 u}{\partial x^2}+\frac{\partial^2 u}{\partial y^2}\right)\right]+f_x=\rho\frac{\partial^2 u}{\partial t^2}$$

$$\frac{E}{2(1+\upsilon)}\left[\frac{1}{1-2\upsilon}\left(\frac{\partial^2 u}{\partial x\partial y}+\frac{\partial^2 v}{\partial y^2}\right)+\left(\frac{\partial^2 u}{\partial x^2}+\frac{\partial^2 u}{\partial y^2}\right)\right]+f_y=\rho\frac{\partial^2 v}{\partial t^2}$$

(2.36a)

It is clearly seen again that it is a set of second order partial differential equations (PDEs).

For static problems, the dynamic inertia term is removed, and the equilibrium equations can be written as:

$$\mathbf{L}^T\boldsymbol{\sigma}+\mathbf{f}_b=0 \tag{2.37}$$

Equations (2.35) or (2.37) will be much easier to solve and computationally less expensive as compared with equations for the 3D solids.

2.4 Equations for truss members

A typical truss structure is shown in Figure 2.7. Each truss member in a truss structure is a solid whose dimension in one direction is much larger than in the other two directions, as shown in Figure 2.8. The force is applied only in the direction with larger dimension, usually the x direction. Therefore a truss member is actually a one-dimensional (1D) solid. The equations for 1D solids can be obtained by further omitting the stress related to the y direction, σ_{yy}, σ_{xy}, from the 2D case.

FIGURE 2.7

A typical structure made up of truss members. The entrance of the faculty of Engineering, National University of Singapore.

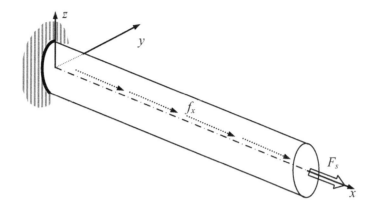

FIGURE 2.8

Truss member. The cross-sectional dimension of the solid is much smaller than that in the axial (x) directions, and the external forces are applied in the x direction, hence the axial displacement is a function of x only.

2.4.1 Stress and strain

Omitting the stress terms in the y direction, the stress in a truss member is only σ_{xx}, which is often simplified as σ_x. The corresponding strain in a truss member is ε_{xx}, which is simplified as ε_x. The strain–displacement relationship is simply given by:

$$\varepsilon_x = \frac{\partial u}{\partial x} \tag{2.38}$$

2.4.2 Constitutive equations

Hooke's law for 1D solids has the following simple form, with the exclusion of the y dimension and hence the Poisson effect:

$$\sigma = E \varepsilon \tag{2.39}$$

This is actually the original Hooke's law. The Young's module E can be obtained using a simple tensile test.

2.4.3 Dynamic equilibrium equations

By eliminating the y dimension term from Eq. (2.33), the dynamic equilibrium equation for 1D solids is:

$$\frac{\partial \sigma_x}{\partial x} + f_x = \rho \ddot{u} \tag{2.40}$$

Substituting Eqs. (2.38) and (2.39) into Eq. (2.40), we obtain the governing equation for elastic and homogenous (E is independent of x) trusses as follows:

$$E\frac{\partial^2 u}{\partial x^2} + f_x = \rho\ddot{u} \tag{2.41}$$

The static equilibrium equation for trusses is obtained by eliminating the inertia term in Eq. (2.40):

$$\frac{\partial \sigma_x}{\partial x} + f_x = 0 \tag{2.42}$$

The static equilibrium equation in terms of displacement for elastic and homogenous trusses is obtained by eliminating the inertia term in Eq. (2.41):

$$E\frac{\partial^2 u}{\partial x^2} + f_x = 0 \tag{2.43}$$

For bars of constant cross-sectional area A, the above equation can be written as:

$$EA\frac{\partial^2 u}{\partial x^2} + F_x = 0 \tag{2.43a}$$

where $F_x = f_x A$ is the external force applied in the axial direction of the bar, in which A is the cross-sectional area of the bar.

EXAMPLE 2.1

A uniform bar subjected to an axial force at the free end

Consider a bar of uniform cross-sectional area A and length l, as shown in Figure 2.8. The bar is fixed at one end and is subjected to a horizontal load of P at the free end and the body force is zero. The bar is made of an isotropic material with Young's modulus E. Determine the displacement and stress in the bar.

Solution

We can derive the solution via analytical methods, as this problem is very simple. From the strong form of the governing equation, Eq. (2.43), we have:

$$\frac{\partial^2 u}{\partial x^2} = 0$$

This is because the bar is free of body forces, or $f_x = 0$. The *general solution* that satisfies the previous governing equation can be obtained very easily as:

$$u(x) = c_0 + c_1 x$$

where c_0 and c_1 are unknown constants to be determined by boundary conditions. We need now to find the *particular solution* that satisfies the boundary condition of the physical (mechanics) problem. The displacement boundary condition for this problem can be given as:

$$u = 0, \quad \text{at } x = 0$$

Therefore, we have $c_0 = 0$. The displacement now becomes:

$$u(x) = c_1 x$$

Using Eqs. (2.38) and (2.39), we obtain:

$$\sigma_x = E\frac{\partial u}{\partial x} = c_1 E$$

The force boundary condition at $x = l$ for this bar can be given as:

$$\sigma_x|_{x=l} = \frac{P}{A}$$

Equating the right-hand side of the previous two equations, we obtain:

$$c_1 = \frac{P}{EA}$$

The stress at any point in the bar is obtained by:

$$\sigma_x = E\frac{\partial u}{\partial x} = c_1 E = \frac{P}{EA}E = \frac{P}{A}$$

which is uniform throughout the bar. We finally obtain the particular solution of the displacement of the bar:

$$u(x) = \frac{P}{EA}x$$

At $x = l$, we have the maximum displacement.

$$u(x = l) = \frac{Pl}{EA}$$

Although this problem is very simple, it shows a typical approach to solving a differential equation analytically for exact solutions:

- Obtain the general solution that satisfies the governing equations containing unknown constants.
- Use the boundary conditions for the problem to determine these constants, leading to a particular solution that satisfies both the governing equation and the boundary conditions.

2.5 Equations for beams

A beam possesses geometrically similar dimensional characteristics to a truss member, as shown in Figure 2.9. The difference is that the forces applied on beams are transversal, meaning the direction of the force is perpendicular to the axis of the beam. Therefore, a beam experiences bending, which is the deflection in the y direction as a function of x.

2.5.1 Stress and strain

The stresses on the cross-section of a beam are the normal stress, σ_{xx}, and shear stress, σ_{xz}. There are several theories for analyzing beam deflections. These theories can be basically divided into two major categories: a theory for *thin* beams and a theory for *thick* beams. This book focuses on the thin beam theory, which is often referred to as the *Euler–Bernoulli* beam theory. The Euler–Bernoulli beam theory assumes that the plane cross-sections, which are normal to the undeformed centroidal axis, remain plane after bending and remain normal to the deformed axis, as shown in Figure 2.10. With this assumption, one can first have:

$$\gamma_{xy} = 0 \tag{2.44}$$

which simply means that the shear stress is assumed to be negligible. Secondly, assume that the centroidal axis of the beam is not stretched or compressed in the x direction. Using the 3rd equation in Eq. (2.27), the axial displacement, u, of a fiber at a distance y from the centroidal axis can be expressed by:

$$u = -y\theta \tag{2.45}$$

where θ is the rotation of the cross-section of the beam in the x–y plane measured counterclockwise. The rotation can be obtained from the deflection of the centroidal axis of the beam, v, in the y direction:

$$\theta = \frac{\partial v}{\partial x} \tag{2.46}$$

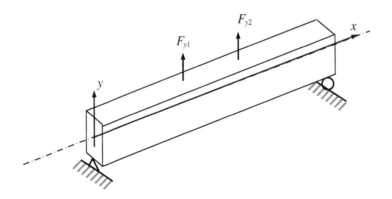

FIGURE 2.9

Simply supported beam. The cross-sectional dimensions of the solid are much smaller than those in the axial (x) directions, and the external forces are applied in the transverse (y) direction, hence the deflection of the beam is a function of x only.

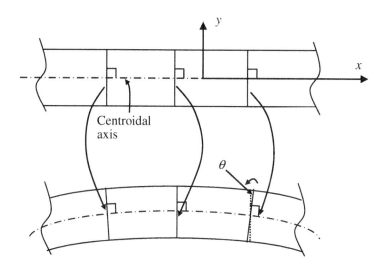

FIGURE 2.10

Euler–Bernoulli assumption for thin beams. The plane cross-sections that are normal to the undeformed centroidal axis remain plane and normal to the deformed axis after bending deformation. We hence have $u = -y\theta$.

The relationship between the normal strain and the deflection can be given by:

$$\varepsilon_{xx} = \frac{\partial u}{\partial x} = \frac{\partial(-y\theta)}{\partial x} = -y\frac{\partial\theta}{\partial x} = -y\frac{\partial^2 v}{\partial x^2} = -yLv \tag{2.47}$$

where L is the differential operator, given by:

$$L = \frac{\partial^2}{\partial x^2} \tag{2.48}$$

2.5.2 Constitutive equations

Similar to the equation for truss members, the original Hooke's law is applicable for beams:

$$\sigma_{xx} = E\varepsilon_{xx} \tag{2.49}$$

2.5.3 Moments and shear forces

Because the loading on the beam is in the transverse direction, there will be moments and corresponding shear forces resulted in the cross-sectional plane of the beam. In addition, bending of the beam can also be achieved if pure moments are applied instead of transverse loading. Figure 2.11 shows a small

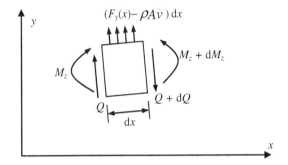

FIGURE 2.11

Isolated beam cell of length dx. Moments and shear forces are obtained by integration of stresses over the cross-section of the beam.

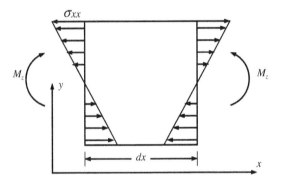

FIGURE 2.12

Normal stress that results in a moment.

representative cell of length dx of the beam. The beam cell is subjected to external force, $F_y = Af_y$, moment, M, shear force, Q, and inertial force, $\rho A \ddot{v}$, where ρ is the density of the material and A is the area of the cross-section. The moment on the cross-section at x results from the distributed normal stress σ_{xx}, as shown in Figure 2.12. The normal stress can be calculated by substituting Eq. (2.47) into Eq. (2.49):

$$\sigma_{xx} = -yELv \qquad (2.50)$$

It can be seen from the above equation that the normal stress σ_{xx} varies linearly with respect to y in the vertical direction on the cross-section of the beam. The moments resulting from the normal stress on the cross-section can be calculated by the following integration over the area of the cross-section:

$$M_z = -\int_A \sigma_{xx} y \, dA = E\left(\int_A y^2 dA\right) Lv = EILv = EI\frac{\partial^2 v}{\partial x^2} \qquad (2.51)$$

where I is the second moment of area (or moment of inertia) of the cross-section with respect to the z axis, which can be calculated for a given shape of the cross-section using the following equation:

$$I = \int_A y^2 \mathrm{d}A \tag{2.52}$$

Using Eqs. (2.50–2.52), we shall have:

$$\sigma = -My/I.$$

We now consider the force equilibrium of the small beam cell in the z direction:

$$\mathrm{d}Q + (F_y(x) - \rho A\ddot{v})\mathrm{d}x = 0 \tag{2.53}$$

or

$$F_y(x) = -\frac{\mathrm{d}Q}{\mathrm{d}x} + \rho A\ddot{v} \tag{2.54}$$

We would also need to consider the moment equilibrium of the small beam cell with respect to any point at the right surface of the cell:

$$\mathrm{d}M_z + Q\mathrm{d}x - \frac{1}{2}(F_y - \rho A\ddot{v})(\mathrm{d}x)^2 = 0 \tag{2.55}$$

Neglecting the small second order term containing $(\mathrm{d}x)^2$ leads to:

$$\frac{\mathrm{d}M_z}{\mathrm{d}x} = -Q \tag{2.56}$$

Finally, substituting Eq. (2.51) into Eq. (2.56) gives:

$$Q = -EI\frac{\partial^3 v}{\partial x^3} \tag{2.57}$$

Equations (2.56) and (2.57) give the relationships between the moments, shear forces, and the deflection of the Euler–Bernoulli beam.

2.5.4 Dynamic equilibrium equations

The dynamic equilibrium equation for beams can be obtained simply by substituting Eq. (2.57) into Eq. (2.54):

$$EI\frac{\partial^4 v}{\partial x^4} + \rho A\ddot{v} = F_y \tag{2.58}$$

The static equilibrium equation for beams can be obtained similarly by dropping the dynamic term in Eq. (2.58):

$$EI \frac{\partial^4 v}{\partial x^4} = F_y \tag{2.59}$$

The displacement boundary conditions for beams will be given in terms of deflection v and rotation θ, respectively, which are complementary to the force boundary conditions in terms of shear force Q and moments M (see Timoshenko and Gere, 1972).

2.6 Equations for plates

The wings of an aircraft can sometimes be simplified as a plate structure carrying transverse loads in the form of the weight of the engines or other components, as shown in Figure 2.13. A plate possesses a geometrically similar dimensional characteristic to that of a 2D solid, as shown in Figure 2.14.

FIGURE 2.13

An aircraft wing can be considered as a plate structure.

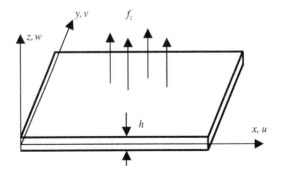

FIGURE 2.14

A plate subjected to transverse load that results in bending deformation.

The difference is that the direction of the forces applied on a plate are perpendicular to the plane of the plate. A plate can also be viewed as a 2D analogy of a beam. Therefore, a plate experiencing bending results in deflection w in the z direction, which is a function of x and y.

2.6.1 Stress and strain

The stress σ_{zz} in a plate is assumed to be zero. Similar to beams, there are several theories for analyzing deflection in plates. These theories can also be basically divided into two major categories: theory for *thin* plates and theory for *thick* plates. This chapter addresses a thin plate theory, often called the *Classical Plate Theory* (CPT), or the *Kirchhoff plate* theory, as well as the *first order shear deformation theory* for thick plates known as the *Reissner–Mindlin* plate theory (Reissner, 1945; Mindlin, 1951). Note that pure plate elements are usually not available in most commercial finite element packages, since most people would use the more general shell elements, which will be discussed in later chapters. Notice also that the formulation for elements for CPT plates is quite similar to that for beams (with just an extension for one more dimension). Hence, in this book, only finite element equations of plate elements based on the Reissner–Mindlin plate theory will be formulated (Chapter 8). Nevertheless, the CPT will also be briefed in this chapter for completeness of this introduction to mechanics for solids and structures.

The CPT assumes that normals to the middle (neutral) plane of the undeformed plate remain straight and orthogonal to the middle plane during deformation or bending. This assumption results in:

$$\gamma_{xz} = 0, \quad \gamma_{yz} = 0 \tag{2.60}$$

Secondly, assume that the middle plan of the plate is not stretched or compressed in the x-y plane. Using the last two equations in Eq. (2.4), the displacements parallel to the—middle plane, u and v, at a distance z from the middle plane, can be expressed by:

$$u = -z\frac{\partial w}{\partial x} \tag{2.61}$$

$$v = -z\frac{\partial w}{\partial y} \tag{2.62}$$

where w is the deflection of the middle plane of the plate in the z direction. The relationship between the components of strain and the deflection can be given by:

$$\varepsilon_{xx} = \frac{\partial u}{\partial x} = -z\frac{\partial^2 w}{\partial x^2} \tag{2.63}$$

$$\varepsilon_{yy} = \frac{\partial v}{\partial y} = -z\frac{\partial^2 w}{\partial y^2} \tag{2.64}$$

$$\gamma_{xy} = \frac{\partial u}{\partial y} + \frac{\partial v}{\partial x} = -2z\frac{\partial^2 w}{\partial x \partial y} \tag{2.65}$$

or in the matrix form

$$\varepsilon = -z\mathbf{L}w \tag{2.66}$$

where ε is the vector of in-plane strains defined by Eq. (2.26), and $\mathbf{L}$ is the differential operator matrix given, in this case, by:

$$\mathbf{L} = \begin{bmatrix} \partial^2/\partial x^2 \\ \partial^2/\partial y^2 \\ 2\partial^2/\partial x\partial y \end{bmatrix} \tag{2.67}$$

2.6.2 Constitutive equations

The original Hooke's law is applicable for plates:

$$\sigma = \mathbf{c}\varepsilon \tag{2.68}$$

where $\mathbf{c}$ has the same form for 2D solids defined by Eq. (2.31) for the plane stress case, since σ_{zz} is assumed to be zero.

2.6.3 Moments and shear forces

Figure 2.15 shows a small representative cell of $dx \times dy$ from a plate of thickness h. The plate cell is subjected to external force f_z (force per unit area) and inertial force $\rho h \ddot{w}$, where ρ is the density of the material. Figure 2.16 shows the moments M_x, M_y, M_z, and M_{xy}, and shear forces Q_x and Q_y present. The moments and shear forces result from the distributed normal and shear stresses σ_{xx}, σ_{yy}, and σ_{xy}, shown in Figure 2.15. The stresses can be obtained by substituting Eq. (2.66) into Eq. (2.68):

$$\sigma = -z\mathbf{c}\mathbf{L}w \tag{2.69}$$

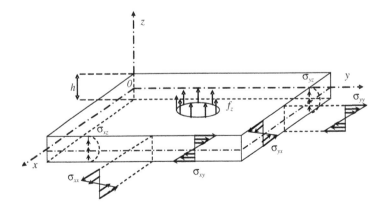

FIGURE 2.15

Stresses on an isolated plate cell. Integration of these stresses results in corresponding moments and shear forces.

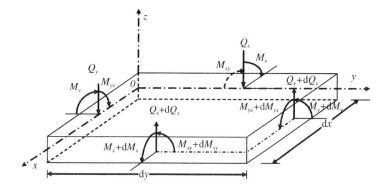

FIGURE 2.16

Shear forces and moments on an isolated plate cell of dx × dy. The equilibrium system equations are established based on this state of forces and moments.

It can be seen from the above equation that the normal stresses vary linearly with respect to z in the thickness direction of the plate. The moments on the cross-section can be calculated in a similar way as for beams via the following integration:

$$\mathbf{M}_p = \begin{Bmatrix} M_x \\ M_y \\ M_{xy} \end{Bmatrix} = \int_A \boldsymbol{\sigma} z \, dA = -\mathbf{c} \left(\int_A z^2 dA \right) \mathbf{L} w = -\frac{h^3}{12} \mathbf{c} \mathbf{L} w \tag{2.70}$$

Consider first the equilibrium of the small plate cell in the z direction, and note that $dQ_x = (\partial Q_x / \partial x)dx$ and $dQ_y = (\partial Q_y / \partial y)dy$, we have:

$$\left(\frac{\partial Q_x}{\partial x} dx \right) dy + \left(\frac{\partial Q_y}{\partial y} dy \right) dx + (f_z - \rho h \ddot{w}) dx \, dy = 0 \tag{2.71}$$

or

$$\frac{\partial Q_x}{\partial x} + \frac{\partial Q_y}{\partial y} + f_z = \rho h \ddot{w} \tag{2.72}$$

Consider then the moment equilibrium of the plate cell with respect to the y axis, and neglecting the small second order term, this leads to a formula for shear force Q_x:

$$Q_x = \frac{\partial M_x}{\partial x} + \frac{\partial M_{xy}}{\partial y} \tag{2.73}$$

Finally, consider the moment equilibrium of the plate cell with respect to the *x axis*, and neglecting the small second order term, this gives:

$$Q_y = \frac{\partial M_{xy}}{\partial x} + \frac{\partial M_y}{\partial y} \tag{2.74}$$

in which we implied that $M_{yx} = M_{xy}$.

2.6.4 Dynamic equilibrium equations

To obtain the dynamic equilibrium equation for plates, we first substitute Eq. (2.70) into Eqs. (2.73) and (2.74), after which Q_x and Q_y are substituted into Eq. (2.72):

$$D\left(\frac{\partial^4 w}{\partial x^4} + 2\frac{\partial^4 w}{\partial x^2 \partial y^2} + \frac{\partial^4 w}{\partial y^4}\right) + \rho h\ddot{w} = f_z \tag{2.75}$$

where $D = Eh^3/(12(1 - v^2))$ is the bending stiffness of the plate. The static equilibrium equation for plates can again be obtained by dropping the dynamic term in Eq. (2.75):

$$D\left(\frac{\partial^4 w}{\partial x^4} + 2\frac{\partial^4 w}{\partial x^2 \partial y^2} + \frac{\partial^4 w}{\partial y^4}\right) = f_z \tag{2.76}$$

Similarly to the beam case, the displacement boundary conditions for beams will be given in terms of deflection v and rotation θ, respectively, complementary to the force boundary conditions in terms of shear force Q and moments M (see Timoshenko, 1940).

2.6.5 Reissner–Mindlin plate

The Reissner–Mindlin plate theory (Reissner, 1945; Mindlin, 1951) is applied for thick plates, where the shear deformation and rotary inertia effects are included. The Reissner–Mindlin theory does not require the cross-section to be perpendicular to the axial axes after deformation, as shown in Figure 2.17. Therefore, $\gamma_{xz} \neq 0$ and $\gamma_{yz} \neq 0$. The displacements parallel to the undeformed middle surface, u and v, at a distance z from the middle plane can be expressed as:

$$u = z\theta_y \tag{2.77}$$

$$v = -z\theta_x \tag{2.78}$$

where θ_x and θ_y are, respectively, the rotations about the x and y axes of lines normal to the middle plane before deformation.

The in-plane strains defined by Eq. (2.26) are given by:

$$\varepsilon = -z\mathbf{L}\theta \tag{2.79}$$

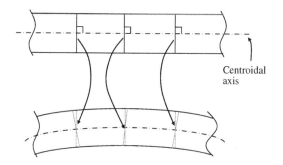

Centroidal
axis

FIGURE 2.17

Shear deformation in a Mindlin plate. The rotations of the cross-sections are treated as independent variables.

where **L,** in this case, is given by:

$$\mathbf{L} = \begin{bmatrix} 0 & -\partial/\partial x \\ \partial/\partial y & 0 \\ \partial/\partial x & -\partial/\partial y \end{bmatrix} \tag{2.80}$$

and

$$\theta = \left\{ \begin{matrix} \theta_x \\ \theta_y \end{matrix} \right\} \tag{2.81}$$

Using Eq. (2.4), the transverse shear strains γ_{xz} and γ_{yz} can be obtained as:

$$\gamma = \left\{ \begin{matrix} \gamma_{xz} \\ \gamma_{yz} \end{matrix} \right\} = \left\{ \begin{matrix} \theta_y + \partial w/\partial x \\ -\theta_x + \partial w/\partial y \end{matrix} \right\} \tag{2.82}$$

Note that if the transverse shear strains are negligible, the above equation will lead to:

$$\theta_x = \frac{\partial w}{\partial y} \tag{2.83}$$

$$\theta_y = -\frac{\partial w}{\partial x} \tag{2.84}$$

and Eq. (2.79) becomes Eq. (2.66) of the CPT. The transverse average shear stress τ relates to the transverse shear strain in the form:

$$\tau = \left\{ \begin{matrix} \sigma_{xz} \\ \sigma_{yz} \end{matrix} \right\} = \kappa \begin{bmatrix} G & 0 \\ 0 & G \end{bmatrix} \gamma = \kappa[\mathbf{D}_s]\gamma \tag{2.85}$$

where G is the shear modulus, and κ is a constant of shear correction factor to account for the assumption that the shear strain is constant across the thickness of the plate. κ is usually taken to be $\pi^2/12$ or 5/6.

The equilibrium equations for a Reissner–Mindlin plate can also be similarly obtained as that of a thin plate. Equilibrium of forces and moments can be carried out but this time taking into account the transverse shear stress and rotary inertia. For the purpose of this book, the above concepts for the Reissner–Mindlin plate will be sufficient and the equilibrium equations will not be shown here. Chapter 8 will detail the derivation of the discrete finite element equations by using energy principles.

2.7 Remarks

Having shown how the equilibrium equations for various types of geometrical structures are obtained, it is noted that all the equilibrium equations are just special cases of the general equilibrium equation for 3D solids. The use of proper assumptions and theories can lead to a dimension reduction, and hence simplify the problem. These kinds of simplification can significantly reduce the size of finite element models.

2.8 Review questions

1. Consider the problem of a 1D bar of uniform cross-section, as shown in Figure 2.18. The bar is fixed at the left end and is of length $l=1$ m and section area $A=0.0001$ m^2. It is subjected to a uniform body force f_x and a concentrated force F_s at the right end. The Young's modulus of the material is $E = 2.0 \times 10^{10}$ N/m^2. Using the analytical (exact) method, obtain solutions in terms of the distribution and the maximum value of the displacement, strain, and stress, for the following cases:
   a. $f_x=0$ and $F_s=1000$ N.
   b. $f_x=1000$ N/m and $F_s=1000$ N.
   c. $f_x=(100x+1000)$ N/m and $F_s=0$.
2. Consider a cantilever beam of uniform cross-section, as shown in Figure 2.19. The beam is clamped at the left end and is of length $l=1$ m and with a square section area of $A=0.001$ m^2. It is subjected to a uniform body force f_b and a concentrated force F_s at the right end. The Young's modulus of the material is $E = 200.0 \times 10^9$ N/m^2. Using the analytical (exact) method, obtain

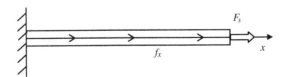

FIGURE 2.18

A bar with uniform cross-section area fixed at the left end.

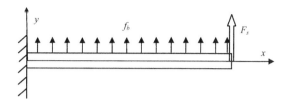

FIGURE 2.19

A cantilever beam with uniform cross-section area.

solutions in terms of the distribution and the maximum value of the deflection, moment, shear force, and normal stresses, for the following cases:

a. $f_b = 0$ and $F_s = 1000\,\text{N}$.

b. $f_b = 1000\,\text{N/m}$ and $F_s = 1000\,\text{N}$.

Fundamentals for Finite Element Method

3

3.1 Introduction

As mentioned in Chapter 1, when using the finite element method (FEM) to solve mechanics problems governed by a set of partial differential equations, the problem domain is first discretized (in a proper manner) into a set of small elements. In each of these elements, the variation/profile/pattern of the displacements is assumed in simple forms to obtain element equations. The equations obtained for each element are then *assembled* together with adjoining elements to form the global finite element equation for the whole problem domain. Equations thus created for the global problem domain can be solved easily for the entire displacement field.

The above-mentioned FEM process does not seem to be a difficult task. However, upon close examination of the process, one would naturally ask a series of questions. How can one simply assume the profile of the solution of the displacement in any simple form? How can one ensure that the governing partial differential equations will be satisfied by the assumed displacement? What should one do when using the assumed profile of the displacement to determine the final displacement field? Yes, one can just simply assume the profile of the solution of the displacements, but a principle has to be followed in order to obtain discretized system equations that can be solved routinely for the final displacement field. The use of such a principle guarantees the *best* (may not be exact) satisfaction of the governing system equation under certain conditions. The following details one of the most important principles, which will be employed to establish the FEM equations for mechanics problems of solids and structures.

Perhaps, for a beginner, full understanding of the details of the equations in this chapter may prove to be a challenging task. It is thus advised that the novice reader just understands the basic concepts involved without digging too much into the equations. It is then recommended to review this chapter again, after going through subsequent chapters, with examples to fully understand the equations. Advanced readers are referred to the FEM handbook (Kardestuncer, 1987) for more complete topics related to FEM.

3.2 Strong and weak forms: problem formulation

The partial differential system equations developed in Chapter 2, such as Eqs. (2.20) and (2.21) with boundary conditions, Eqs. (2.22) and (2.23), are *strong forms* of the governing system of equations for solid mechanics problems. The strong form, in contrast to a *weak form*, requires *strong* continuity on the dependent field variables (the displacements u, v, and w in this case). By "strong," we mean that the functions used to approximate the field variables have to be differentiable up to the order of the partial differential equations (the strong form of the system equations). Obtaining the exact solution for a strong form of the system equation is usually very difficult for practical engineering problems. The finite difference method (FDM) can be used to solve system equations of the strong form to obtain an approximated solution. However, the method usually works well only for problems with simple and regular geometry and boundary conditions.

A weak form of the system equations is usually created using one of the following widely used methods:

- Energy principles (see, for example, Washizu, 1974; Reddy, 1984; Liu, 2009).
- Weighted residual methods (see, for example, Crandall, 1956; Finlayson and Scriven, 1966; Finlayson, 1972; Zienkiewicz and Taylor, 2000; Liu, 2009; Liu and Zhang, 2013).

The energy principle can be categorized as a special form of the variational principle which is particularly suited to problems of the mechanics of solids and structures. The weighted residual method is a more general mathematical tool applicable, in principle, for solving all kinds of partial differential equations. Both methods are very easy to understand and apply. This book will demonstrate both methods for creating FEM equations. An energy principle will be used for mechanics problems of solids and structures, and the weighted residual method will be used for formulating the heat transfer problems. It is also equally applicable to use the energy principle for heat transfer problems, and the weighted residual method for solid mechanics problems, and the procedure is very much the same. A comprehensive discussion on the differences and terminologies on energy methods, variation methods, functional analysis approaches, weak form methods, and weakened weak form methods can be found in the mesh-free book by Liu (2009), and the recent book on the smoothed point interpolation method (S-PIM) by Liu and Zhang (2013).

The weak form is often an integral form and requires a weaker continuity on the field variables. Due to the weaker requirement on the field variables, and the integral operation, a formulation based on a weak form usually produces a set of discretized system equations that give much more stable (and hence reliable) and accurate results, especially for problems of complex geometry. Hence, the weak form is preferred by many for obtaining an approximate solution. The FEM is a typical example of successfully using weak form formulations. Using the weak form usually leads to a set of well-behaved algebraic system equations, if the problem domain is discretized properly into elements. As the problem domain can be discretized into different types of elements, the FEM can be applied for many practical engineering problems with most kinds of complex geometry and boundary conditions.

In the following section, Hamilton's principle, which is one of the most powerful energy principles, is introduced for FEM formulation of problems of mechanics of solids and structures. Hamilton's principle is chosen because it is simple and can be used for dynamic problems. The approach adopted in this book is to directly work out the dynamic system equations, after which the static system equations can be easily obtained by simply dropping out the dynamic terms. This can be done because of the simple fact that the dynamic system equations are the general system equations, and the static case can be considered to be just a special case of the dynamic equations.

3.3 Hamilton's principle: A weak formulation
3.3.1 Hamilton's principle

Hamilton's principle is a simple yet powerful tool that can be used to derive discretized dynamic system equations. It states simply that,

> Of all the **admissible** time histories of displacement the most accurate solution makes the Lagrangian functional stationary.

An *admissible* displacement must satisfy the following conditions:

a. Compatibility equations
b. Essential or the kinematic boundary conditions
c. Conditions at initial (t_1) and final time (t_2).

Condition (a) ensures that the displacements are compatible (continuous) in the problem domain. As will be seen in Chapter 11, there are situations when incompatibility can occur at the edges between elements. Condition (b) ensures that the displacement constraints are satisfied; and condition (c) requires the displacement history to satisfy the constraints at the initial and final times.

Mathematically, Hamilton's principle states:

$$\delta \int_{t_1}^{t_2} L dt = 0 \tag{3.1}$$

The Langrangian functional, L, is obtained using a set of admissible time histories of displacements, and it consists of

$$L = \Pi - T - W_f \tag{3.2}$$

where T is the kinetic energy, Π is the potential energy (for our purposes, it is the elastic strain energy), and W_f is the work done by the external forces. It is clear that the Langrangian functional L is some kind of potential energy of the entire system, and thus the kinetic energy and the work done by the external forces results in deduction in paternal energy.

The kinetic energy of the entire problem domain is defined in the integral form

$$T = \frac{1}{2} \int_V \rho \dot{\mathbf{U}}^T \dot{\mathbf{U}} dV \tag{3.3}$$

where V represents the whole volume of the solid, and $\mathbf{U}$ is the set of admissible time histories of displacements.

The strain energy in the entire domain of elastic solids and structures can be expressed as

$$\Pi = \frac{1}{2} \int_V \boldsymbol{\varepsilon}^T \boldsymbol{\sigma} dV = \frac{1}{2} \int_V \boldsymbol{\varepsilon}^T \mathbf{c} \boldsymbol{\varepsilon} dV \tag{3.4}$$

where $\boldsymbol{\varepsilon}$ are the strains obtained using the set of admissible time histories of displacements. The work done by the external forces over the set of admissible time histories of displacements can be obtained by

$$W_f = \int_V \mathbf{U}^T \mathbf{f}_b dV + \int_{S_f} \mathbf{U}^T \mathbf{f}_s dS_f \tag{3.5}$$

where S_f represents the surface of the solid on which the surface forces are prescribed (see Figure 2.2).

Hamilton's principle allows one to simply assume any set of displacements, as long as it satisfies the three admissible conditions. The assumed set of displacements will not usually satisfy the strong form of governing system equations unless we are extremely lucky, or the problem is extremely simple and we know the exact solution (in this case we do not need to solve the problem). Application of Hamilton's principle will conveniently guarantee a combination of this assumed set of displacements to produce the most accurate (in the energy norm measure) solution for the system that is governed by the strong form of the system equations.

The power of Hamilton's principle (or any other variational principle) is that it provides freedom of choice, opportunity, and possibility. For practical engineering problems, one usually does not have to pursue the exact solution, which in most cases is usually unobtainable, because we now have a choice of quite conveniently obtaining a good approximation using Hamilton's principle, by assuming the likely form, pattern or *shape* of the solutions. Hamilton's principle thus provides the foundation for the FEMs. Furthermore, the simplicity of Hamilton's principle (or any other energy principle) manifests itself in the use of scalar energy quantities. Engineers and scientists like working with scalar quantities when it comes to numerical calculations, as they do not need to worry about the direction. All the mathematical tools required to derive final discrete system equations are basic operations of integration, differentiation, and variation, all of which are standard linear operations. Another plus point of Hamilton's principle is that the final discrete system equations produced are usually a set of linear algebraic equations that can be solved using conventional methods and standard computational routines. The following demonstrates how the finite element equations can be established using Hamilton's principle and its simple operations.

3.3.2 Minimum total potential energy principle

For static problems, Hamilton's principle reduces to the well-known minimum total potential energy principle, which may be stated as,

> *Of all the **admissible** displacements the most accurate solution makes the total potential energy minimum.*

An admissible displacement must satisfy the following conditions:

a. Compatibility equations
b. Essential or the kinematic boundary conditions.

Mathematically, the "most" accurate solution is found using

$$\delta \Pi_t = 0 \tag{3.1a}$$

The total potential energy, Π_t, is obtained using the assume admissible displacement, and it consists of

$$\Pi_t = \Pi - W_f \tag{3.2a}$$

It can be proven for stable materials of a well-posed solid mechanics problem that the stationary point given in (3.1a) is a minimum point of the total potential energy for the solid (see, e.g., Liu, 2009).

3.4 FEM procedure

The standard FEM procedure can be briefly summarized in the following sections.

3.4.1 Domain discretization

The solid or structure is divided into N_e elements with a set of N_n nodes. The procedure is often called *meshing*, which is usually performed using so-called pre-processors. This is especially true for complex geometries. Figure 3.1 shows an example of a mesh for a two-dimensional solid.

The pre-processor generates unique numbers for all the elements and nodes for the solid or structure in a proper manner. An element is formed by connecting its nodes in a pre-defined consistent fashion to

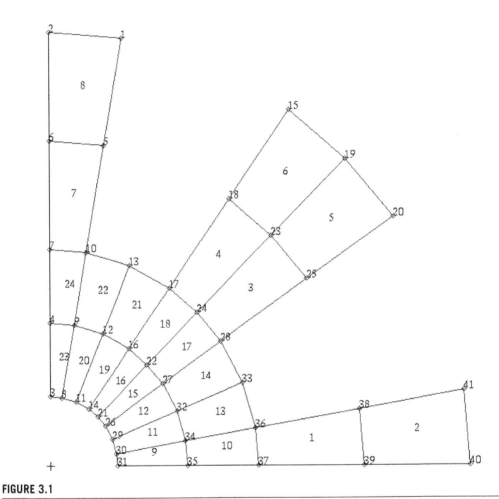

FIGURE 3.1

Example of a mesh with elements and node properly numbered.

create the *connectivity* of the element. All the elements together form the entire domain of the problem without any gap or overlapping. It is possible to have different types of elements with different numbers of nodes, as long as they are *compatible* (no gaps and overlapping; the admissible condition (a) required by Hamilton's principle) on the boundaries between different elements. In fact, meshes of mixed triangular and quadrilateral elements are frequently used to overcome the difficulties of creating preferable pure quadrilateral elements. The density of the mesh depends upon the accuracy requirement of the analysis and the computational resources available. Generally, a finer mesh will yield results that are more accurate, but will increase the computational cost. As such, the mesh is usually not uniform, with a finer mesh being used in the areas where the displacement gradient is larger or where the accuracy is critical to the analysis. The purpose of the domain discretization is to make it easier in assuming the pattern of the displacement field.

3.4.2 Displacement interpolation

The FEM formulation has to be based on a coordinate system. In formulating FEM equations for elements, it is often convenient to use a *local coordinate system* that is defined for an element in reference to the global coordination system that is usually defined for the entire structure, as shown in Figure 3.4. Based on the *local coordinate system* attached on an element, the displacement within the element is now assumed simply by polynomial interpolation using the displacements at its nodes (or *nodal displacements*) as

$$\mathbf{U}^h(x, y, z) = \sum_{i=1}^{n_d} \mathbf{N}_i(x, y, z)\mathbf{d}_i = \mathbf{N}(x, y, z)\mathbf{d}_e \qquad (3.6)$$

where the superscript h stands for approximation, n_d is the number of nodes forming the element, and $\mathbf{d}_i$ is the nodal displacement at the ith node, which is the unknown the analyst wants to find out, and can be expressed in a general form of

$$\mathbf{d}_i = \begin{Bmatrix} d_1 \\ d_2 \\ \vdots \\ d_{n_f} \end{Bmatrix} \begin{matrix} \rightarrow \text{ displacement component 1} \\ \rightarrow \text{ displacement component 2} \\ \vdots \\ \rightarrow \text{ displacement component } n_f \end{matrix} \qquad (3.7)$$

where n_f is the number of Degrees Of Freedom (DOF) at a node. For 3D solids, $n_f = 3$, and

$$\mathbf{d}_i = \begin{Bmatrix} u_i \\ v_i \\ w_i \end{Bmatrix} \begin{matrix} \rightarrow \text{ displacement in the } x\text{-direction} \\ \rightarrow \text{ displacement in the } y\text{-direction} \\ \rightarrow \text{ displacement in the } x\text{-direction} \end{matrix} \qquad (3.8)$$

Note that the displacement components can also consist of rotations for structures of beams and plates. The vector $\mathbf{d}_e$ in Eq. (3.6) is the displacement vector for the entire element, and has the form of

$$\mathbf{d}_e = \begin{Bmatrix} \mathbf{d}_1 \\ \mathbf{d}_2 \\ \vdots \\ \mathbf{d}_{n_d} \end{Bmatrix} \begin{matrix} \rightarrow \text{ displacements at node 1} \\ \rightarrow \text{ displacements at node 2} \\ \vdots \\ \rightarrow \text{ displacements at node } n_d \end{matrix} \qquad (3.9)$$

Therefore, the total DOF for the entire element is $n_d \times n_f$.

In Eq. (3.6), **N** is a matrix of *shape functions* for the nodes in the element, which are pre-defined to assume the shapes of the displacement variations with respect to the coordinates. It has the general form of

$$\mathbf{N}(x, y, z) = [\mathbf{N}_1(x, y, z) \ \mathbf{N}_2(x, y, z) \ \cdots \ \mathbf{N}_{nd}(x, y, z)]$$

$$\downarrow \qquad \qquad \downarrow \qquad \cdots \qquad \downarrow \qquad \qquad \text{(3.10)}$$

$$\text{for node 1} \quad \text{for node 2} \ \cdots \ \text{for node } n_d$$

where $\mathbf{N}_i$ is a sub-matrix of shape functions for displacement components, which is arranged as

$$\mathbf{N}_i = \begin{bmatrix} N_{i1} & 0 & 0 & 0 \\ 0 & N_{i2} & 0 & 0 \\ 0 & 0 & \ddots & 0 \\ 0 & 0 & 0 & N_{in_f} \end{bmatrix} \qquad \text{(3.11)}$$

where N_{ik} is the shape function for the kth displacement component, DOF at the ith node. For 3D solids, $n_f = 3$, and often we use $N_{i1} = N_{i2} = N_{i3} = N_i$. Note that it is not always necessary to use the same shape function for all the displacement components at a node. For example, we often use different shape functions for translational and rotational displacements.

Note that this approach of assuming the displacements is often called the *displacement method*. Most of the commercially available FEM packages use this approach. There are, however, FEM approaches that assume the stresses instead, but they will not be covered in this book.

3.4.3 Standard procedure for constructing shape functions

Consider an element with n_d nodes at $\mathbf{x}_i$; $(i = 1, 2, ..., n_d)$, where $\mathbf{x}^T = \{x\}$ for one-dimensional problems, $\mathbf{x}^T = \{x \ y\}$ for two-dimensional problems, and $\mathbf{x}^T = \{x \ y \ z\}$ for three-dimensional problems. We should have n_d shape functions for each displacement component for an element. In the following, we consider only one displacement component in the explanation of the standard procedure for constructing the shape functions. The standard procedure is applicable for any other displacement components. First, the displacement component is approximated in the form of a linear combination of n_d linearly-independent basis functions $p_i(\mathbf{x})$, i.e.

$$u^h(\mathbf{x}) = \sum_{i=1}^{n_d} p_i(\mathbf{x})\alpha_i = \mathbf{p}^T(\mathbf{x})\boldsymbol{\alpha} \qquad \text{(3.12)}$$

where u^h is the approximation of the displacement component, $p_i(\mathbf{x})$ is the basis function of monomials in the space coordinates $\mathbf{x}$, and α_i is the coefficient for the monomial $p_i(\mathbf{x})$. Vector $\boldsymbol{\alpha}$ is defined as

$$\boldsymbol{\alpha}^T = \{\alpha_1 \quad \alpha_2 \quad \alpha_3 \ ... \ \alpha_{n_d}\} \qquad \text{(3.13)}$$

The $p_i(\mathbf{x})$ in Eq. (3.12) is built with n_d terms of one-dimensional monomials; based on the Pascal's triangle shown in Figure 3.2 for two-dimensional problems; or the well-known Pascal's pyramid shown in Figure 3.3 for three-dimensional problems. A basis of complete order of p in the one-dimensional domain has the form

$$\mathbf{p}^T(x) = \{1 \quad x \quad x^2 \quad x^3 \quad x^4 \dots x^p\} \tag{3.14}$$

A basis of complete order of p in the two-dimensional domain is provided by

$$\mathbf{p}^T(\mathbf{x}) = \mathbf{p}^T(x,y) = \{1 \quad x \quad y \quad xy \quad x^2 \quad y^2 \dots x^p \quad y^p\} \tag{3.15}$$

and that in the three-dimensional domain can be written as

$$\begin{aligned}
\mathbf{p}^T(\mathbf{x}) &= \mathbf{p}^T(x,y,z) \\
&= \{1 \quad x \quad y \quad z \quad xy \quad yz \quad zx \quad x^2 \quad y^2 \quad z^2 \dots x^p \quad y^p \quad z^p\}
\end{aligned} \tag{3.16}$$

As a general rule, the n_d terms of $p_i(\mathbf{x})$ used in the basis should be selected from the constant term to higher orders symmetrically from the Pascal triangle shown in Figure 3.2 or Figure 3.3. Some higher order terms can be selectively included in the polynomial basis if there is a need in specific circumstances.

The coefficients α_i in Eq. (3.12) can be determined by enforcing the displacements calculated using Eq. (3.12) to be equal to the nodal displacements at the n_d nodes of the element. At node i we can have

$$d_i = \mathbf{p}^T(\mathbf{x}_i)\alpha \quad i = 1, 2, 3, \dots, n_d \tag{3.17}$$

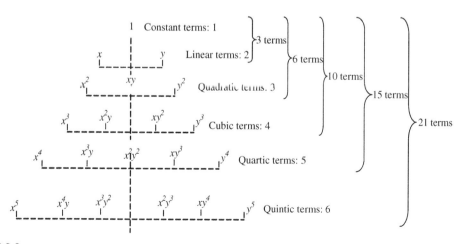

FIGURE 3.2

Pascal triangle of monomials. Two-dimensional case.

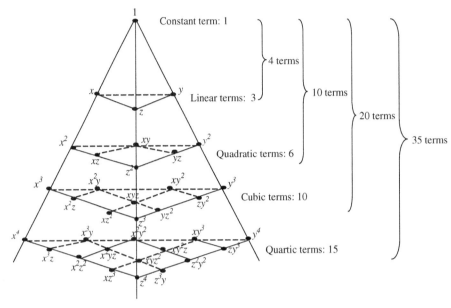

FIGURE 3.3

Pascal pyramid of monomials. Three-dimensional case.

where d_i is the nodal value of u^h at $\mathbf{x} = \mathbf{x}_i$. Equation (3.17) can be written in the following matrix form:

$$\mathbf{d}_e = \mathbf{P}\boldsymbol{\alpha} \tag{3.18}$$

where $\mathbf{d}_e$ is the vector that includes the values of the displacement component at all the n_d nodes in the element:

$$\mathbf{d}_e = \begin{Bmatrix} d_1 \\ d_2 \\ \vdots \\ d_{n_d} \end{Bmatrix} \tag{3.19}$$

and **P** is given by

$$\mathbf{P} = \begin{bmatrix} \mathbf{p}^T(\mathbf{x}_1) \\ \mathbf{p}^T(\mathbf{x}_2) \\ \vdots \\ \mathbf{p}^T(\mathbf{x}_{n_d}) \end{bmatrix} \tag{3.20}$$

which is called the *moment matrix*. The expanded form of **P** is

$$
\mathbf{P} = \begin{bmatrix}
p_1(\mathbf{x}_1) & p_2(\mathbf{x}_1) & \cdots & p_{n_d}(\mathbf{x}_1) \\
p_1(\mathbf{x}_2) & p_2(\mathbf{x}_2) & \cdots & p_{n_d}(\mathbf{x}_2) \\
\vdots & \vdots & \ddots & \vdots \\
p_1(\mathbf{x}_{n_d}) & p_2(\mathbf{x}_{n_d}) & \cdots & p_{n_d}(\mathbf{x}_{n_d})
\end{bmatrix}
\tag{3.21}
$$

For two-dimensional polynomial basis functions, we may have

$$
\mathbf{P} = \begin{bmatrix}
1 & x_1 & y_1 & x_1 y_1 & x_1^2 & y_1^2 & x_1^2 y_1 & x_1 y_1^2 & x_1^3 & \cdots \\
1 & x_2 & y_2 & x_2 y_2 & x_2^2 & y_2^2 & x_2^2 y_2 & x_2 y_2^2 & x_2^3 & \cdots \\
\vdots & \vdots & \vdots & \vdots & \vdots & \vdots & \vdots & \vdots & \vdots & \vdots \\
1 & x_{n_d} & y_{n_d} & x_{n_d} y_{n_d} & x_{n_d}^2 & y_{n_d}^2 & x_{n_d}^2 y_{n_d} & x_{n_d} y_{n_d}^2 & x_{n_d}^3 & \cdots
\end{bmatrix}
\tag{3.22}
$$

Using Eq. (3.18), and assuming that the inverse of the moment matrix **P** exists, we can then have

$$
\boldsymbol{\alpha} = \mathbf{P}^{-1} \mathbf{d}_e
\tag{3.23}
$$

Substituting Eq. (3.23) into Eq. (3.12), we then obtain

$$
u^h(\mathbf{x}) = \sum_{i=1}^{n_d} N_i(\mathbf{x}) d_i
\tag{3.24}
$$

or in matrix form

$$
u^h(\mathbf{x}) = \mathbf{N}(\mathbf{x}) \mathbf{d}_e
\tag{3.25}
$$

where $\mathbf{N}(\mathbf{x})$ is a matrix of shape functions $N_i(\mathbf{x})$ defined by

$$
\mathbf{N}(\mathbf{x}) = \mathbf{p}^T(\mathbf{x}) \mathbf{P}^{-1} = \left[\underbrace{\mathbf{p}^T(\mathbf{x}) P_1^{-1}}_{N_1(\mathbf{x})} \quad \underbrace{\mathbf{p}^T(\mathbf{x}) P_2^{-1}}_{N_2(\mathbf{x})} \quad \cdots \quad \underbrace{\mathbf{p}^T(\mathbf{x}) P_n^{-1}}_{N_n(\mathbf{x})} \right]
\tag{3.26}
$$

$$
= [N_1(\mathbf{x}) \quad N_2(\mathbf{x}) \quad \cdots \quad N_n(\mathbf{x})]
$$

where $\mathbf{P}_i^{-1}$ is the ith column of matrix $\mathbf{P}^{-1}$, and

$$N_i(\mathbf{x}) = \mathbf{p}^T(\mathbf{x})\mathbf{P}_i^{-1} \tag{3.27}$$

The derivatives of the shape functions can be obtained very easily, as all the functions involved are polynomials. The lth derivative of the shape functions is simply given by

$$N_i^{(l)}(\mathbf{x}) = [\mathbf{p}^{(l)}(\mathbf{x})]^T \mathbf{P}_i^{-1} \tag{3.28}$$

3.4.3.1 *On the inverse of the moment matrix*

In obtaining Eq. (3.23), we have assumed the existence of the inverse of $\mathbf{P}$. There could be cases where $\mathbf{P}^{-1}$ does not exist, and the construction of shape functions will fail. The existence of $\mathbf{P}^{-1}$ depends upon (1) the basis function used, and (2) the nodal distribution of the element. The basis functions have to be chosen first from a linearly-independent set of bases, and then the inclusion of the basis terms should be based on the nodal distribution in the element. The discussion in this direction is more involved, and interested readers are referred to a monograph on mesh-free methods by Liu (2009). In this book, we shall only discuss elements whose corresponding moment matrices are invertible.

3.4.3.2 *On the compatibility of the shape functions*

The other possible failure of the general procedure of shape function construction is the compatibility of the shape functions created. This is often the case for higher order elements created on physical coordinates. This is the main reason why most of the higher order elements are created in the so-called natural coordinate system (see topics on isoparametric elements). The issues related to the compatibility of element shape functions will be addressed in greater detail in Chapter 11. Note that the compatibility conditions can be relaxed if the S-PIM (Liu and Zhang, 2013) is used. This is because the S-PIM is based on the so-called weakened weak form, and the stability and convergence are ensured by the G-space theory.

3.4.3.3 *On other means of construct shape functions*

There are many other methods for creating shape functions which do not necessarily follow the standard procedure described above. Some of these often used *shortcut* methods will be discussed in later chapters, when we develop different types of elements. These shortcut methods need to make use of the properties of shape functions detailed in the next section.

3.4.4 **Properties of the shape functions**

PROPERTY 1

Reproduction property and consistency
The consistency of the shape function within the element depends upon the complete orders of the monomial $p_i(\mathbf{x})$ used in Eq. (3.12), and hence is also dependent upon the number of nodes of the element. If the complete order of monomial is k, the shape function is said to possess C^k consistency. To demonstrate, we consider a field given by

$$f(\mathbf{x}) = \sum_{j}^{k} p_j(\mathbf{x})\beta_j, \quad ; k \le n_d \tag{3.29}$$

where $p_j(\mathbf{x})$ are monomials that are included in Eq. (3.12). Such a given field can always be written using Eq. (3.12) using all the basis terms, including those in Eq. (3.29):

$$f(\mathbf{x}) = \sum_{j}^{n_d} p_j(\mathbf{x})\beta_j = \mathbf{p}^T(\mathbf{x})\alpha \tag{3.30}$$

where

$$\alpha^T = [\beta_1 \quad \beta_2 \ \ldots \ \beta_k \quad 0 \ \ldots \ 0] \tag{3.31}$$

Using n nodes in the support domain of $\mathbf{x}$, we can obtain the vector of nodal function value $\mathbf{d}_e$ as:

$$\mathbf{d}_e = \begin{Bmatrix} f_1 \\ f_2 \\ \vdots \\ f_k \\ f_{k+1} \\ \vdots \\ f_n \end{Bmatrix} = \begin{bmatrix} p_1(\mathbf{x}_1) & p_2(\mathbf{x}_1) & \cdots & p_k(\mathbf{x}_1) & p_{k+1}(\mathbf{x}_1) & p_{n_d}(\mathbf{x}_1) \\ p_1(\mathbf{x}_2) & p_2(\mathbf{x}_2) & \cdots & p_k(\mathbf{x}_2) & p_{k+1}(\mathbf{x}_2) & p_{n_d}(\mathbf{x}_2) \\ \vdots & \vdots & \cdots & \vdots & \vdots & \vdots \\ p_1(\mathbf{x}_k) & p_2(\mathbf{x}_k) & \cdots & p_k(\mathbf{x}_k) & p_{k+1}(\mathbf{x}_k) & p_{n_d}(\mathbf{x}_k) \\ p_1(\mathbf{x}_{k+1}) & p_2(\mathbf{x}_{k+1}) & \cdots & p_k(\mathbf{x}_{k+1}) & p_{k+1}(\mathbf{x}_{k+1}) & p_{n_d}(\mathbf{x}_{k+1}) \\ \vdots & \vdots & \cdots & \vdots & \vdots & \vdots \\ p_1(\mathbf{x}_{n_d}) & p_2(\mathbf{x}_{n_d}) & \cdots & p_k(\mathbf{x}_{n_d}) & p_{k+1}(\mathbf{x}_{n_d}) & p_{n_d}(\mathbf{x}_{n_d}) \end{bmatrix} \begin{Bmatrix} \beta_1 \\ \beta_2 \\ \vdots \\ \beta_k \\ 0 \\ \vdots \\ 0 \end{Bmatrix} \tag{3.32}$$

$$= \mathbf{P}\alpha$$

Substituting Eq. (3.32) into Eq. (3.25), we have the approximation of

$$u^h(\mathbf{x}) = \mathbf{p}^T(\mathbf{x})\mathbf{P}^{-1}\mathbf{d}_e = \mathbf{p}^T(\mathbf{x})\mathbf{P}^{-1}\mathbf{P}\alpha = \mathbf{p}^T(\mathbf{x})\alpha = \sum_{j}^{k} p_j(\mathbf{x})\alpha_j = f(\mathbf{x}) \tag{3.33}$$

which is exactly what is given in Eq. (3.30). This proves that any field given by Eq. (3.29) will be exactly reproduced in the element by the approximation using the shape functions, as long as the given field function is included in the basis functions used for constructing the shape functions. This feature of the shape function is in fact also very easy to understand by intuition: Any function given in the form of $f(\mathbf{x}) = \sum_j^k p_j(\mathbf{x})\beta_j$ can be produced exactly by letting $\alpha_j = \beta_j$ ($j = 1, 2, ..., k$) and $\alpha_j = 0$ ($j = k+1, ..., n_d$). This can always be done as long as the moment matrix $\mathbf{P}$ is invertible so as to ensure the uniqueness of the solution for α.

The proof of the consistency of the shape function implies another important feature of the shape function: That is the *reproduction property,* which states that any function that appears in the basis can be reproduced exactly. This important property can be used for creating fields of special features. To ensure that the shape functions have linear consistency, all one needs to do is include the constant (unit) and linear monomials into the basis. We can make use of the feature of the shape function to compute accurate results for problems by including terms in the basis functions that are good approximations of the problem solution. The difference between consistency and reproduction is:

- Consistency depends upon the complete order of the basis functions
- Reproduction depends upon whatever is included in the basis functions.

PROPERTY 2

Linear independence

Shape functions are linearly-independent. This is because basis functions are of linear independence and $\mathbf{P}^{-1}$ is assumed to exist. The existence of $\mathbf{P}^{-1}$ implies that the shape functions are equivalent to the basis functions in the function space, as shown in Eq. (3.26). Because the basis functions are linearly-independent, the shape functions are hence also linearly-independent. Many FEM users do not pay much attention to this linear independence property; however, it is the foundation for the shape functions to have the delta function property stated below.

Since the elements are independently geometrical (it occupies its own domain), all the nodal shape functions for the entire FEM domain will be independent, which is essential to any FEM model.

PROPERTY 3

Delta function properties

$$N_i(\mathbf{x}_j) = \delta_{ij} = \begin{cases} 1 & i = j, \ j = 1, 2, \ldots, n_d \\ 0 & i \neq j, \ i,j = 1, 2, \ldots, n_d \end{cases} \qquad (3.34)$$

where δ_{ij} is the delta function. The delta function property implies that the shape function N_i of an element should be unit at its *home node* i, and vanishes at the *remote nodes* $j = i$ of the element.

The delta function property can be proven easily as follows: because the shape functions $N_i(\mathbf{x})$ are linearly-independent, any vector of length n_d should be uniquely produced by linear combination of these n_d shape functions. Assume that the displacement at node i is $d_i \neq 0$ and the displacements at other nodes are zero, i.e.

$$\mathbf{d}_e = \{0 \quad 0 \ldots d_i \ldots 0\}^T \qquad (3.35)$$

This vector must be uniquely produced by linear combination of these n_d shape functions. Substituting the above equation into Eq. (3.24), we have at $\mathbf{x} = \mathbf{x}_j$, that

$$u^h(\mathbf{x}_j) = \sum_{k=1}^{n_d} N_k(\mathbf{x}_j)d_k = N_i(\mathbf{x}_j)d_i \qquad (3.36)$$

and when $i = j$, we must have

$$u_i = d_i = N_i(\mathbf{x}_i)d_i \tag{3.37}$$

which (since $d_i \neq 0$) implies that

$$N_i(\mathbf{x}_i) = 1 \tag{3.38}$$

This proves the first row of Eq. (3.34). When $i \neq j$, we must have

$$u_j = 0 = N_i(\mathbf{x}_j)d_i \tag{3.39}$$

which (since again $d_i \neq 0$) requires

$$N_i(\mathbf{x}_j) = 0 \tag{3.40}$$

This proves the second row of Eq. (3.34). We can then conclude that the shape functions possess the delta function property, as depicted by Eq. (3.34). Note that there are elements, such as the thin beam and plate elements, whose shape functions may not possess the delta function property (see Section 5.2.1 for details).

PROPERTY 4

Partitions of unity property
Shape functions are partitions of unity:

$$\sum_{i=1}^{n_d} N_i(\mathbf{x}) = 1 \tag{3.41}$$

if the constant is included in the basis. This can be proven easily from the reproduction feature of the shape function. Let $u(\mathbf{x}) = c$, where c is an arbitrary non-zero constant; we should have

$$\mathbf{d}_e = \begin{Bmatrix} d_1 \\ d_2 \\ \vdots \\ d_{n_d} \end{Bmatrix} = \begin{Bmatrix} c \\ c \\ \vdots \\ c \end{Bmatrix} \tag{3.42}$$

which implies the same constant displacement for all the nodes. Substituting the above equation into Eq. (3.24), we obtain

$$u(\mathbf{x}) = c \underset{\text{reproduction}}{=} u^h(\mathbf{x}) \underset{\text{approximation}}{=} \sum_{i=1}^{n_d} N_i(\mathbf{x})d_i = \sum_{i=1}^{n_d} N_i(\mathbf{x})c = c \sum_{i=1}^{n_d} N_i(\mathbf{x}) \tag{3.43}$$

which gives Eq. (3.41). This shows that the partitions of unity of the shape functions in the element allow a constant field or rigid body movement to be reproduced. Note that Eq. (3.41) does not require $0 \leq N_i(\mathbf{x}) \leq 1$.

PROPERTY 5

Linear field reproduction
If the first order monomial is included in the basis, the shape functions constructed reproduce the linear field, i.e.

$$\sum_{i=1}^{n_d} N_i(x)x_i = x \tag{3.44}$$

where x_i is the coordinate at node i (and similarly for y and z). This can be proven easily from the reproduction feature of the shape function in exactly the same manner for proving Property 4. Let $u(x) = x$, we should have

$$\mathbf{d}_e = \{x_1 \quad x_2 \quad \cdots \quad x_{n_d}\}^T \tag{3.45}$$

Substituting the above equation into Eq. (3.24), we obtain

$$u^h(x) = x = \sum_{1}^{n_d} N_i(x)x_i \tag{3.46}$$

which is Eq. (3.44).

LEMMA 1

Condition for shape functions being partitions of unity
For a set of shape functions in the general form

$$N_i(\mathbf{x}) = c_{1i} + c_{2i}p_2(\mathbf{x}) + c_{3i}p_3(\mathbf{x}) + \cdots + c_{n_d i}p_{n_d}(\mathbf{x}) \tag{3.47}$$

where $p_i(\mathbf{x})$ $(p_1(\mathbf{x}) = 1, i = 2, n_d)$ is a set of independent base functions, the sufficient and necessary condition for this set of shape functions being partitions of unity is

$$\begin{aligned} C_1 &= 1 \\ C_2 &= C_3 = \cdots = C_{n_d} = 0 \end{aligned} \tag{3.48}$$

where

$$C_k = \sum_{i=1}^{n_d} c_{ki} \tag{3.49}$$

PROOF

Using Eq. (3.47), the summation of the shape functions is

$$\sum_{i=1}^{n_d} N_i(\mathbf{x}) = \sum_{i=1}^{n_d} c_{1i} + p_2(\mathbf{x}) \sum_{i=1}^{n_d} c_{2i} + p_3(\mathbf{x}) \sum_{i=1}^{n_d} c_{3i} + \cdots + p_{n_d}(\mathbf{x}) \sum_{i=1}^{n_d} c_{n_d i}$$
$$= \underbrace{C_1}_{1} + \underbrace{C_2}_{0} \, p_2(\mathbf{x}) + \underbrace{C_2}_{0} \, p_3(\mathbf{x}) + \cdots + \underbrace{C_{n_d}}_{0} \, p_{n_d}(\mathbf{x}) = 0 \tag{3.50}$$

which proves the sufficient condition. To prove the necessary condition, we argue that, to have the partitions of unity, we have

$$\sum_{i=1}^{n_d} N_i(\mathbf{x}) = C_1 + C_2 \, p_2(\mathbf{x}) + C_2 \, p_3(\mathbf{x}) + \cdots + C_{n_d} p_{n_d}(\mathbf{x}) = 1 \tag{3.51}$$

or

$$(C_1 - 1) + C_2 \, p_2(\mathbf{x}) + C_2 \, p_3(\mathbf{x}) + \cdots + C_{n_d} p_{n_d}(\mathbf{x}) = 0 \tag{3.52}$$

Because $p_i(\mathbf{x})$ ($p_1(\mathbf{x}) = 1$, $i = 2$, n_d) is a set of independent base functions, the necessary condition for Eq. (3.52) to be satisfied is Eq. (3.48). $\square$

LEMMA 2

Condition for shape functions being partitions of unity
Any set of n_d shape functions will automatically satisfy the partitions of unity property if it satisfies:

- *Condition* 1: It is given in a linear combination of the same linearly-independent set of n_d basis functions that contain the constant basis, and the moment matrix defined by Eq. (3.21) is of full rank.
- *Condition* 2: It possesses the delta function property.

PROOF

From Eq. (3.26), we can see that all the n_d shape functions are formed via a combination of the same basis function $p_i(\mathbf{x})$ ($i = 1, 2, \ldots, n_d$). This feature, together with the delta function property, can ensure the property of partitions of unity. To prove this, we write a set of shape functions in the general form

$$N_i(\mathbf{x}) = c_{1i} + c_{2i} p_2(\mathbf{x}) + c_{3i} p_3(\mathbf{x}) + \cdots + c_{n_d i} p_{n_d}(\mathbf{x}) \tag{3.53}$$

where we ensured the inclusion of the constant basis of $p_1(\mathbf{x}) = 1$. The other basis function $p_i(\mathbf{x})$ ($i = 2, \ldots, n_d$) in Eq. (3.53) can be monomials or any other type of basis functions as long as all the basis functions (including $p_1(\mathbf{x})$) are linearly-independent.

From the Condition 2, the shape functions possess a delta function property that leads to

$$\sum_{i=1}^{n_d} N_i(\mathbf{x}_j) = 1 \quad \text{for } j = 1, 2, \ldots, n_d \tag{3.54}$$

Substituting Eq. (3.53) into the previous equations, we have

$$\sum_{i=1}^{n_d} c_{1i} + p_2(\mathbf{x}_j)\sum_{i=1}^{n_d} c_{2i} + p_3(\mathbf{x}_j)\sum_{i=1}^{n_d} c_{3i} + \cdots + p_{n_d}(\mathbf{x}_j)\sum_{i=1}^{n_d} c_{n_d i} = 1 \tag{3.55}$$

$$\text{for} \quad j = 1, 2, \ldots, n_d$$

or

$$C_1 + p_2(\mathbf{x}_j)C_2 + p_3(\mathbf{x}_j)C_3 + \cdots + p_{n_d}(\mathbf{x}_j)C_{n_d} = 1 \quad \text{for } j = 1, 2, \ldots, n_d \tag{3.56}$$

Expanding Eq. (3.56) gives

$$C_1 + p_2(\mathbf{x}_1)C_2 + p_3(\mathbf{x}_1)C_3 + \cdots + p_{n_d}(\mathbf{x}_1)C_{n_d} = 1$$
$$C_1 + p_2(\mathbf{x}_2)C_2 + p_3(\mathbf{x}_2)C_3 + \cdots + p_{n_d}(\mathbf{x}_2)C_{n_d} = 1$$
$$\vdots \tag{3.57}$$
$$C_1 + p_2(\mathbf{x}_{n_d})C_2 + p_3(\mathbf{x}_{n_d})C_3 + \cdots + p_{n_d}(\mathbf{x}_{n_d})C_{n_d} = 1$$

or in the matrix form

$$\begin{bmatrix} 1 & p_2(\mathbf{x}_2) & p_3(\mathbf{x}_2) & \cdots & p_{n_d}(\mathbf{x}_2) \\ 1 & p_2(\mathbf{x}_3) & p_3(\mathbf{x}_3) & \cdots & p_{n_d}(\mathbf{x}_3) \\ \vdots & \vdots & \vdots & \ddots & \vdots \\ 1 & p_2(\mathbf{x}_{n_d}) & p_3(\mathbf{x}_{n_d}) & \cdots & p_{n_d}(\mathbf{x}_{n_d}) \end{bmatrix} \begin{Bmatrix} C_1 - 1 \\ C_2 \\ C_3 \\ \vdots \\ C_{n_d} \end{Bmatrix} = 0 \tag{3.58}$$

Note that the coefficient matrix of Eq. (3.58) is the moment matrix that has a full rank (Condition 1); we then have

$$C_1 = 1 \tag{3.59}$$
$$C_2 = C_3 = \cdots = C_{n_d} = 0$$

The use of Lemma 1 proves the partitions of the unity property of shape functions. $\square$

LEMMA 3

Condition for shape functions being linear field reproduction

Any set of n_d shape functions will automatically satisfy the linear reproduction property, if it satisfies

- *Condition* 1: It is given in a linear combination of the same linearly-independent set of n_d basis functions that contain the linear basis function, and the moment matrix defined by Eq. (3.21) is of full rank.
- *Condition* 2: It possesses the delta function property.

To prove this, we write a set of shape functions in the following general form of

$$N_i(\mathbf{x}) = c_{1i}p_1(\mathbf{x}) + c_{2i}x + c_{3i}p_3(\mathbf{x}) + \cdots + c_{ni}p_{n_d}(\mathbf{x}) \tag{3.60}$$

where we ensure inclusion of the complete linear basis functions of $p_2(\mathbf{x}) = x$. The other basis function $p_i(\mathbf{x})$ $(i = 1, 3, \ldots, n_d)$ in Eq. (3.53) can be monomials or any other type of basis function as long as all the basis functions are linearly-independent.

Consider a linear field of $u(\mathbf{x}) = x$, we should have the nodal vector as follows:

$$\mathbf{d}_e = \{x_1 \; x_2 \; \ldots \; x_{n_d}\}^T \tag{3.61}$$

Substituting the above equation into Eq. (3.24), we obtain

$$\begin{aligned}
u^h(\mathbf{x}) &= \sum_{i=1}^{n_d} N_i(\mathbf{x})x_i \\
&= \sum_{i=1}^{n_d} [c_{1i}p_1(\mathbf{x}) + c_{2i}x + c_{3i}p_3(\mathbf{x}) + \cdots + c_{n_d i}p_{n_d}(\mathbf{x})]x_i \\
&= \sum_{i=1}^{n_d} c_{1i}p_1(\mathbf{x})x_i + \sum_{i=1}^{n_d} c_{2i}xx_i + \sum_{i=1}^{n_d} c_{3i}p_3(\mathbf{x})x_i + \cdots + \sum_{i=1}^{n_d} c_{n_d i}p_{n_d}(\mathbf{x})x_i \qquad (3.62) \\
&= \sum_{i-1}^{n_d} c_{1i}p_1(\mathbf{x})x_i + x\sum_{i=1}^{n_d} c_{2i}x_i + p_3(\mathbf{x})\sum_{l=1}^{n_d} c_{3i}x_i + \cdots + p_{n_d}(\mathbf{x})\sum_{l=1}^{n_d} c_{n_d i}x_i \\
&= p_1(\mathbf{x})C_{x1} + xC_{x2} + p_3(\mathbf{x})C_{x3} + \cdots + p_{n_d}(\mathbf{x})C_{xn_d}
\end{aligned}$$

At the n_d nodes of the element, we have n_d equations:

$$\begin{aligned}
u^h(\mathbf{x}_1) &= p_1(\mathbf{x})C_{x1} + x_1 C_{x2} + p_3(\mathbf{x}_1)C_{x3} + \cdots + p_{n_d}(\mathbf{x}_1)C_{xn_d} \\
u^h(\mathbf{x}_2) &= p_1(\mathbf{x})C_{x1} + x_2 C_{x2} + p_3(\mathbf{x}_2)C_{x3} + \cdots + p_{n_d}(\mathbf{x}_2)C_{xn_d} \\
&\vdots \qquad\qquad\qquad\qquad\qquad\qquad\qquad\qquad\qquad (3.63) \\
u^h(\mathbf{x}_{n_d}) &= p_1(\mathbf{x})C_{x1} + x_{n_d} C_{x2} + p_3(\mathbf{x}_{n_d})C_{x3} + \cdots + p_{n_d}(\mathbf{x}_{n_d})C_{xn_d}
\end{aligned}$$

Using the delta function property of the shape functions, we have

$$u^h(\mathbf{x}_j) = \sum_{i=1}^{n_d} N_i(\mathbf{x}_j) x_i$$

$$= \underbrace{N_1(\mathbf{x}_j) x_1}_{0} + \underbrace{N_2(\mathbf{x}_j) x_2}_{0} + \cdots + \underbrace{N_j(\mathbf{x}_j) x_j}_{1} + \cdots + \underbrace{N_{n_d}(\mathbf{x}_j) x_{n_d}}_{0}$$

$$= x_j$$

(3.64)

Hence, Eq. (3.63) becomes

$$0 = p_1(\mathbf{x})C_{x1} + x_1(C_{x2} - 1) + p_3(\mathbf{x}_2)C_{x3} + \cdots + p_{n_d}(\mathbf{x}_2)C_{xn_d}$$

$$0 = p_1(\mathbf{x}_3)C_{x1} + x_2(C_{x2} - 1) + p_3(\mathbf{x}_3)C_{x3} + \cdots + p_{n_d}(\mathbf{x}_3)C_{xn_d}$$

$$\vdots$$

$$0 = p_1(\mathbf{x}_{n_d})C_{x1} + x_{n_d}(C_{x2} - 1) + p_3(\mathbf{x}_{n_d})C_{x3} + \cdots + p_{n_d}(\mathbf{x}_{n_d})C_{xn_d}$$

(3.65)

Or in matrix form,

$$
\begin{bmatrix}
p_1(\mathbf{x}_1) & x_1 & p_3(\mathbf{x}_1) & \cdots & p_{n_d}(\mathbf{x}_1) \\
p_1(\mathbf{x}_1) & x_2 & p_3(\mathbf{x}_2) & \cdots & p_{n_d}(\mathbf{x}_2) \\
\vdots & \vdots & \vdots & \ddots & \vdots \\
p_1(\mathbf{x}_1) & x_{n_d} & p_3(\mathbf{x}_{n_d}) & \cdots & p_{n_d}(\mathbf{x}_{n_d})
\end{bmatrix}
\begin{Bmatrix}
C_{x1} \\
C_{x2} - 1 \\
C_{x3} \\
\vdots \\
Cxn_d
\end{Bmatrix} = 0
$$

(3.66)

Note that the coefficient matrix of Eq. (3.66) is the moment matrix that has a full rank. We thus have

$$C_{x1} = 0$$

$$(C_{x2} - 1) = 0$$

$$C_{x3} = \cdots = C_{n_d} = 0$$

(3.67)

Substituting the previous equation back into Eq. (3.62), we obtain

$$u^h(x) = \sum_{i=1}^{n_d} N_i(x)x_i = 0$$

(3.68)

This proves the property of linear field reproduction.

The delta function property (Property 3) ensures convenient imposition of the essential boundary conditions (admissible condition (b) required by Hamilton's principle), because the nodal displacement at a node is independent of that at any other nodes. The constraints can often be written in the form of a so-called *Single Point Constraint* (SPC). If the displacement at a node is fixed, all one needs to do is to remove corresponding rows and columns without affecting the other rows and columns.

The proof of Property 4 gives a convenient way to confirm the partitions of unity property of shape functions. As long as the constant (unit) basis is included in the basis functions, the shape functions constructed are partitions of unity. Properties 4 and 5 are essential for the FEM to pass the standard *patch test*, used for decades in the FEM for validating the elements. In the standard patch test, the patch is meshed with a number of elements, with at least one interior node. Linear displacements are then enforced on the boundary (edges) of the patch. A successful patch test requires the FEM solution to produce the linear displacement (or constant strain) field at any interior node. Therefore, the property of reproduction of a linear field of shape function provides the foundation for passing the patch test. Note that the property of reproducing the linear field of the shape function does not guarantee successful patch tests, as there could be other sources of numerical error, such as numerical integration, which can cause failure.

Lemma 1 seems to be redundant, since we already have Property 4. However, Lemma 1 is a very convenient property to use for checking the property of partitions of unity of shape functions that are constructed using other *shortcut* methods, rather than the standard procedure. Using Lemma 1, one only needs to make sure whether the shape functions satisfy Eq. (3.48).

Lemma 2 is another very convenient property to use for checking the property of partitions of unity of shape functions. Using Lemma 2, we only need to make sure that the constructed n_d shape functions are of the delta function property, and they are linear combinations of the same n_d basis functions that are linearly-independent and contain the constant basis function. The conformation of full rank of the moment matrix of the basis functions can sometimes be difficult. In this book, we usually assume that the rank is full for the normal elements, as long as the basis functions are linearly-independent. In usual situations, one will not be able to obtain the shape functions if the rank of the moment matrix is not full. If we somehow obtained the shape functions successfully, we can usually be sure that the rank of the corresponding moment matrix is full.

Lemma 3 is a very convenient property to use for checking the property of linear field reproduction of shape functions. Using Lemma 3, we only need to make sure that the constructed n_d shape functions are of the delta function property, and they are linear combinations of the same n_d basis functions that are linearly-independent and contain the linear basis function.

3.4.5 Formulation of finite element equations in local coordinate system

Once the shape functions are constructed, the finite element (FE) equation for an element can be formulated using the following process. By substituting the interpolation of the nodes, Eq. (3.6), and the strain–displacement equation, say Eq. (2.5), into the strain energy term (Eq. (3.4)), we have

$$\Pi = \frac{1}{2} \int_{V_e} \boldsymbol{\varepsilon}^T \mathbf{c} \boldsymbol{\varepsilon} \, dV = \frac{1}{2} \int_{V_e} \mathbf{d}_e^T \mathbf{B}^T \mathbf{c} \underbrace{\mathbf{B} \mathbf{d}_e}_{\boldsymbol{\varepsilon}} \, dV = \frac{1}{2} \mathbf{d}_e^T \left(\int_{V_e} \mathbf{B}^T \mathbf{c} \mathbf{B} d_e dV \right) \mathbf{d}_e \qquad (3.69)$$

where the subscript e stands for the element. Note that the volume integration over the global domain has been changed to that over the elements. This can be done because we assume that the assumed displacement field satisfies the compatibility condition (see Section 3.3) on all the edges between the elements. Otherwise, techniques discussed in Chapter 11 are needed. In Eq. (3.69), **B** gives the strain in the element using the element nodal displacements, and is often called the strain–displacement matrix or simply *strain matrix*. It is defined by

$$\mathbf{B} = \mathbf{LN} \tag{3.70}$$

where **L** is the differential operator that is defined for different problems in Chapter 2. This implies that the strain matrix is determined once the shape functions are obtained. For 3D solids, **L** is given by Eq. (2.7). By denoting

$$\mathbf{k}_e = \int_{V_e} \mathbf{B}^T \mathbf{c} \mathbf{B} dV \tag{3.71}$$

which is called the element stiffness matrix, Eq. (3.69) can be rewritten as

$$\Pi = \frac{1}{2} \mathbf{d}_e^T \mathbf{k}_e \mathbf{d}_e \tag{3.72}$$

Note that the stiffness matrix $\mathbf{k}_e$ is symmetrical, because

$$[\mathbf{k}_e]^T = \int_{V_e} [\mathbf{B}^T \mathbf{c} \mathbf{B}]^T dV = \int_{V_e} \mathbf{B}^T \mathbf{c}^T [\mathbf{B}^T]^T dV = \int_{V_e} \mathbf{B}^T \mathbf{c} \mathbf{B} dV = \mathbf{k}_e \tag{3.73}$$

which shows that the transpose of matrix $\mathbf{k}_e$ is itself. In deriving the above equation, $\mathbf{c} = \mathbf{c}^T$ has been employed. Making use of the symmetry of the stiffness matrix, only half of the terms in the matrix need to be evaluated and stored.

By substituting Eq. (3.6) into Eq. (3.3), the kinetic energy can be expressed as

$$T = \frac{1}{2} \int_{V_e} \rho \dot{\mathbf{U}}^T \dot{\mathbf{U}} dV = \frac{1}{2} \int_{V_e} \rho \dot{\mathbf{d}}_e^T \mathbf{N}^T \mathbf{N} \dot{\mathbf{d}}_e dV = \frac{1}{2} \dot{\mathbf{d}}_e^T \left(\int_{V_e} \rho \mathbf{N}^T N dV \right) \dot{\mathbf{d}}_e \tag{3.74}$$

By denoting

$$\mathbf{m}_e = \int_{V_e} \rho \mathbf{N}^T \mathbf{N} dV \tag{3.75}$$

which is called the mass matrix of the element, Eq. (3.74) can be rewritten as

$$T = \frac{1}{2} \dot{\mathbf{d}}_e^T \mathbf{m}_e \dot{\mathbf{d}}_e \tag{3.76}$$

It is obvious that the element mass matrix is also symmetrical.

Finally, to obtain the work done by the external forces, Eq. (3.6) is substituted into Eq. (3.5):

$$W_f = \int_{V_e} \mathbf{d}_e^T \mathbf{N}^T \mathbf{f}_b dV + \int_{S_e} \mathbf{d}_e^T \mathbf{N}^T \mathbf{f}_s dS$$
$$= \mathbf{d}_e^T \underbrace{\left(\int_{V_e} \mathbf{N}^T \mathbf{f}_b dV \right)}_{\mathbf{F}_b} + \mathbf{d}_e^T \underbrace{\left(\int_{S_e} \mathbf{N}^T \mathbf{f}_s dS \right)}_{\mathbf{F}_S}$$

(3.77)

where the surface integration is performed only for elements on the force boundary of the problem domain. By denoting

$$\mathbf{F}_b = \int_{V_e} \mathbf{N}^T \mathbf{f}_b dV$$

(3.78)

and

$$\mathbf{F}_S = \int_{S_e} \mathbf{N}^T \mathbf{f}_s dS$$

(3.79)

Eq. (3.77) can then be rewritten as

$$W_f = \mathbf{d}_e^T \mathbf{F}_b + \mathbf{d}_e^T \mathbf{F}_s = \mathbf{d}_e^T \mathbf{f}_e$$

(3.80)

$\mathbf{F}_b$ and $\mathbf{F}_s$ are the nodal forces acting on the nodes of the elements, which are equivalent to the body forces and surface forces applied on the element in terms of the work done on a virtual displacement. These two nodal force vectors can then be added up to form the total nodal force vector $\mathbf{f}_e$:

$$\mathbf{f}_e = \mathbf{F}_b + \mathbf{F}_s$$

(3.81)

Substituting Eqs. (3.72), (3.76), and (3.80) into Lagrangian functional L (Eq. (3.2)), we have

$$L = -\frac{1}{2} \dot{\mathbf{d}}_e^T \mathbf{m}_e \dot{\mathbf{d}}_e + \frac{1}{2} \mathbf{d}_e^T \mathbf{k}_e \mathbf{d}_e - \mathbf{d}_e^T \mathbf{f}_e$$

(3.82)

Applying Hamilton's principle (Eq. (3.1)), we have

$$\delta \int_{t_1}^{t_2} \left(-\frac{1}{2} \dot{\mathbf{d}}_e^T \mathbf{m}_e \dot{\mathbf{d}}_e + \frac{1}{2} \mathbf{d}_e^T \mathbf{k}_e \mathbf{d}_e - \mathbf{d}_e^T \mathbf{f}_e \right) dt = 0$$

(3.83)

Note that the variation and integration operators are interchangeable, hence we obtain

$$\int_{t_1}^{t_2} (-\delta \dot{\mathbf{d}}_e^T \mathbf{m}_e \dot{\mathbf{d}}_e + \delta \mathbf{d}_e^T \mathbf{k}_e \mathbf{d}_e - \delta \mathbf{d}_e^T \mathbf{f}_e) dt = 0 \tag{3.84}$$

To explicitly illustrate the process of deriving Eq. (3.84) from Eq. (3.83), we use a two-degree of freedom system as an example. Here, we show the procedure for deriving the second term in Eq. (3.84):

$$\delta\left(\frac{1}{2}\mathbf{d}_e^T \mathbf{k}_e \mathbf{d}_e\right) = \delta\left(\frac{1}{2}\{d_1\, d_2\} \begin{bmatrix} k_{11} & k_{12} \\ k_{12} & k_{22} \end{bmatrix} \begin{Bmatrix} d_1 \\ d_2 \end{Bmatrix}\right)$$

$$= \frac{1}{2}\delta\left(\{d_1 k_{11} + d_2 k_{12} \quad d_1 k_{12} + d_2 k_{22}\} \begin{Bmatrix} d_1 \\ d_2 \end{Bmatrix}\right)$$

$$= \frac{1}{2}\delta\left(d_1^2 k_{11} + 2d_1 d_2 k_{12} + d_2^2 k_{22}\right)$$

$$= \frac{1}{2}\left[\frac{\partial(d_1^2 k_{11} + 2d_1 d_2 k_{12} + d_2^2 k_{22})}{\partial d_1}\delta d_1 + \frac{\partial(d_1^2 k_{11} + 2d_1 d_2 k_{12} + d_2^2 k_{22})}{\partial d_2}\delta d_2\right]$$

$$= (d_1 k_{11} + d_2 k_{12})\delta d_1 + (d_1 k_{12} + d_2 k_{22})\delta d_2$$

$$= \{\delta d_1\, \delta d_2\} \begin{Bmatrix} d_1 k_{11} + d_2 k_{12} \\ d_1 k_{12} + d_2 k_{22} \end{Bmatrix} = \{\delta d_1 \quad \delta d_2\} \begin{bmatrix} k_{11} & k_{12} \\ k_{12} & k_{22} \end{bmatrix} \begin{Bmatrix} d_1 \\ d_2 \end{Bmatrix} = \delta \mathbf{d}_e^T \mathbf{k}_e \mathbf{d}_e$$

In Eq. (3.84), the variation and differentiation with time are also interchangeable, i.e.,

$$\delta \dot{\mathbf{d}}_e^T = \delta\left(\frac{d\mathbf{d}_e^T}{dt}\right) = \frac{d}{dt}\left(\delta \mathbf{d}_e^T\right) \tag{3.85}$$

Hence, by substituting Eq. (3.85) into Eq. (3.84), and integrating the first term by parts, we obtain

$$\int_{t_1}^{t_2} \delta \dot{\mathbf{d}}_e^T \mathbf{m}_e \dot{\mathbf{d}}_e \, dt = \underbrace{\delta \mathbf{d}_e^T \mathbf{m}_e \dot{\mathbf{d}}_e|_{t_1}^{t_2}}_{=0} - \int_{t_1}^{t_2} \delta \mathbf{d}_e^T \mathbf{m}_e \ddot{\mathbf{d}}_e dt = -\int_{t_1}^{t_2} \delta \mathbf{d}_e^T \mathbf{m}_e \ddot{\mathbf{d}}_e dt \tag{3.86}$$

Note that in deriving Eq. (3.86) as above, the condition $\delta \mathbf{d}_e = 0$ at t_1 and t_2 have been used, which leads to the vanishing of the first term on the right-hand side. This is because the initial condition at t_1 and final condition at t_2 have to be satisfied for any $\mathbf{d}_e$ (admissible conditions (c) required by Hamilton's principle), and no variation at t_1 and t_2 is allowed. Substituting Eq. (3.86) into Eq. (3.84) leads to

$$\int_{t_1}^{t_2} \delta \mathbf{d}_e^T (\mathbf{m}_e \ddot{\mathbf{d}}_e + \mathbf{k} \mathbf{d}_e - \mathbf{f}_e) dt = 0 \tag{3.87}$$

To have the integration in Eq. (3.87) as zero for an arbitrary integrand, the integrand itself has to vanish, i.e,

$$\delta \mathbf{d}_e^T (\mathbf{m}_e \ddot{\mathbf{d}}_e + \mathbf{k} \mathbf{d}_e - \mathbf{f}_e) = 0 \tag{3.88}$$

Due to the arbitrary nature of the variation of the displacements, the only insurance for Eq. (3.88) to be satisfied is

$$\mathbf{k}_e \mathbf{d}_e + \mathbf{m}_e \ddot{\mathbf{d}}_e = \mathbf{f}_e \tag{3.89}$$

Equation (3.89) is the FEM equation for an element, while $\mathbf{k}_e$ and $\mathbf{m}_e$ are the *stiffness* and *mass* matrices for the element, and $\mathbf{f}_e$ is the element force vector of the total external forces acting on the nodes of the element. All these element matrices and vectors can be obtained simply by integration for the given shape functions of displacements.

3.4.6 Coordinate transformation

The element equation given by Eq. (3.89) is often formulated based on the local coordinate system defined on an element. In general, the structure is divided into many elements with different orientations (see Figure 3.4). To assemble all the element equations to form the global system equations, a coordinate transformation has to be performed for each element in order to convert its element equation in reference to the *global coordinate system* which is defined on the whole structure.

The coordinate transformation gives the relationship between the displacement vector $\mathbf{d}_e$ based on the local coordinate system and the displacement vector $\mathbf{D}_e$ for the same element, but based on the global coordinate system. It is therefore a transformation of a vector between two (local and global) coordinate systems, which can be written in general in the form of

$$\mathbf{d}_e = \mathbf{T} \mathbf{D}_e \tag{3.90}$$

where $\mathbf{T}$ is the transformation matrix that can have different forms depending upon the type of element, and will be discussed in detail in later chapters when different elements are formulated. The transformation matrix can also be applied to the force vectors between the local and global coordinate systems:

$$\mathbf{f}_e = \mathbf{T} \mathbf{F}_e \tag{3.91}$$

in which $\mathbf{F}_e$ stands for the force vector at node i on the global coordinate system. Substitution of Eq. (3.90) into Eq. (3.89) leads to the element equation based on the global coordinate system:

$$\mathbf{K}_e \mathbf{D}_e + \mathbf{M}_e \ddot{\mathbf{D}}_e = \mathbf{F}_e \tag{3.92}$$

where

$$\mathbf{K}_e = \mathbf{T}^T \mathbf{k}_e \mathbf{T} \tag{3.93}$$

$$\mathbf{M}_e = \mathbf{T}^T \mathbf{m}_e \mathbf{T} \tag{3.94}$$

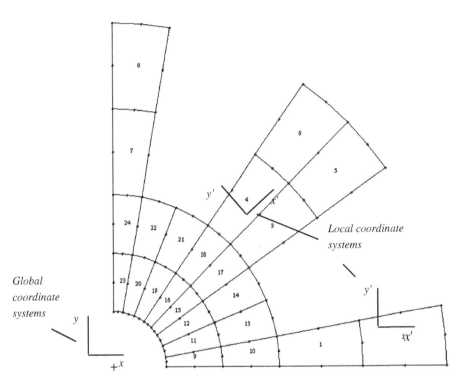

FIGURE 3.4

Local and global coordinate systems.

$$F_e = T^T f_e \tag{3.95}$$

In the cases when the element equations are formulated based on the global coordinate system (for example for the linear triangular or tetrahedral elements), the above coordinate transformation is not necessary, or the T matrix becomes an identity matrix.

3.4.7 Assembly of global FE equation

The FE equations for all the individual elements can be assembled together to form the global FE system equation:

$$KD + M\ddot{D} = F \tag{3.96}$$

where K and M are the global stiffness and mass matrix, D is a vector of all the displacements at all the nodes in the entire problem domain, and F is a vector of all the equivalent nodal force vectors. The process of assembly is one of simply adding up the contributions from all the elements connected at a node. The detailed process will be demonstrated in Chapter 4 using example problems. It may be noted here that the assembly of the global matrices can be skipped by combining assembling with the procedure of equation solving. This means that the assembling of a term in the global matrix is done only when the equation solver is operating on this term. Such a procedure can drastically reduce the need for storage during the computation.

3.4.8 **Imposition of displacement constraints**

The global stiffness matrix $\mathbf{K}$ in Eq. (3.96) does not usually have a full rank, because displacement constraints (supports) are not yet imposed, and it is non-negative definite or positive semi-definite. Physically, an unconstrained solid or structure is capable of performing rigid movements. Therefore, if the solid or structure is free of support, Eq. (3.96) gives the behavior that includes the rigid body dynamics, if it is subjected to dynamic forces. If the external forces applied are static, the displacements cannot be uniquely determined from Eq. (3.96) for any given force vector. It is physically meaningless to try to determine the static displacements of an unconstrained solid or structure that can move freely.

For constrained solids and structures, the constraints can be imposed by simply removing the rows and columns corresponding to the constrained nodal displacements. We shall demonstrate this method in example problems in later chapters. After such treatments of constraints (and if the constraints are sufficient), the stiffness matrix $\mathbf{K}$ in Eq. (3.96) will be of full rank, and will be Positive Definite (PD). Since we have already proven that $\mathbf{K}$ is symmetric, $\mathbf{K}$ is of a symmetric positive definite property.

3.4.9 **Solving the global FE equation**

By solving the global FE equation, displacements at the nodes can be obtained. The strain and stress in any element can then be retrieved using Eq. (3.6) in Eqs. (2.5) and (2.8).

3.5 **Static analysis**

Static analysis involves the solving of Eq. (3.96) without the term with the global mass matrix, $\mathbf{M}$. Hence, as discussed, the static system of equations takes the form

$$\mathbf{KD} = \mathbf{F} \tag{3.97}$$

There are numerous methods and algorithms to solve the above matrix equation. The methods often used are Gauss elimination or LU decompositions for small systems, and iterative methods for large systems. These methods are all routinely available in any math library of any computer system.

3.6 **Analysis of free vibration (eigenvalue analysis)**

For a structural system with a total DOF of N, the stiffness matrix $\mathbf{K}$ and mass matrix $\mathbf{M}$ in Eq. (3.96) have a dimension of $N \times N$. By solving the above equation we can obtain the displacement field, and the stress and strain can then be calculated. The question now is how to solve this equation, as N is usually very large for practical engineering structures. One way to solve this equation is by using the so-called direct integration method discussed in the next section. An alternative way of solving Eq. (3.96) is by using the so-called modal analysis (or mode superposition) technique. In this technique, we first have to solve the homogenous equation of Eq. (3.96). The homogeneous equation is when we consider the case of $\mathbf{F}=0$, therefore it is also called *free vibration* analysis, as the system is free of external forces. For a solid or structure that undergoes a free vibration, the discretized system equation Eq. (3.96) becomes

$$\mathbf{KD} + \mathbf{M\ddot{D}} = 0 \tag{3.98}$$

The solution for the free vibration problem can be assumed as

$$\mathbf{D} = \phi\exp(i\omega t) \tag{3.99}$$

where ϕ is the amplitude of the nodal displacement, ω is the (angular) frequency of the free vibration, and t is the time. By substituting Eq. (3.99) into Eq. (3.98), we obtain

$$[K - \omega^2\mathbf{M}]\boldsymbol{\phi} = 0 \tag{3.100}$$

or

$$[\mathbf{K} - \lambda\mathbf{M}]\boldsymbol{\phi} = 0 \tag{3.101}$$

where

$$\lambda = \omega^2 \tag{3.102}$$

Equation (3.100), (or (3.101)) is called the *eigenvalue equation.* To have a non-zero solution for ϕ, the determinate of the matrix must vanish:

$$\det[\mathbf{K} - \lambda\mathbf{M}] = |\mathbf{K} - \lambda\mathbf{M}| = 0 \tag{3.103}$$

The expansion of the above equation will lead to a polynomial of λ of order N. This polynomial equation will have N roots, $\lambda_1, \lambda_2, ..., \lambda_N$, called *eigenvalues,* which relate to the *natural frequencies* of the system by Eq. (3.100). The natural frequency is a very important characteristic of the structure carrying dynamic loads. It has been found that if a structure is excited by a load with a frequency of one of the structure's natural frequencies, the structure can undergo extremely violent vibration, which often leads to catastrophic failure of the structural system. Such a phenomenon is called *resonance.* Therefore, an eigenvalue analysis has to be performed in designing a structural system that is to be subjected to dynamic loadings.

By substituting an eigenvalue λ_i back into the eigenvalue equation, Eq. (3.101), we have

$$[\mathbf{K} - \lambda_i\mathbf{M}]\boldsymbol{\phi} = 0 \tag{3.104}$$

which is a set of algebraic equations. Solving the above equation for ϕ, a vector denoted by ϕ_i can then be obtained. This vector corresponding to the ith eigenvalue λ_i is called the ith *eigenvector* that satisfies the following equation:

$$[\mathbf{K} - \lambda_i\mathbf{M}]\boldsymbol{\phi}_i = 0 \tag{3.105}$$

An eigenvector ϕ_i corresponds to a *vibration mode* that gives the shape of the vibrating structure of the *i*th mode. Therefore, analysis of the eigenvalue equation also gives very important information on possible vibration modes experienced by the structure when it undergoes a vibration. Vibration modes of a structure are therefore another important characteristic of the structure. Mathematically, the eigenvectors can be used to construct the displacement fields. It has been found that using a few of the lowest modes can obtain very accurate results for many engineering problems. Modal analysis techniques have been developed to take advantage of these properties of natural modes.

In Eq. (3.101), since the mass matrix **M** is symmetric positive definite and the stiffness matrix **K** is symmetric and either positive or positive semi-definite, the eigenvalues are all real and either positive or zero. It is possible that some of the eigenvalues may coincide. The corresponding eigenvalue equation is said to have multiple eigenvalues. If there are *m* coincident eigenvalues, the eigenvalue is said to be of a multiplicity of *m*. For an eigenvalue of multiplicity *m*, there are *m* vectors satisfying Eq. (3.105).

Methods for the effective computation of the eigenvalues and eigenvectors for an eigenvalue equations system like Eq. (3.100) or Eq. (3.101) are outside the scope of this book. Intensive research has been conducted to date, and many sophisticated algorithms are already well established and readily available in the open literature, and routinely in computational libraries. The commonly used methods are (see, for example, Petyt, 1990):

- Jacobi's method;
- Given's method and householder's method;
- the bisection method (using Sturm sequences);
- inverse iteration;
- QR method;
- subspace iteration;
- Lanczos' method.

3.7 Transient response

Structural systems are very often subjected to *transient* excitation. A transient excitation is a highly dynamic time-dependent force exerted on the solid or structure, such as earthquake, impact, and shocks. The discrete governing equation system for such a structure is still Eq. (3.96), but it often requires a different solver from that used in the eigenvalue analysis. The widely used method is the so-called *direct integration method*.

The direct integration method basically uses the *finite difference method* (FDM) for time stepping to solve Eq. (3.96). There are two main types of direct integration method: *implicit* and *explicit*. Implicit methods are generally more efficient for a relatively slow phenomenon, and explicit methods are more efficient for a very fast phenomenon, such as impact and explosion. The literature on the various algorithms available for solving transient problems is vast. This section introduces the basic idea of time stepping used in FDMs, which are employed in solving transient problems.

Before discussing the equations used for the time stepping techniques, the general system equation for a structure considering damping effects is written as

$$\mathbf{KD} + \mathbf{C\dot{D}} + \mathbf{M\ddot{D}} = \mathbf{F} \tag{3.106}$$

where $\dot{\mathbf{D}}$ is the vector of velocity components, and $\mathbf{C}$ is the matrix of damping coefficients for the solid/ structure that are determined experimentally. The damping coefficients are often expressed simply as proportions of the mass and stiffness matrices, called *proportional damping* (e.g. Petyt, 1990; Clough and Penzien, 1975). For a proportional damping system, $\mathbf{C}$ can be simply determined in the form

$$\mathbf{C} = c_K \mathbf{K} + c_M \mathbf{M} \tag{3.107}$$

where c_K and c_M are determined by experiments.

3.7.1 Central difference algorithm

We first re-write the system equation (3.106) in the form

$$\mathbf{M}\ddot{\mathbf{D}} = \mathbf{F} - \underbrace{[\mathbf{C}\dot{\mathbf{D}} + \mathbf{K}\mathbf{D}]}_{\mathbf{F}^{\text{int}}} = \mathbf{F} - \mathbf{F}^{\text{int}} = \mathbf{F}^{\text{residual}} \tag{3.108}$$

where $\mathbf{F}^{\text{residual}}$ is the *residual* force vector, and

$$\mathbf{F}^{\text{int}} = [\mathbf{C}\dot{\mathbf{D}} + \mathbf{K}\mathbf{D}] \tag{3.109}$$

is defined as the *int*ernal force at time t. The acceleration, $\ddot{\mathbf{D}}$, can be simply obtained by

$$\ddot{\mathbf{D}} = \mathbf{M}^{-1}\mathbf{F}^{\text{residual}} \tag{3.110}$$

In practice, the above equation does not usually require the solving of the matrix equation, since lumped masses are usually used which form a diagonal mass matrix (Petyt, 1990). The solution to Eq. (3.110) is thus trivial, and the matrix equation is the set of independent equations for each degree of freedom i as follows:

$$d_i = \frac{f_i^{\text{residual}}}{m_i} \tag{3.111}$$

where f_i^{residual} is the residual force, and m_i is the lumped mass corresponding to the ith DOF. We now introduce the following finite central difference equations:

$$\mathbf{D}_{t+\Delta t} = 2(\Delta t)\dot{\mathbf{D}}_t + \mathbf{D}_{t-\Delta t} \tag{3.112}$$

$$\dot{\mathbf{D}}_{t+\Delta t} = 2(\Delta t)\ddot{\mathbf{D}}_t + \dot{\mathbf{D}}_{t-\Delta t} \tag{3.113}$$

$$\ddot{\mathbf{D}}_t = \frac{1}{(\Delta t)^2}(\mathbf{D}_{t+\Delta t} - 2\mathbf{D}_t + \mathbf{D}_{t-\Delta t}) \tag{3.114}$$

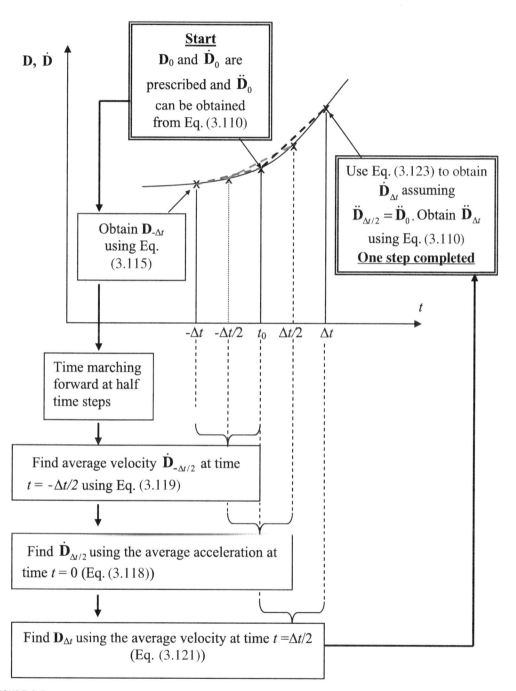

FIGURE 3.5

Time marching in the central difference algorithm: explicitly advancing in time.

By eliminating $\mathbf{D}_{t+\Delta t}$ from Eqs. (3.112) and (3.114), we have

$$\mathbf{D}_{t-\Delta t} = \mathbf{D}_t - (\Delta t)\dot{\mathbf{D}}_t + \frac{(\Delta t)^2}{2}\ddot{\mathbf{D}}_t \qquad (3.115)$$

To explain the time stepping procedure, refer to Figure 3.5, which shows an arbitrary plot of either displacement or velocity against time. The time stepping/marching procedure in the central difference method starts at $t=0$, and computes the acceleration $\ddot{\mathbf{D}}_0$ using Eq. (3.110):

$$\ddot{\mathbf{D}}_0 = \mathbf{M}^{-1}\mathbf{F}_0^{\text{residual}} \qquad (3.116)$$

For given initial conditions, $\mathbf{D}_0$ and $\dot{\mathbf{D}}_0$ are known. Substituting $\mathbf{D}_0$, $\dot{\mathbf{D}}_0$, and $\ddot{\mathbf{D}}_0$ into Eq. (3.115), we find $\mathbf{D}_{-\Delta t}$. Considering a half of the time step and using the central difference Eqs. (3.112) and (3.113), we have

$$\mathbf{D}_{t+\Delta t/2} = (\Delta t)\dot{\mathbf{D}}_t + \mathbf{D}_{t-\Delta t/2} \qquad (3.117)$$

$$\dot{\mathbf{D}}_{t+\Delta t/2} = (\Delta t)\ddot{\mathbf{D}}_t + \dot{\mathbf{D}}_{t-\Delta t/2} \qquad (3.118)$$

The velocity, $\dot{\mathbf{D}}_{-\Delta t/2}$ at $t=-\Delta t/2$ can be obtained by Eq. (3.117) by performing the central differencing at $t=-\Delta t/2$ and using values of $\mathbf{D}_{-\Delta t}$ and $\mathbf{D}_0$:

$$\dot{\mathbf{D}}_{-\Delta t/2} = \frac{\mathbf{D}_0 - \mathbf{D}_{-\Delta t}}{(\Delta t)} \qquad (3.119)$$

After this, Eq. (3.118) is used to compute $\dot{\mathbf{D}}_{\Delta t/2}$ using $\ddot{\mathbf{D}}_0$ and $\dot{\mathbf{D}}_{-\Delta t/2}$:

$$\dot{\mathbf{D}}_{\Delta t/2} = (\Delta t)\ddot{\mathbf{D}}_0 + \dot{\mathbf{D}}_{-\Delta t/2} \qquad (3.120)$$

Then, Eq. (3.117) is used once again to compute $\mathbf{D}_{\Delta t}$ using $\dot{\mathbf{D}}_{-\Delta t/2}$ and $\mathbf{D}_0$:

$$\mathbf{D}_{\Delta t} = (\Delta t)\dot{\mathbf{D}}_{\Delta t/2} + \mathbf{D}_0 \qquad (3.121)$$

Once $\mathbf{D}_{\Delta t}$ is determined, Eq. (3.118) at $t=\Delta t/2$ can be used to obtain $\dot{\mathbf{D}}_{\Delta t}$ by assuming that the acceleration is constant over the step Δt

$$\ddot{\mathbf{D}}_{t+\Delta t/2} \approx \ddot{\mathbf{D}}_t \qquad (3.122)$$

and using $\dot{\mathbf{D}}_0$ (that is the prescribed initial velocity). We have,

$$\dot{\mathbf{D}}_{\Delta t} = (\Delta t)\underbrace{\ddot{\mathbf{D}}_{\Delta t/2}}_{\approx \ddot{\mathbf{D}}_0} + \dot{\mathbf{D}}_0 = (\Delta t)\ddot{\mathbf{D}}_0 + \dot{\mathbf{D}}_0 \qquad (3.123)$$

At the next step in time, $\ddot{\mathbf{D}}_{\Delta t}$ is again computed using Eq. (3.110). The above process is then repeated. The time marching is continued until it reaches the final desired time.

Note that in the above process, the solutions (displacement, velocity, and acceleration) are obtained without solving any matrix form of system equation, but repeatedly using Eqs. (3.110), (3.117), (3.118), and (3.122). The central difference method is therefore an *explicit method.* The time marching in explicit methods is therefore extremely fast, and the coding is also very straightforward. It is particularly suited for simulating highly nonlinear, large deformation, contact, and extremely fast events of mechanics.

The central difference method, like most explicit methods, is *conditionally stable.* This means that if the time step, Δt, becomes too large to exceed a critical time step, Δt_{cr}, then the computed solution will become unstable and might grow without limit. The critical time step Δt_{cr} should be the time taken for the fastest stress wave in the solids/structure to across the smallest element in the mesh. Therefore, the time steps used in the explicit methods are typically 100 to 1000 times smaller than those used with implicit methods (that are outlined in the next subsection). The need to use a small time step, and especially its dependence on the smallest element size, makes the explicit codes lose out to implicit codes for some of the problems, especially for those of slow dynamic events.

3.7.2 Newmark's method (Newmark, 1959)

Newmark's method is the most widely used implicit algorithm. The example software used in this book, ABAQUS, also uses the Newmark's method as its implicit solver except that the equilibrium equation defined in Eq. (3.106) is modified with the introduction of an operator defined by Hilber et al. (1978). In this book, we will introduce the standard Newmark's method as follows. It is first assumed that

$$\mathbf{D}_{t+\Delta t} = \mathbf{D}_t + (\Delta t)\dot{\mathbf{D}}_t + (\Delta t)^2 \left[\left(\frac{1}{2} - \beta \right) \ddot{\mathbf{D}}_t + \beta \ddot{\mathbf{D}}_{t+\Delta t} \right] \tag{3.124}$$

$$\dot{\mathbf{D}}_{t+\Delta t} = \dot{\mathbf{D}}_t + (\Delta t) \left[(1 - \gamma)\ddot{\mathbf{D}}_t + \gamma \ddot{\mathbf{D}}_{t+\Delta t} \right] \tag{3.125}$$

where β and γ are constants chosen by the analyst. Equations (3.124) and (3.125) are then substituted into the system equation, Eq. (3.106) to give

$$\begin{aligned} \mathbf{K} &\left\{ \mathbf{D}_t + (\Delta t)\dot{\mathbf{D}}_t + (\Delta t)^2 \left[\left(\frac{1}{2} - \beta \right) \ddot{\mathbf{D}}_t + \beta \ddot{\mathbf{D}}_{t+\Delta t} \right] \right\} \\ &+ \mathbf{C} \left\{ \dot{\mathbf{D}}_t + (\Delta t) \left[(1 - \gamma)\ddot{\mathbf{D}}_t + \gamma \ddot{\mathbf{D}}_{t+\Delta t} \right] \right\} + \mathbf{M}\ddot{\mathbf{D}}_{t+\Delta t} = \mathbf{F}_{t+\Delta t} \end{aligned} \tag{3.126}$$

If we group all the terms involving $\ddot{\mathbf{D}}_{t+\Delta t}$ on the left and shift the remaining terms to the right, we can write

$$\mathbf{K}_{cm}\ddot{\mathbf{D}}_{t+\Delta t} = \mathbf{F}_{t+\Delta t}^{residual} \tag{3.127}$$

where

$$\mathbf{K}_{cm} = [\mathbf{K}\beta(\Delta t)^2 + \mathbf{C}\gamma\,\Delta t + \mathbf{M}] \tag{3.128}$$

and

$$
\begin{aligned}
\mathbf{F}_{t+\Delta t}^{\text{residual}} = \mathbf{F}_{t+\Delta t} &- \mathbf{K}\left[\mathbf{D}_t + (\Delta t)\dot{\mathbf{D}}_t + (\Delta t)^2\left(\frac{1}{2} - \beta\right)\ddot{\mathbf{D}}_t\right] \\
&- \mathbf{C}\left\{\dot{\mathbf{D}}_t + (\Delta t)(1 - \gamma)\ddot{\mathbf{D}}_t\right\}
\end{aligned}
\tag{3.129}
$$

The accelerations $\ddot{\mathbf{D}}_{t+\Delta t}$ can then be obtained by solving the matrix system equation, Eq. (3.127):

$$\ddot{\mathbf{D}}_{t+\Delta t} = \mathbf{K}_{cm}^{-1}\mathbf{F}_{t+\Delta t}^{\text{residual}} \tag{3.130}$$

Note that the above equation involves matrix inversion, and hence it is analogous to solving a matrix equation. This makes it an implicit method, meaning that one needs to solve a set of linear algebraic equations to obtain a solution at the current time step.

The algorithm normally starts with a prescribed initial velocity and displacements, $\mathbf{D}_0$ and $\dot{\mathbf{D}}_0$. The initial acceleration $\ddot{\mathbf{D}}_0$ can then be obtained by substituting $\mathbf{D}_0$ and $\dot{\mathbf{D}}_0$ into Eq. (3.106), if $\ddot{\mathbf{D}}_0$ is not prescribed initially. Then Eq. (3.130) can be used to obtain the acceleration at the next time step, $\ddot{\mathbf{D}}_{\Delta t}$. The displacements $\mathbf{D}_{\Delta t}$ and velocities $\dot{\mathbf{D}}_{\Delta t}$ can then be calculated using Eqs. (3.124) and (3.125), respectively. The procedure then repeats to march forward in time until arriving at the final desired time. Because at each time step, the matrix system Eq. (3.127) has to be solved, which can be very time consuming, the implicit algorithm is a very slow time stepping process.

Newmark's method, like most implicit methods, is *unconditionally stable* if $\gamma \geq 0.5$ and $\beta \geq (2\gamma + 1)^2/16$. Unconditionally stable methods are those in which the size of the time step, Δt, will not affect the stability of the solution, but rather it is governed by accuracy considerations. The unconditionally stable property allows the implicit algorithms to use significantly larger time steps when the external excitation is of a slow time variation. When the external excitation changes too fast with time, the time step needs to be sufficiently small to capture the time variation features of the excitation.

3.8 Remarks

3.8.1 Summary of shape function properties

The properties of the shape functions are summarized in Table 3.1.

3.8.2 Sufficient requirements for FEM shape functions

Properties 3 and 4 are the minimum requirements for shape functions workable for the FEM. In meshfree methods (Liu, 2009), Property 3 is not a necessary condition for shape functions. Property 5 is a sufficient requirement for shape functions workable for the FEM for solid mechanics problems. It is possible for shape functions that do not possess Property 5 to produce convergent FEM solutions. In this book, however, we generally require all the FEM shape functions to satisfy Properties 3, 4, and 5.

Table 3.1   List of properties of shape functions.

Item	Name	Significance
Property 1	Reproduction property and consistency	Ensures shape functions produce all the functions that can be formed using basis functions used to create the shape functions. It is useful for constructing shape functions with desired accuracy and consistency in displacement field approximation.
Property 2	Shape functions are linearly-independent	Ensures the shape functions are qualified for a unique approximation of the (displacement) field function. It is often viewed as the default property for any FEM model.
Property 3	Delta function properties	Facilitates an easy imposition of essential boundary conditions. This is a minimum requirement for shape functions workable for the standard FEM.
Property 4	Partitions of unity property	Enables the shape functions to produce the rigid body movements. This is a minimum requirement for shape functions for the standard FEM models.
Property 5	Linear field reproduction	Ensures shape functions produce the linear displacement field. This is a sufficient requirement for shape functions capable of passing the patch test, and hence enabling FEM solution to converge to any smooth displacement solution for solid mechanics problems.
Lemma 1	Condition for shape functions being partitions of unity	Provides an alternative tool for checking the property of partitions of unity of shape functions.
Lemma 2	Condition for shape functions being partitions of unity	Provides an alternative tool for checking the property of partitions of unity of shape functions.
Lemma 3	Condition for shape functions being linear field reproduction	Provides an alternative tool for checking the property of linear field reproduction of shape functions.

Property 2 is the default property for any FEM model;
Property 3, 4, and 5 are the sufficient requirements for FEM shape functions.

These three requirements are called the sufficient requirements in this book for FEM shape functions; they are the *delta function property*, *partitions of unity*, and *linear field reproduction*.

3.8.3 **Recap of FEM procedure**

In FEMs, the displacement field **U** is expressed by displacements at *nodes* using *shape functions* defined over *elements*. Once the shape functions are found, the mass matrix and force vector can be obtained using Eqs. (3.75), (3.78), (3.79), and (3.81). The stiffness matrix can also be obtained using Eq. (3.71), once the shape functions and the strain matrix have been found. Therefore, to develop FE equations for any type of structure components, all one needs to do is formulate the shape function N and then establish the strain matrix **B**. The other procedures are very much the same. Hence, in the following chapters, the focus will be mainly on the derivation of the shape function and strain matrix for various types of elements for solids and structural components.

3.9 **Review questions**

1. What is the difference between the strong and weak forms of system equations?
2. What are the conditions that a summed displacement has to satisfy in order to apply the Hamilton's principle?
3. Briefly describe the standard steps involved in the FEM.
4. Do we have to discretize the problem domain in order to apply the Hamilton's principle?
5. What is the purpose of dividing the problem domain into elements?
6. For a function defined in $[0, l]$ as

$$f(x) = a_0 + a_1 x + a_2 x^2$$

where l is a given constant, x is an independent variable, and a_i ($i=0,1,2$) are variable constants (but independent of x). Show that:
   a. The variation operator and definite integral operator are exchangeable, i.e.,

$$\int_0^l [\delta f(x)] dx = \delta \left[\int_0^l f(x) dx \right]$$

   b. The variation operator and the differential operator are exchangeable, i.e.,

$$\delta \frac{df(x)}{dx} = \frac{d}{dx} [\delta f(x)]$$

7. What are the properties of a shape function? Can we use shape functions that do not possess these properties?

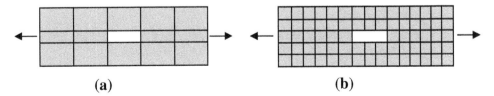

(a)   (b)

8. Answer the following questions with reference to the two meshes, (a) and (b), shown above:
   a. Which mesh will yield more accurate results?
   b. Which will be more computationally expensive?
   c. Suggest a way of meshing which will yield relatively accurate results and at the same time be less computationally expensive than B?
   9. Why is there a need to perform coordinate transformation for each element?

10. Describe how element matrices can be assembled together to form the global system matrix.

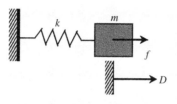

11. Consider a 1 DOF system with a rigid block of mass m that can move only in the horizontal direction, as shown in the above figure. The system is at the static equilibrium:
   a. Give the expressions for the strain potential energy, work done by external forces, and the total potential energy.
   b. Derive the equilibrium equations for the system using the minimum potential energy principle.
   c. Show that the total potential energy of the 1DOF system is at a minimum at the equilibrium position.
12. Consider a 1DOF system with a rigid block of mass m that can move only in the horizontal direction, as shown in the previous figure. The system is at a dynamic equilibrium. Derive the dynamic equilibrium equation for this system.

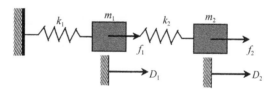

13. Consider a 2DOF system with two rigid blocks that can move only in the horizontal direction, as shown in the above figure. The system is at the static equilibrium:
   a. Give the expressions for the strain potential energy, work done by external forces, and the total potential energy.
   b. Drive the equilibrium equations for the system using the minimum potential energy principle.
14. Consider again a 2DOF system with two rigid blocks that can move only in the horizontal direction, as shown in the previous figure. The system is at a dynamic equilibrium.
   a. Give the expressions for the strain potential energy, work done by external forces, and the total potential energy.
   b. Drive the equilibrium equations for the system using the minimum potential energy principle.
15. Consider again the simplest problem of a continuum 1D bar of uniform cross-section, as shown in Figure 2.18. The bar is fixed at the left end and is of length l and section area A. It is subjected to a uniform body force f_x and a concentrated force F_s at the right end. The Young's modulus of the material is E. Using the method of minimum potential energy, obtain the distribution of the displacement for the following cases:
   a. $f_x = 0$ and $F_s = $ constant;
   b. $f_x = $ constant and $F_s = $ constant.

FEM for Trusses

4.1 Introduction

A truss is one of the simplest and most widely used structural members. It is a straight *bar* that is designed to take only axial forces, and therefore deforms only along its axial direction. An example of a structure consisting of an assemblage of trusses can be seen in Figure 2.7. The cross-section of the truss can be arbitrary, but the dimensions of the cross-section should be much smaller than that in the axial direction—a characteristic of a truss or bar structure.

Finite element equations for such truss members will be developed in this chapter. The one-dimensional element developed is commonly known as the *truss element* or *bar element*. Such elements are applicable for analysis of skeletal-type truss structural systems both in two-dimensional planes and in three-dimensional space. The basic concepts, procedures, and formulations can also be found in many existing textbooks (see, e.g. Reddy, 1993; Rao, 1999; Zienkiewicz and Taylor, 2000).

In planar trusses there are two components in the x and y directions for the displacements as well as for the forces. For spatial trusses, however, there will be three components in the x, y, and z directions for both displacements and forces. In skeletal structures consisting of truss members, the truss members are joined together by pins or hinges (as opposed to welding), so that only forces (not moments) are transmitted between the bars. For the purpose of explaining the concepts more clearly, this book will assume that the truss elements have a uniform cross-section. The concepts described will apply equally and easily to the formulation of bars with varying cross-sections. Note, however, that there is no reason from a mechanics viewpoint to use bars with a varying cross-section, because the force in a bar is uniform.

4.2 FEM equations
4.2.1 Shape function construction

Consider a structure consisting of a number of trusses or bar members. Each of the members can be considered as a truss/bar element of a uniform cross-section bounded by two nodes ($n_d = 2$). Consider a bar element with nodes 1 and 2 at each end of the element, as shown in Figure 4.1. The length of the element is l_e. The local x axis is taken in the axial direction of the element with the origin at node 1. In the local coordinate system, there is only one DOF at each node of the element, and that is the *axial displacement*. Therefore, there is a total of two DOFs for the element, i.e. $n_e = 2$. In the FEM discussed in the previous chapter, the displacement in an element should be written in the form

$$u^h(x) = \mathbf{N}(x)\mathbf{d}_e \tag{4.1}$$

where u^h is the *approximation* of the axial displacement within the element, $\mathbf{N}$ is a matrix of shape functions that possess the properties described in Chapter 3, and $\mathbf{d}_e$ should be the vector of the displacements at the two nodes of the element:

$$\mathbf{d}_e = \begin{Bmatrix} u_1 \\ u_2 \end{Bmatrix} \tag{4.2}$$

The question now is how we can determine the shape functions for the truss element.

We can follow the standard procedure described in Section 3.4.3 for constructing shape functions, and assume that the axial displacement in the truss element can be given in a general form

$$u^h(x) = \alpha_0 + \alpha_1 x = \underbrace{\begin{Bmatrix} 1 & x \end{Bmatrix}}_{\mathbf{p}^T} \underbrace{\begin{Bmatrix} \alpha_0 \\ \alpha_1 \end{Bmatrix}}_{\boldsymbol{\alpha}} = \mathbf{p}^T \boldsymbol{\alpha} \tag{4.3}$$

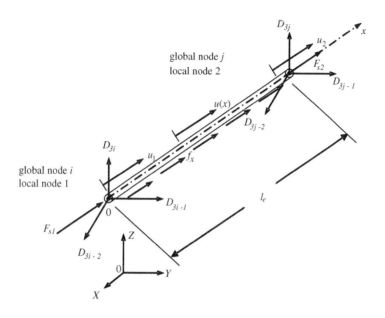

FIGURE 4.1

Truss element and the coordinate system.

where u^h is the approximation of the displacement, $\boldsymbol{\alpha}$ is the vector of two unknown constants, α_0 and α_1, and $\mathbf{p}$ is the vector of polynomial basis functions (or monomials). Since we have two nodes with a total of two DOFs in the element, we choose to have two terms of basis functions as in Eq. (4.3). Notice that we choose to use matrix expression in Eq. (4.3), despite the simplicity of the equation. This is to intentionally help readers get familiar with matrix expression and manipulation, which is essential in FEM formulations. For this particular problem, the polynomial basis functions consist of terms up to the first order. Depending upon the problem, we could use higher order polynomial basis functions. The order of polynomial basis functions up to the nth order can be given by

$$\mathbf{p}^T = \left\{1 \ x \ \cdots \ x^n\right\} \tag{4.4}$$

In this case, vector $\boldsymbol{\alpha}$ shall have n unknown constants, $\alpha_0, \alpha_1, \ldots \alpha_n$, and Eq. (4.3) can still be written as,

$$u^h(x) = \mathbf{p}^T \boldsymbol{\alpha} \tag{4.3a}$$

The number of terms of basis functions or monomials we should use depends upon the number of nodes and degrees of freedom in the element.

Note that we usually use polynomial basis functions of complete orders, meaning we do not skip any lower terms in constructing Eq. (4.3a). This is to ensure that the shape functions constructed will be able to reproduce complete polynomials up to an order of n. If a polynomial basis of the kth order is skipped, the shape function constructed will only be able to ensure a consistency of $(k-1)$th order, regardless of how many higher orders of monomials are included in the basis. This is because of the consistency property of the shape function (Property 1), discussed in Section 3.4.4. From Properties 3, 4, and 5 discussed in Chapter 3, we can expect that the complete linear basis functions used in Eq. (4.3) guarantee that the shape function to be constructed satisfies the *sufficient requirements* for the FEM shape functions: the delta function property, partition of unity, and linear field reproduction.

In deriving the shape function, we use the fact that

$$\text{at } x = 0, \quad u(x = 0) = u_1$$
$$\text{at } x = l_e, \quad u(x = l_e) = u_2 \tag{4.5}$$

Using Eqs. (4.3) and (4.5), we then have

$$\underbrace{\begin{Bmatrix} u_1 \\ u_2 \end{Bmatrix}}_{\mathbf{d}_e} = \underbrace{\begin{bmatrix} 1 & 0 \\ 1 & l_e \end{bmatrix}}_{\mathbf{P}} \underbrace{\begin{Bmatrix} \alpha_0 \\ \alpha_1 \end{Bmatrix}}_{\alpha} = \mathbf{P}\alpha \tag{4.6}$$

For $l_e > 0$, the *moment matrix* $\mathbf{P}$ is clearly not singular. Thus, solving the above equation for α, we have

$$\underbrace{\begin{Bmatrix} \alpha_0 \\ \alpha_1 \end{Bmatrix}}_{\alpha} = \underbrace{\begin{bmatrix} 1 & 0 \\ -1/l_e & 1/l_e \end{bmatrix}}_{\mathbf{P}^{-1}} \underbrace{\begin{Bmatrix} u_1 \\ u_2 \end{Bmatrix}}_{\mathbf{d}_e} \tag{4.7}$$

Substituting the above equation back into Eq. (4.3), we obtain

$$u^h(x) = \mathbf{p}^T\alpha = \mathbf{p}^T \underbrace{\mathbf{P}^{-1}\mathbf{d}_e}_{\alpha} = \underbrace{\underbrace{\begin{Bmatrix} 1 & x \end{Bmatrix}}_{\mathbf{p}^T} \underbrace{\begin{bmatrix} 1 & 0 \\ -1/l_e & 1/l_e \end{bmatrix}}_{\mathbf{P}^{-1}}}_{\mathbf{N}(x)} \underbrace{\begin{Bmatrix} u_1 \\ u_2 \end{Bmatrix}}_{\mathbf{d}_e} \tag{4.8}$$

$$= \underbrace{\underbrace{\begin{Bmatrix} \underbrace{1 - x/l_e}_{N_1(x)} & \underbrace{x/l_e}_{N_2(x)} \end{Bmatrix}}_{\mathbf{N}(x)} \underbrace{\begin{Bmatrix} u_1 \\ u_2 \end{Bmatrix}}_{\mathbf{d}_e}$$

$$= \mathbf{N}(x)\mathbf{d}_e$$

which is in the form of Eq. (4.1). The matrix of shape functions is then obtained in the form

$$\mathbf{N}(x) = \begin{bmatrix} N_1(x) & N_2(x) \end{bmatrix} \tag{4.9}$$

where the shape functions for a truss element can be finally written as

$$N_1(x) = 1 - \frac{x}{l_e}$$
$$N_2(x) = \frac{x}{l_e} \tag{4.10}$$

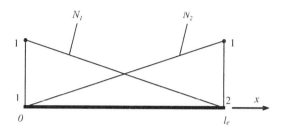

FIGURE 4.2

Linear shape functions for a truss element with two nodes.

Notice that we obtained two shape functions because we have two DOFs in the truss elements.

It is easy to verify that these two shape functions satisfy the delta function property defined by Eq. (3.34), and the partitions of unity in Eq. (3.41). We leave this verification to the reader as a simple exercise. The graphical representation of the linear shape functions is shown in Figure 4.2. It is clearly shown that N_i gives the *shape* of the contribution from nodal displacement at node i, and that is why it is called a shape function. In this case, the shape functions vary linearly across the element, and they are termed *linear shape functions*. Substituting Eqs. (4.9), (4.10), and (4.2) into Eq. (4.1), we have

$$u(x) = N_1(x)u_1 + N_2(x)u_2 = u_1 + \frac{u_2 - u_1}{l_e}x \tag{4.11}$$

which clearly states that the displacement within the element varies linearly for any given nodal displacements u_1 and u_2. The element is therefore called a *linear element*.

4.2.2 Strain matrix

As discussed in Chapter 2, there is only one stress component σ_x in a truss, and the corresponding strain can be obtained by

$$\varepsilon_x = \frac{\partial u}{\partial x} = \frac{u_2 - u_1}{l_e} \tag{4.12}$$

which is a direct result from differentiating Eq. (4.11) with respect to x. Note that the strain in Eq. (4.12) is a constant value (does not vary with x) in the element for any given nodal displacements u_1 and u_2.

It was mentioned in the previous chapter that we would need to first obtain the strain matrix, **B**, after which we can obtain the stiffness and mass matrices. In this two-node element case, this can be easily done. Equation (4.12) can be re-written in a matrix form as

$$\varepsilon_x = \frac{\partial u}{\partial x} = \underbrace{\frac{\partial}{\partial x}}_{L} \underbrace{u}_{\mathbf{N}\mathbf{d}_e} = \underbrace{L\mathbf{N}}_{\mathbf{B}} \, \mathbf{d}_e = \mathbf{B}\mathbf{d}_e \tag{4.13}$$

where the strain matrix **B** has the following form for the truss element:

$$\mathbf{B} = L\mathbf{N} = \frac{\partial}{\partial x}\left[1 - x/l_e \;\; x/l_e\right] = \left[-1/l_e \;\; 1/l_e\right] \tag{4.14}$$

Substituting **B** back to Eq. (4.13) gives Eq. (4.12).

4.2.3 **Element matrices in the local coordinate system**

Once the strain matrix **B** has been obtained, the stiffness matrix for truss elements can be obtained using Eq. (3.71) from the previous chapter:

$$\mathbf{k}_e = \int_{V_e} \mathbf{B}^T \mathbf{c}\mathbf{B}\, dV = A \int_0^{l_e} \underbrace{\begin{bmatrix} -1/l_e \\ 1/l_e \end{bmatrix}}_{\mathbf{B}^T} E \underbrace{\left[-1/l_e \;\; 1/l_e\right]}_{\mathbf{B}}\, dx$$

$$\underbrace{\qquad\qquad\qquad}_{\text{Constants: not functions of } x} \tag{4.15}$$

$$= AE \underbrace{\begin{bmatrix} -1/l_e \\ 1/l_e \end{bmatrix}}_{\mathbf{B}^T} \underbrace{\left[-1/l_e \;\; 1/l_e\right]}_{\mathbf{B}} \underbrace{\int_0^{l_e} dx}_{l_e} = \frac{AE}{l_e}\begin{bmatrix} 1 & -1 \\ -1 & 1 \end{bmatrix}$$

where A is the area of the cross-section of the truss element. Note that the material constant matrix **c** reduces to the elastic modulus, E, for the one-dimensional truss element (see Eq. (2.39)). It is noted that the element stiffness matrix as shown in Eq. (4.15) is symmetrical. This confirms the proof given in Eq. (3.73). Making use of the symmetry of the stiffness matrix, only half of the terms in the matrix need to be evaluated and stored during computation.

The mass matrix for truss elements can be obtained using Eq. (3.75):

$$\mathbf{m}_e = \int_{V_e} \rho \mathbf{N}^T \mathbf{N}\, dV = A\rho \int_0^{l_e} \begin{bmatrix} N_1 N_1 & N_1 N_2 \\ N_2 N_1 & N_2 N_2 \end{bmatrix} dx$$

$$= \begin{bmatrix} \underbrace{A\rho \int_0^{l_e} N_1 N_1\, dx}_{m_{11}} & \underbrace{A\rho \int_0^{l_e} N_1 N_2\, dx}_{m_{12}} \\ \underbrace{A\rho \int_0^{l_e} N_2 N_1\, dx}_{m_{21}} & \underbrace{A\rho \int_0^{l_e} N_2 N_2\, dx}_{m_{22}} \end{bmatrix} = \begin{bmatrix} m_{11} & m_{12} \\ m_{21} & m_{22} \end{bmatrix} = \frac{A\rho l_e}{6}\begin{bmatrix} 2 & 1 \\ 1 & 2 \end{bmatrix} \tag{4.16}$$

Similarly, the mass matrix is found to be symmetrical.

The nodal force vector for truss elements comprises contributions from the body force and the surface force and can be obtained using Eqs. (3.78), (3.79), and (3.81). Suppose, as an example, the element is loaded by an evenly distributed force f_x (N/m) along the x axis, and two concentrated forces F_{s1} (N) and F_{s2} (N), respectively, at two nodes, 1 and 2, as shown in Figure 4.1. Note that for a 1D truss element, a force distributed along the length of the truss is considered a body force (like for example, the self weight of the truss if that is to be taken into consideration). On the other hand, surface forces are the forces applied to the end nodes. The total nodal force vector can therefore be written as

$$
\mathbf{f}_e = \int_{V_e} \mathbf{N}^T f_b \, dV + \int_{S_e} \mathbf{N}^T f_s \, dS = \underbrace{A f_b}_{f_x} \int_0^{l_e} \begin{bmatrix} N_1 \\ N_2 \end{bmatrix} dx + \begin{Bmatrix} F_{s1} \\ F_{s2} \end{Bmatrix}
$$

$$
= \begin{Bmatrix} f_x l_e/2 + F_{s1} \\ f_x l_e/2 + F_{s2} \end{Bmatrix}
$$

(4.17)

Notice that an evenly distributed force acting along the length of the truss element contributes equally at the two nodes, as shown in Eq. (4.17). Note also that the "surface" force for our 1D problem is just the two concentrated point forces at the endpoints of the bar.

4.2.4 Element matrices in the global coordinate system

Element matrices in Eqs. (4.15), (4.16), and (4.17) were formulated based on the *local coordinate system*, where the x axis coincides with the mid axis of the bar 1–2, shown in Figure 4.1. In a typical assemblage of trusses, there are many bars of different orientations and at different locations. To assemble all the element matrices to form the global system matrices, a coordinate transformation has to be performed for each element to formulate its element matrix based on the *global coordinate system* for the entire truss structure. The following performs the transformation for both spatial and planar trusses.

4.2.4.1 *Spatial trusses*

Assume that the local nodes 1 and 2 of the element correspond to the global nodes i and j, respectively, as shown in Figure 4.1. The displacement at a global node in space should have three components in the X, Y, and Z directions, and numbered sequentially. For example, the three components at the ith node are denoted by D_{3i-2}, D_{3i-1}, and D_{3i}. The coordinate transformation gives the relationship between the displacement vector $\mathbf{d}_e$ based on the local coordinate system and the displacement vector $\mathbf{D}_e$ for the same element, but based on the global coordinate system XYZ:

$$
\mathbf{d}_e = \mathbf{T}\mathbf{D}_e
$$

(4.18)

where

$$
\mathbf{D}_e = \begin{Bmatrix} D_{3i-2} \\ D_{3i-1} \\ D_{3i} \\ D_{3j-2} \\ D_{3j-1} \\ D_{3j} \end{Bmatrix}
$$

(4.19)

and **T** is the *transformation matrix* for the truss element, given by

$$\mathbf{T} = \begin{bmatrix} l_{ij} & m_{ij} & n_{ij} & 0 & 0 & 0 \\ 0 & 0 & 0 & l_{ij} & m_{ij} & n_{ij} \end{bmatrix}_e \tag{4.20}$$

in which

$$l_{ij} = \cos(x, X) = \frac{X_j - X_i}{l_e}$$

$$m_{ij} = \cos(x, Y) = \frac{Y_j - Y_i}{l_e} \tag{4.21}$$

$$n_{ij} = \cos(x, Z) = \frac{Z_j - Z_i}{l_e}$$

are the *direction cosines* of the axial axis of the element, which are the projections of the local axial axis onto the global X, Y, and Z axes. It is easy to confirm that

$$\mathbf{T}\mathbf{T}^T = \mathbf{I} \tag{4.22}$$

where **I** is an identity matrix of 2×2. Therefore, matrix **T** is an *orthogonal matrix*. The length of the element, l_e, can be calculated using the global coordinates of the two nodes of the element by

$$l_e = \sqrt{(X_j - X_i)^2 + (Y_j - Y_i)^2 + (Z_j - Z_i)^2} \tag{4.23}$$

Equation (4.18) can be easily verified, as it simply says that at node i, d_1 equals the summation of all the projections of D_{3i-2}, D_{3i-1}, and D_{3i} onto the local x axis, and the same can be said for node j. The matrix **T** for a truss element transforms a 6×1 vector in the global coordinate system into a 2×1 vector in the local coordinate system.

The transformation matrix also applies to the force vectors between the local and global coordinate systems:

$$\mathbf{f}_e = \mathbf{T}\mathbf{F}_e \tag{4.24}$$

where

$$\mathbf{F}_e = \begin{Bmatrix} F_{3i-2} \\ F_{3i-1} \\ F_{3i} \\ F_{3j-2} \\ F_{3j-1} \\ F_{3j} \end{Bmatrix} \tag{4.25}$$

in which F_{3i-2}, F_{3i-1}, and F_{3i} represent the three components of the force vector at node i based on the global coordinate system.

Substitution of Eq. (4.18) into Eq. (3.89) leads to the element equation based on the global coordinate system:

$$\mathbf{k}_e \mathbf{T} \mathbf{D}_e + \mathbf{m}_e \mathbf{T} \ddot{\mathbf{D}}_e = \mathbf{f}_e \tag{4.26}$$

Pre-multiply $\mathbf{T}^T$ to both sides in the above equation to obtain:

$$\left(\mathbf{T}^T \mathbf{k}_e \mathbf{T}\right) \mathbf{D}_e + \left(\mathbf{T}^T \mathbf{m}_e \mathbf{T}\right) \ddot{\mathbf{D}}_e = \mathbf{T}^T \mathbf{f}_e \tag{4.27}$$

or

$$\mathbf{K}_e \mathbf{D}_e + \mathbf{M}_e \ddot{\mathbf{D}}_e = \mathbf{F}_e \tag{4.28}$$

where

$$\mathbf{K}_e = \mathbf{T}^T \mathbf{k}_e \mathbf{T}$$

$$= \frac{AE}{l_e}
\begin{bmatrix}
l_{ij}^2 & l_{ij}m_{ij} & l_{ij}n_{ij} & -l_{ij}^2 & -l_{ij}m_{ij} & -l_{ij}n_{ij} \\
l_{ij}m_{ij} & m_{ij}^2 & m_{ij}n_{ij} & -l_{ij}m_{ij} & -m_{ij}^2 & -m_{ij}n_{ij} \\
l_{ij}n_{ij} & m_{ij}n_{ij} & n_{ij}^2 & -l_{ij}n_{ij} & -m_{ij}n_{ij} & -n_{ij}^2 \\
-l_{ij}^2 & -l_{ij}m_{ij} & -l_{ij}n_{ij} & l_{ij}^2 & l_{ij}m_{ij} & l_{ij}n_{ij} \\
-l_{ij}m_{ij} & -m_{ij}^2 & -m_{ij}n_{ij} & l_{ij}m_{ij} & m_{ij}^2 & m_{ij}n_{ij} \\
-l_{ij}n_{ij} & -m_{ij}n_{ij} & -n_{ij}^2 & l_{ij}n_{ij} & m_{ij}n_{ij} & n_{ij}^2
\end{bmatrix} \tag{4.29}$$

and

$$\mathbf{M}_e = \mathbf{T}^T \mathbf{m}_e \mathbf{T}$$

$$= \frac{A\rho l_e}{6}
\begin{bmatrix}
2l_{ij}^2 & 2l_{ij}m_{ij} & 2l_{ij}n_{ij} & l_{ij}^2 & l_{ij}m_{ij} & l_{ij}n_{ij} \\
2l_{ij}m_{ij} & 2m_{ij}^2 & 2m_{ij}n_{ij} & l_{ij}m_{ij} & m_{ij}^2 & m_{ij}n_{ij} \\
2l_{ij}n_{ij} & 2m_{ij}n_{ij} & 2n_{ij}^2 & l_{ij}n_{ij} & m_{ij}n_{ij} & n_{ij}^2 \\
l_{ij}^2 & l_{ij}m_{ij} & l_{ij}n_{ij} & 2l_{ij}^2 & 2l_{ij}m_{ij} & 2l_{ij}n_{ij} \\
l_{ij}m_{ij} & m_{ij}^2 & m_{ij}n_{ij} & 2l_{ij}m_{ij} & 2m_{ij}^2 & 2m_{ij}n_{ij} \\
l_{ij}n_{ij} & m_{ij}n_{ij} & n_{ij}^2 & 2l_{ij}n_{ij} & 2m_{ij}n_{ij} & 2n_{ij}^2
\end{bmatrix} \tag{4.30}$$

Note that the coordinate transformation preserves the symmetrical properties of both stiffness and mass matrices.

For the corresponding transformation of the force vector given in Eq. (4.17), we obtain

$$\mathbf{F}_e = \mathbf{T}^T \mathbf{f}_e$$

$$= \begin{Bmatrix} (f_x l_e/2 + F_{s1})l_{ij} \\ (f_x l_e/2 + F_{s1})m_{ij} \\ (f_x l_e/2 + F_{s1})n_{ij} \\ (f_y l_e/2 + F_{s2})l_{ij} \\ (f_y l_e/2 + F_{s2})m_{ij} \\ (f_y l_e/2 + F_{s2})n_{ij} \end{Bmatrix} \tag{4.31}$$

Note that the element stiffness matrix $\mathbf{K}_e$ and mass matrix $\mathbf{M}_e$ have a dimension of 6×6 in the three-dimensional global coordinate system, and the displacement $\mathbf{D}_e$ and the force vector $\mathbf{F}_e$ have a dimension of 6×1.

4.2.4.2 *Planar trusses*

For a planar truss, the global coordinates X–Y can be employed to represent the plane of the truss. All the formulations of coordinate transformation can be obtained from that for the spatial trusses counterpart by simply removing the rows and/or columns corresponding to the z- (or Z-) axis. The displacement at the global node i should have two components in the X and Y directions only: D_{2i-1} and D_{2i}. The coordinate transformation, which gives the relationship between the displacement vector $\mathbf{d}_e$ based on the local coordinate system and the displacement vector $\mathbf{D}_e$, has the same form as Eq. (4.18), except that

$$\mathbf{D}_e = \begin{Bmatrix} D_{2i-1} \\ D_{2i} \\ D_{2j-1} \\ D_{2j} \end{Bmatrix} \tag{4.32}$$

and the transformation matrix $\mathbf{T}$ is given by

$$\mathbf{T} = \begin{bmatrix} l_{ij} & m_{ij} & 0 & 0 \\ 0 & 0 & l_{ij} & m_{ij} \end{bmatrix} \tag{4.33}$$

The force vector in the global coordinate system is

$$\mathbf{F}_e = \begin{Bmatrix} F_{2i-1} \\ F_{2i} \\ F_{2j-1} \\ F_{2j} \end{Bmatrix} \tag{4.34}$$

All the other equations for a planar truss have the same form as the corresponding equations for a spatial truss. The $\mathbf{K}_e$ and $\mathbf{M}_e$ for the planar truss have a dimension of 4×4 in the global coordinate system. They are listed as follows:

$$
\mathbf{K}_e = \mathbf{T}^T \mathbf{k}_e \mathbf{T} =
\begin{bmatrix}
K_{11}^e & K_{12}^e & K_{13}^e & K_{14}^e \\
K_{12}^e & K_{22}^e & K_{23}^e & K_{24}^e \\
K_{13}^e & K_{23}^e & K_{33}^e & K_{34}^e \\
K_{14}^e & K_{24}^e & K_{34}^e & K_{44}^e
\end{bmatrix}
$$

$$
= \frac{AE}{l_e}
\begin{bmatrix}
l_{ij}^2 & l_{ij}m_{ij} & -l_{ij}^2 & -l_{ij}m_{ij} \\
l_{ij}m_{ij} & m_{ij}^2 & -l_{ij}m_{ij} & -m_{ij}^2 \\
-l_{ij}^2 & -l_{ij}m_{ij} & l_{ij}^2 & l_{ij}m_{ij} \\
-l_{ij}m_{ij} & -m_{ij}^2 & l_{ij}m_{ij} & m_{ij}^2
\end{bmatrix}
\tag{4.35}
$$

$$
\mathbf{M}_e = \mathbf{T}^T \mathbf{m}_e \mathbf{T} =
\begin{bmatrix}
M_{11}^e & M_{12}^e & M_{13}^e & M_{14}^e \\
M_{12}^e & M_{22}^e & M_{23}^e & M_{24}^e \\
M_{13}^e & M_{23}^e & M_{33}^e & M_{34}^e \\
M_{14}^e & M_{24}^e & M_{34}^e & M_{44}^e
\end{bmatrix}
$$

$$
= \frac{A\rho l_e}{6}
\begin{bmatrix}
2l_{ij}^2 & 2l_{ij}m_{ij} & l_{ij}^2 & l_{ij}m_{ij} \\
2l_{ij}m_{ij} & 2m_{ij}^2 & l_{ij}m_{ij} & m_{ij}^2 \\
l_{ij}^2 & l_{ij}m_{ij} & 2l_{ij}^2 & 2l_{ij}m_{ij} \\
l_{ij}m_{ij} & m_{ij}^2 & 2l_{ij}m_{ij} & 2m_{ij}^2
\end{bmatrix}
\tag{4.36}
$$

4.2.5 Boundary conditions

The stiffness matrix $\mathbf{K}_e$ in Eq. (4.28) is usually singular, because the whole structure can perform rigid body movements. There are two DOFs of rigid movements for planer trusses and three DOFs for space trusses. These rigid body movements are constrained by supports or displacement constraints. In practice, truss structures are fixed somehow to the ground or to a fixed main structure at a number of the nodes. When a node is fixed, the displacement at the node must be zero, while the structure takes external loadings. This fixed displacement boundary condition can be imposed on Eq. (4.28). The imposition leads to a cancelation of the corresponding rows and columns in the stiffness matrix. The reduced stiffness matrix becomes Symmetric Positive Definite (SPD), if sufficient displacements are constrained to remove all the possible rigid movements.

4.2.6 **Recovering stress and strain**

Equation (4.28) can be solved using standard routines and the displacements at all the nodes can be obtained after imposing sufficient boundary conditions. The displacements at any position other than the nodal positions can also be obtained by interpolation using the shape functions. The stress in a truss element can then be recovered using the following equation:

$$\sigma_x = E\mathbf{B}\mathbf{d}_e = E\mathbf{B}\mathbf{T}\mathbf{D}_e \tag{4.37}$$

In deriving the above equation, Hooke's law in the form of $\sigma = E\varepsilon$ is used, together with Eqs. (4.13) and (4.18).

4.3 **Worked examples**

EXAMPLE 4.1

A uniform bar subjected to an axial force
Consider a bar of uniform cross-sectional area, shown in Figure 4.3. The bar is fixed at one end and is subjected to a horizontal load of P at the free end. The dimensions of the bar are shown in the figure, and the bar is made of an isotropic material with Young's modulus E.

Exact solution

From Example 2.1, we have the exact solutions for the stress and displacements as,

$$\sigma_x = \frac{P}{A}; \quad u(x) = \frac{P}{EA}x; \quad u(x = l) = \frac{Pl}{EA} \tag{4.38}$$

FEM solution

Using one element, the bar is modeled as shown in Figure 4.4. From Eq. (4.15), the stiffness matrix of the bars is given by

$$\mathbf{K} = \mathbf{k}_e = \frac{AE}{l}\begin{bmatrix} 1 & -1 \\ -1 & 1 \end{bmatrix}$$

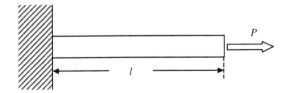

FIGURE 4.3

Clamped bar of uniform cross-section subject to a static axial load.

node 1   node 2

u_1 u_2

FIGURE 4.4

One truss element is used to model the lamped bar subject to a static load.

In this case, coordinate transformation is not required since both local and global coordinates are the same. There is also no need to perform assembly, because only one element is used. The finite element equation becomes

$$\frac{AE}{l}\begin{bmatrix} 1 & -1 \\ -1 & 1 \end{bmatrix}\begin{Bmatrix} u_1 \\ u_2 \end{Bmatrix} = \begin{Bmatrix} F_1 =? \\ F_2 = P \end{Bmatrix} \tag{4.39}$$

where F_1 is the reaction force applied at node 1, which is unknown at this stage. Instead, what we do know is the displacement boundary is condition at node 1:

$$u(x = 0) = 0; \quad \text{or} \quad u_1 = 0 \tag{4.40}$$

We can then simply remove the first equation in Eq. (4.39), i.e.

$$\frac{AE}{l}\begin{bmatrix} 1 & -1 \\ -1 & 1 \end{bmatrix}\begin{Bmatrix} u_1 \\ u_2 \end{Bmatrix} = \begin{Bmatrix} F_1 =? \\ F_2 = P \end{Bmatrix} \tag{4.41}$$

which leads to

$$u_2 = \frac{Pl}{AE} \tag{4.42}$$

This is the finite element solution of the bar, which is exactly the same as the exact solution obtained in Eq. (4.38). The distribution of the displacement in the bar can be obtained by substituting Eqs. (4.40) and (4.42) into Eq. (4.1),

$$u(x) = \mathbf{N}(x)\mathbf{d}_e = \left\{1 - \frac{x}{l} \quad \frac{x}{l}\right\}\begin{Bmatrix} u_1 \\ u_2 \end{Bmatrix} = \left\{1 - \frac{x}{l} \quad \frac{x}{l}\right\}\begin{Bmatrix} 0 \\ \frac{Pl}{EA} \end{Bmatrix} = \frac{P}{EA}x \tag{4.43}$$

which gives the same solution as the exact solution obtained in Eq. (4.38). Using Eqs. (4.37) and (4.14), we obtain the stress in the bar

$$\sigma_x = E\mathbf{B}\mathbf{d}_e = E\begin{bmatrix} -\frac{1}{l} & \frac{1}{l} \end{bmatrix}\begin{Bmatrix} 0 \\ u_2 \end{Bmatrix} = \frac{P}{A} \tag{4.44}$$

which is again the exact solution given in Eq. (4.38).

4.3.1 Properties of the FEM

4.3.1.1 Reproduction property of the FEM

Using the FEM, one can usually expect only an approximated solution. In Example 4.1, however, we obtained the exact solution. Why? This is because the exact solution of the deformation for the bar is a first order polynomial (see Eq. (4.38)). The shape functions used in our FEM analysis are also first order polynomials that are constructed using complete monomials up to the first order. Therefore, the exact solution of the problem is included in the set of assumed displacements in FEM shape functions. In Chapter 3, we understand that the FEM based on Hamilton's principle guarantees to choose the best possible solution that can be produced by the shape functions. In Example 4.1, the best possible solution that can be produced by the shape function is the exact solution, due to the reproduction property of the shape functions, and the FEM has indeed reproduced it exactly. We therefore verified the *reproduction* property of the FEM that if the exact solution can be formed by the basis functions used to construct the FEM shape function, the FEM will always produce the exact solution, provided there is no numerical error involved in computation of the FEM solution.

Making use of this property, one may try to deliberately include additional basis functions that form the exact solution or part of the exact solution, if that is possible, so as to achieve better accuracy in the FEM solution. This technique is known as enrichment and is frequently used in simulating singular stress fields (e.g., Chapter 10 in Liu and Nguyen, 2010).

4.3.1.2 Convergence property of the FEM

For complex problems in general, the exact or analytical solution (if it exists) cannot be written in the form of a combination of monomials. Therefore, the FEM using polynomial shape functions will not be able to produce the exact solution for such a problem. The question now is, how can one ensure that the FEM can produce a *good* approximation of the solution for a complex problem? The insurance is given by the *convergence* property of the FEM, which states that the FEM solution will converge to the exact solution that is continuous at arbitrary accuracy when the element size becomes infinitely small, and as long as the complete linear polynomial basis is included in the basis to form the FEM shape functions. The theoretical background for this convergence feature of the FEM is due to the fact that any continuous function can always be approximated by a first order polynomial with a second order refinement error. This fact can be revealed by using the local Taylor expansion, based on which a continuous (displacement) function $u(x)$ can always be approximated using the following equation:

$$u = u_i + \frac{\partial u}{\partial x}\bigg|_i (x - x_i) + O(h^2) \tag{4.45}$$

where h is the characteristic size that relates to $(x - x_i)$, or the size of the element.

According to Eq. (4.45), one may argue that the use of a constant can also reproduce the function u, but with an accuracy of $O(h^1)$. However, the constant displacements produced by the elements will not be continuous in between elements, unless the entire displacement field is constant (rigid movement), which is trivial. Therefore, to guarantee the convergence of a continuous solution, a complete polynomial up to at least the first order is used.

4.3.1.3 **Rate of convergence of FEM results**

The Taylor expansion up to the order of p can be given as

$$u = u_i + \left.\frac{\partial u}{\partial x}\right|_i (x - x_i) + \frac{1}{2!}\left.\frac{\partial^2 u}{\partial x^2}\right|_i (x - x_i)^2 + \cdots$$
$$+ \frac{1}{p!}\left.\frac{\partial^p u}{\partial x^p}\right|_i (x - x_i)^p + O\left(h^{p+1}\right) \qquad (4.46)$$

If the complete polynomials up to the pth order are used for constructing the shape functions, the first $(p+1)$ terms in Eq. (4.46) will be reproduced by the FEM shape function. The error is of the order of $O(h^{p+1})$; the order of the rate of convergence is therefore $O(h^{p+1})$. For linear elements we have $p=1$, and the order of the rate of convergence for the displacement is therefore $O(h^2)$. This implies that if the element size is halved, the error of the results in displacement will be reduced to one quarter: known as *quadratic convergence.*

These properties of the FEM, reproduction and convergence, are key for the FEM to provide reliable numerical results for mechanics problems, because we are assured as to what kind of results we are going to get. For simple problems whose exact solutions are of polynomial types, the FEM is capable of reproducing the exact solution using a minimum number of elements, as long as complete order of basis functions up to the order of the exact solution is used. In Example 4.1, one element of first order is sufficient. For complex problems whose exact solution is of a very high order of polynomial type, or often a non-polynomial type, it is then up to the analyst to use a proper density of the element mesh to obtain FEM results of desired accuracy where the error converges at the rate of h^{p+1} for the displacements.

As an extension of this discussion, we would like to introduce the concepts of so-called h-adaptivity and p-adaptivity that are intensively used in the recent development of FEM analyses. We conventionally use h to present the characteristic size of the elements, and p to represent the order of the polynomial basis functions. h-adaptive analysis uses finer element meshes (reducing h), and p-adaptivity analysis uses a higher order of shape functions (increasing p) to achieve the desired accuracy of FEM results .

EXAMPLE 4.2

A triangular truss structure subjected to a vertical force

Consider the plane truss structure shown in Figure 4.5. The structure is made of three planar truss members as shown, and a vertical downward force of 1000 N is applied at node 2. The figure also shows the numbering of the elements used (in squared numbers), as well as the numbering of the nodes (in circled numbers).

The local coordinates of the three truss elements are shown in Figure 4.6. The figure also shows the numbering of the global degrees of freedom, $D_1, D_2, ..., D_6$, corresponding to the three nodes in the structure. Note that there are six global degrees of freedom altogether, with each node having two degrees of freedom of displacement in the X and Y directions. However, there is actually only one degree of freedom in each node in the local coordinate system for each element. From the figure, it is shown clearly that the degrees of freedom at each node have contributions from more than one element. For example, at node 1, the global degrees of freedom D_1 and D_2 have a contribution from elements 1 and 2. These will play an important role in the assembly of the final finite element matrices. Table 4.1 shows the dimensions and material properties of the truss members in the structure. Calculate the displacements at nodes 2 and 3, as well as the axial stresses in the three truss members.

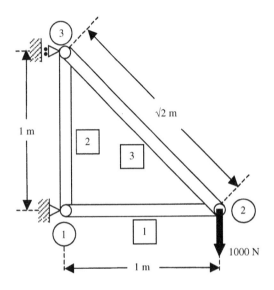

FIGURE 4.5

A three member truss structure.

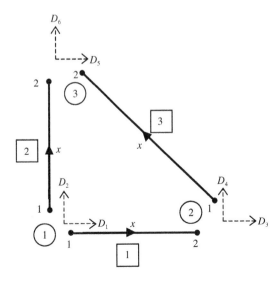

FIGURE 4.6

Local coordinates and degrees of freedom of truss elements.

Table 4.1   Dimensions and properties of truss members.

Element Number	Cross-Sectional Area, A_e (m^2)	Length le (m)	Young's Modulus E (N/m^2)
1	0.1	1	70×10^9
2	0.1	1	70×10^9
3	0.1	$\sqrt{2}$	70×10^9

Table 4.2   Global coordinates of nodes and direction cosines of elements.

Element Number	Global Node Corresponding to		Coordinates in Global Coordinate System		Direction Cosines	
	Local Node 1 (i)	Local Node 2 (j)	X_i, Y_i	X_j, Y_j	l_{ij}	m_{ij}
1	1	2	0, 0	1, 0	1	0
2	1	3	0, 0	0, 1	0	1
3	2	3	1, 0	0, 1	$-1/\sqrt{2}$	$1/\sqrt{2}$

Step 1: Obtaining the direction cosines of the elements

Knowing the coordinates of the nodes in the global coordinate system, the first step would be to take into account the orientation of the elements with respect to the global coordinate system. This can be done by computing the direction cosines using Eq. (4.21). Since this problem is a planar problem, there is no need to compute n_{ij}. The coordinates of all the nodes and the direction cosines of l_{ij} and m_{ij} are shown in Table 4.2.

Step 2: Calculation of element matrices in the global coordinate system

After obtaining the direction cosines, the element matrices in the global coordinate system can be obtained. Note that the problem here is a static one, hence there is no need to compute the element mass matrices. What is required is only the stiffness matrix. Recall that the element stiffness matrix in the local coordinate system is a 2×2 matrix, since the degrees of freedom are two for each element. However, in the transformation to the global coordinate system, the degrees of freedom for each element become four, therefore the element stiffness matrix in the global coordinate system is a 4×4 matrix. The stiffness matrices can be computed using Eq. (4.35), as is shown below:

$$\mathbf{K}^{e1} = \frac{(0.1)(70 \times 10^9)}{1.0} \begin{bmatrix} 1 & 0 & -1 & 0 \\ 0 & 0 & 0 & 0 \\ -1 & 0 & 1 & 0 \\ 0 & 0 & 0 & 0 \end{bmatrix}$$

$$= \begin{bmatrix} 7 & 0 & -7 & 0 \\ 0 & 0 & 0 & 0 \\ -7 & 0 & 7 & 0 \\ 0 & 0 & 0 & 0 \end{bmatrix} \times 10^9 \, \text{Nm}^{-1}$$

(4.47)

$$\mathbf{K}^{e2} = \frac{(0.1)(70 \times 10^9)}{1} \begin{bmatrix} 0 & 0 & 0 & 0 \\ 0 & 1 & 0 & -1 \\ 0 & 0 & 0 & 0 \\ 0 & -1 & 0 & 1 \end{bmatrix}$$

$$= \begin{bmatrix} 0 & 0 & 0 & 0 \\ 0 & 7 & 0 & -7 \\ 0 & 0 & 0 & 0 \\ 0 & -7 & 0 & 7 \end{bmatrix} \times 10^9 \ \mathrm{Nm}^{-1}$$

(4.48)

$$\mathbf{K}^{e3} = \frac{(0.1)(70 \times 10^9)}{\sqrt{2}} \begin{bmatrix} 1/2 & -1/2 & -1/2 & 1/2 \\ -1/2 & 1/2 & 1/2 & -1/2 \\ -1/2 & 1/2 & 1/2 & -1/2 \\ 1/2 & -1/2 & -1/2 & 1/2 \end{bmatrix}$$

$$= \begin{bmatrix} 7/2\sqrt{2} & -7/2\sqrt{2} & -7/2\sqrt{2} & 7/2\sqrt{2} \\ -7/2\sqrt{2} & 7/2\sqrt{2} & 7/2\sqrt{2} & -7/2\sqrt{2} \\ -7/2\sqrt{2} & 7/2\sqrt{2} & 7/2\sqrt{2} & -7/2\sqrt{2} \\ 7/2\sqrt{2} & -7/2\sqrt{2} & -7/2\sqrt{2} & 7/2\sqrt{2} \end{bmatrix} \times 10^9 \ \mathrm{Nm}^{-1}$$

(4.49)

Step 3: Assembly of global FE matrices

The next step after getting the element matrices will be to assemble the element matrices into a global matrix. Since the total global degrees of freedom in the structure are six, the global stiffness matrix will be a 6×6 matrix. The assembly is done by adding up the contributions for each node by the elements that share the node. For example, looking at Figure 4.6, element 1 contributes to the degrees of freedom D_1 and D_2 at node 1, and also to the degrees of freedom D_3 and D_4 at node 2. On the other hand, element 2 also contributes to degrees of freedom D_1 and D_2 at node 1, and also to D_5 and D_6 at node 3. By adding the contributions from the individual element matrices into the respective positions in the global matrix according the contributions to the degrees of freedom, the global matrix can be obtained. This assembly process is termed *direct assembly*.

At the beginning of the assembly, the entire global stiffness matrix is zeroed initially. By adding the element matrix for element $\boxed{1}$ into the global element, we have

$$
\mathbf{K} = 10^9 \times
\begin{array}{cccccc}
D_1 & D_2 & D_3 & D_4 & & \\
\uparrow & \uparrow & \uparrow & \uparrow & &
\end{array}
\left[
\begin{array}{cccccc}
7 & 0 & -7 & 0 & 0 & 0 \\
0 & 0 & 0 & 0 & 0 & 0 \\
-7 & 0 & 7 & 0 & 0 & 0 \\
0 & 0 & 0 & 0 & 0 & 0 \\
0 & 0 & 0 & 0 & 0 & 0 \\
0 & 0 & 0 & 0 & 0 & 0
\end{array}
\right]
\begin{array}{l}
\to D_1 \\
\to D_2 \\
\to D_3 \\
\to D_4 \\
\\
\\
\end{array}
\tag{4.50}
$$

Note that element 1 contributes to DOFs of D_1 to D_4. By adding the element matrix for element 2 on top of the new global element, it becomes

$$
\mathbf{K} = 10^9 \times
\begin{array}{cccccc}
D_1 & D_2 & & & D_5 & D_6 \\
\uparrow & \uparrow & & & \uparrow & \uparrow
\end{array}
\left[
\begin{array}{cccccc}
7+0 & 0+0 & -7 & 0 & 0 & 0 \\
0+0 & 0+7 & 0 & 0 & 0 & -7 \\
-7 & 0 & 7 & 0 & 0 & 0 \\
0 & 0 & 0 & 0 & 0 & 0 \\
0 & 0 & 0 & 0 & 0 & 0 \\
0 & -7 & 0 & 0 & 0 & 7
\end{array}
\right]
\begin{array}{l}
\to D_1 \\
\to D_2 \\
\\
\\
\to D_5 \\
\to D_6
\end{array}
\tag{4.51}
$$

Element $\boxed{2}$ contributes to DOFs of D_1, D_2, D_5, and D_6. Finally, by adding the element matrix for element $\boxed{3}$ on top of the current global element, we obtain

$$
\mathbf{K} = 10^9 \times
\begin{array}{cccccc}
& & D_3 & D_4 & D_5 & D_6 \\
& & \uparrow & \uparrow & \uparrow & \uparrow
\end{array}
\left[
\begin{array}{cccccc}
7 & 0 & -7 & 0 & 0 & 0 \\
0 & 7 & 0 & 0 & 0 & -7 \\
-7 & 0 & 7+7/2\sqrt{2} & -7/2\sqrt{2} & -7/2\sqrt{2} & 7/2\sqrt{2} \\
0 & 0 & -7/2\sqrt{2} & 7/2\sqrt{2} & 7/2\sqrt{2} & -7/2\sqrt{2} \\
0 & 0 & -7/2\sqrt{2} & 7/2\sqrt{2} & 7/2\sqrt{2} & -7/2\sqrt{2} \\
0 & -7 & 7/2\sqrt{2} & -7/2\sqrt{2} & -7/2\sqrt{2} & 7+7/2\sqrt{2}
\end{array}
\right]
\begin{array}{l}
\\
\\
\to D_3 \\
\to D_4 \\
\to D_5 \\
\to D_6
\end{array}
\tag{4.52}
$$

Element $\boxed{3}$ contributes to DOFs of D_3 to D_6. In summary, we have the final global stiffness matrix:

$$
\mathbf{K}=10^9\times
\begin{array}{cccccc}
D_1 & D_2 & D_3 & D_4 & D_5 & D_6 \\
\uparrow & \uparrow & \uparrow & \uparrow & \uparrow & \uparrow
\end{array}
$$

$$
\mathbf{K}=10^9\times
\begin{bmatrix}
7 & 0 & -7 & 0 & 0 & 0 \\
0 & 7 & 0 & 0 & 0 & -7 \\
-7 & 0 & 7+7/2\sqrt{2} & -7/2\sqrt{2} & -7/2\sqrt{2} & 7/2\sqrt{2} \\
0 & 0 & -7/2\sqrt{2} & 7/2\sqrt{2} & 7/2\sqrt{2} & -7/2\sqrt{2} \\
0 & 0 & -7/2\sqrt{2} & 7/2\sqrt{2} & 7/2\sqrt{2} & -7/2\sqrt{2} \\
0 & -7 & 7/2\sqrt{2} & -7/2\sqrt{2} & -7/2\sqrt{2} & 7+7/2\sqrt{2}
\end{bmatrix}
\begin{array}{l}
\to D_1 \\ \to D_2 \\ \to D_3 \\ \to D_4 \\ \to D_5 \\ \to D_6
\end{array}
\tag{4.53}
$$

The direct assembly process shown above is very simple, and can be coded in a computer program very easily. All one needs to do is add entries of the element matrix to the corresponding entries in the global stiffness matrix. The correspondence is usually facilitated using a so-called index that gives the relation between the element number and the global nodal numbers.

One may now ask how one can simply add up element matrices into the global matrix like this, and whether we can prove that this will indeed lead to the global stiffness matrix. The answer is that we can, and the proof can be performed simply using the equilibrium conditions at all of these nodes in the entire problem domain. The following gives such a simple proof.

We choose to prove the assembled result of the entries of the third row in the global stiffness matrix given in Eq. (4.53). The proof process applies exactly to all other rows. For the third row of the equation, we consider the equilibrium of forces in the x-direction at node 2 of elements 1 and 3, which corresponds to the third global DOF of the truss structure that links elements 1 and 3. For static problems, the FE equation for element 1 can be written in the following general form (in the global coordinate system):

$$
\begin{bmatrix}
K_{11}^{e1} & K_{12}^{e1} & K_{13}^{e1} & K_{14}^{e1} \\
K_{12}^{e1} & K_{22}^{e1} & K_{23}^{e1} & K_{24}^{e1} \\
K_{13}^{e1} & K_{23}^{e1} & K_{33}^{e1} & K_{34}^{e1} \\
K_{14}^{e1} & K_{24}^{e1} & K_{34}^{e1} & K_{44}^{e1}
\end{bmatrix}
\begin{Bmatrix}
D_1 \\ D_2 \\ D_3 \\ D_4
\end{Bmatrix}
=
\begin{Bmatrix}
F_1^{e1} \\ F_2^{e1} \\ F_3^{e1} \\ F_4^{e1}
\end{Bmatrix}
\tag{4.54}
$$

The third equation of Eq. (4.54), which corresponds to the third global DOF, is

$$
K_{13}^{e1}D_1 + K_{23}^{e1}D_2 + K_{33}^{e1}D_3 + K_{34}^{e1}D_4 = F_3^{e1}
\tag{4.55}
$$

The FE equation for element 3 can be written in the following general form:

$$
\begin{bmatrix}
K_{11}^{e3} & K_{12}^{e3} & K_{13}^{e3} & K_{14}^{e3} \\
K_{12}^{e3} & K_{22}^{e3} & K_{23}^{e3} & K_{24}^{e3} \\
K_{13}^{e3} & K_{23}^{e3} & K_{33}^{e3} & K_{34}^{e3} \\
K_{14}^{e3} & K_{24}^{e3} & K_{34}^{e3} & K_{44}^{e3}
\end{bmatrix}
\begin{Bmatrix}
D_3 \\ D_4 \\ D_5 \\ D_6
\end{Bmatrix}
=
\begin{Bmatrix}
F_3^{e3} \\ F_4^{e3} \\ F_5^{e3} \\ F_6^{e3}
\end{Bmatrix}
\tag{4.56}
$$

The first equation of Eq. (4.54), which corresponds to the third global DOF, is

$$K_{11}^{e3}D_3 + K_{12}^{e3}D_4 + K_{13}^{e3}D_5 + K_{14}^{e1}D_6 = F_3^{e3} \tag{4.57}$$

Forces in the x-direction applied at node 2 consist of element force F_3^{e3} from element 1 and F_3^{e3} from element 3, and the possible external force F_3. All these forces have to satisfy the following equilibrium equation:

$$F_3^{e1} + F_3^{e3} = F_3 \tag{4.58}$$

Substitution of Eqs. (4.55) and (4.57) into the above equation leads to

$$K_{13}^{e1}D_1 + K_{23}^{e1}D_2 + \left(K_{33}^{e1} + K_{11}^{e3}\right)D_3 + \left(K_{34}^{e1} + K_{12}^{e3}\right)D_4 + K_{13}^{e3}D_5 + K_{14}^{e3}D_6 = F_3 \tag{4.59}$$

This confirms that the coefficients on the left-hand side of the above equations are the entries for the third row of the global stiffness matrix given in Eq. (4.53). The above proof process is also valid for all the other rows of entries in the global stiffness matrix.

Step 4: Applying boundary conditions

The global matrix can normally be reduced in size after applying boundary conditions. In this case, D_1, D_2, and D_5 are constrained, and thus

$$D_1 = D_2 = D_5 = 0 \text{ m} \tag{4.60}$$

This implies that the first, second, and fifth rows and columns will actually have no effect on the solving of the matrix equation. Hence, we can simply remove the corresponding rows and columns:

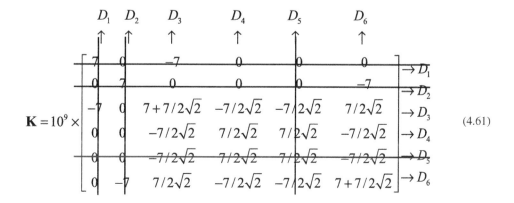

The condensed global matrix becomes a 3×3 matrix, given as follows:

$$\mathbf{K} = \begin{bmatrix} 7 + 7/2\sqrt{2} & -7/2\sqrt{2} & 7/2\sqrt{2} \\ -7/2\sqrt{2} & 7/2\sqrt{2} & -7/2\sqrt{2} \\ 7/2\sqrt{2} & -7/2\sqrt{2} & 7 + 7/2\sqrt{2} \end{bmatrix} \times 10^9 \ \text{Nm}^{-1} \tag{4.62}$$

It can easily be confirmed that this condensed stiffness matrix is SPD. The constrained global FE equation is

$$\mathbf{KD} = \mathbf{F} \tag{4.63}$$

where

$$D^T = [D_3 \quad D_4 \quad D_6] \tag{4.64}$$

and the force vector $\mathbf{F}$ is given as

$$\mathbf{F} = \left\{ \begin{array}{c} 1 \\ -1000 \\ 0 \end{array} \right\} \text{N} \tag{4.65}$$

Note that the only force applied is at node 2 in the downward direction of D_4. Equation (4.63) is actually equivalent to three simultaneous equations involving the three unknowns D_3, D_4, and D_6, as shown below:

$$\begin{aligned} [(7 + 7/2\sqrt{2})D_3 - (7/2\sqrt{2})D_4 + (7/2\sqrt{2})D_6] \times 10^9 &= 0 \\ [(-7/2\sqrt{2})D_3 + (7/2\sqrt{2})D_4 - (7/2\sqrt{2})D_6] \times 10^9 &= -1000 \\ [(7/2\sqrt{2})D_3 - (7/2\sqrt{2})D_4 + (7 + 7/2\sqrt{2})D_6] \times 10^9 &= 0 \end{aligned} \tag{4.66}$$

Step 5: Solving the FE matrix equation

The final step would be to solve the FE equation, Eqs. (4.63), (4.66), to obtain the solution for D_3, D_4, and D_6. Solving this equation manually is possible, since this only involves three unknowns in three equations. To this end, we obtain

$$\begin{aligned} D_3 &= -1.429 \times 10^{-7} \ \text{m} \\ D_4 &= -6.898 \times 10^{-7} \ \text{m} \\ D_6 &= -1.429 \times 10^{-7} \ \text{m} \end{aligned} \tag{4.67}$$

To obtain the stresses in the elements, Eq. (4.37) is used as follows:

$$\sigma_x^1 = E\mathbf{B}\mathbf{T}\mathbf{D}_e = 70 \times 10^9 \begin{bmatrix} -1 & 1 \end{bmatrix} \begin{bmatrix} 1 & 0 & 0 & 0 \\ 0 & 0 & 1 & 0 \end{bmatrix} \begin{bmatrix} 0 \\ 0 \\ -1.429 \times 10^{-7} \\ -6.898 \times 10^{-7} \end{bmatrix} \qquad (4.68)$$

$$= -10003 \text{ Pa}$$

$$\sigma_x^2 = E\mathbf{B}\mathbf{T}\mathbf{D}_e = 70 \times 10^9 \begin{bmatrix} -1 & 1 \end{bmatrix} \begin{bmatrix} 0 & 1 & 0 & 0 \\ 0 & 0 & 0 & 1 \end{bmatrix} \begin{bmatrix} 0 \\ 0 \\ 0 \\ -1.429 \times 10^{-7} \end{bmatrix} \qquad (4.69)$$

$$= -10003 \text{ Pa}$$

$$\sigma_x^3 = E\mathbf{B}\mathbf{T}\mathbf{D}_e = 70 \times 10^9 \begin{bmatrix} \frac{-1}{\sqrt{2}} & \frac{1}{\sqrt{2}} \end{bmatrix} \begin{bmatrix} \frac{-1}{\sqrt{2}} & \frac{-1}{\sqrt{2}} & 0 & 0 \\ 0 & 0 & \frac{-1}{\sqrt{2}} & \frac{1}{\sqrt{2}} \end{bmatrix} \begin{bmatrix} -1.429 \times 10^{-7} \\ -6.898 \times 10^{-7} \\ 0 \\ -1.429 \times 10^{-7} \end{bmatrix} \qquad (4.70)$$

$$= -14140 \text{ Pa}$$

In engineering practice, the problem can be of a much larger scale, and thus the unknowns or number of DOFs will also be very much more. Therefore, efficient equation solvers for FEM equations have to be used. Typical real-life engineering problems might involve hundreds of thousands, and even millions, of DOFs. Many kinds of such solvers are routinely available in math or numerical libraries in computer systems.

4.4 High order one-dimensional elements

For truss members that are free of body forces, there is no need to use higher order elements, as the linear element can already give the exact solution, as shown in Example 4.1. However, for truss members subjected to body forces arbitrarily distributed in the truss elements along their axial direction, higher order elements can be used for more accurate analysis. The procedure for developing such high order one-dimensional elements is the same as for the linear elements. The only difference is the shape function.

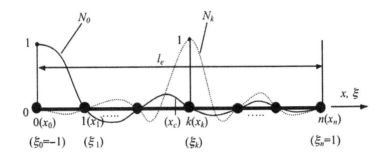

FIGURE 4.7

One-dimensional element of nth order with $(n+1)$ nodes.

In deriving high order shape functions, we usually use the natural coordinate ξ, instead of the physical coordinate x. The natural coordinate ξ is defined as

$$\xi = 2\frac{x - x_c}{l_e} \tag{4.71}$$

where x_c is the physical coordinate of the mid-point of the one-dimensional element. In the natural coordinate system, the element is defined in the range of $-1 \le \xi \le 1$. Figure 4.7 shows a one-dimensional element of nth order with $(n+1)$ nodes. The shape function of the element can be written in the following form using so-called Lagrange interpolants:

$$N_k(\xi) = l_k^n(\xi) \tag{4.72}$$

where $l_k^n(x)$ is the well-known Lagrange interpolant, defined as

$$l_k^n(\xi) = \frac{(\xi - \xi_0)(\xi - \xi_1)\cdots(\xi - \xi_{k-1})(\xi - \xi_{k+1})\cdots(\xi - \xi_n)}{(\xi_k - \xi_0)(\xi_k - \xi_1)\cdots(\xi_k - \xi_{k-1})(\xi_k - \xi_{k+1})\cdots(\xi_k - \xi_n)} \tag{4.73}$$

Note in this definition that the subscript starts from zero. From Eq. (4.73), it is clear that

$$N_k(\xi) = \begin{cases} 1 \text{ at node k where } \xi = \xi_k \\ 0 \text{ at other nodes} \end{cases} \tag{4.74}$$

Therefore, the high order shape functions defined by Eq. (4.72) are of the delta function property.

Using Eq. (4.73), *the quadratic one-dimensional element* with three nodes shown in Figure 4.8a can be obtained explicitly as

$$N_0(\xi) = -\frac{1}{2}\xi(1 - \xi)$$

$$N_1(\xi) = (1 + \xi)(1 - \xi) \tag{4.75}$$

$$N_2(\xi) = \frac{1}{2}\xi(1 + \xi)$$

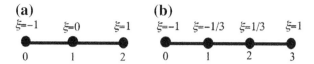

FIGURE 4.8

One-dimensional quadratic and cubic element with evenly distributed nodes. (a) Quadratic element; (b) cubic element.

The *cubic one-dimensional element* with four nodes shown in Figure 4.8b can be obtained as

$$N_0(\xi) = -\frac{1}{16}(1 - \xi)(1 - 9\xi)^2$$

$$N_1(\xi) = \frac{9}{16}(1 - 3\xi)(1 - \xi^2)$$

$$N_2(\xi) = \frac{9}{16}(1 + 3\xi)(1 - \xi)^2 \qquad (4.76)$$

$$N_3(\xi) = -\frac{1}{16}(1 + \xi)(1 - 9\xi)^2$$

4.5 Review questions

1. What are the characteristics of the joints in a truss structure and what are the effects of this on the deformation and stress properties in a truss element?
2. How many DOFs does a 2-nodal planar truss element have in its local coordinate system, and in the global coordinate system? Why is there a difference in DOFs in these two coordinate systems?
3. How many DOFs does a 2-nodal space truss element have in its local coordinate system, and in the global coordinate system? Why is there such a difference?
4. Derive the expression for all the entries for the expression for the element stiffness matrix, ke, with Young's modulus, E, length, le, and cross-sectional area, $A = 0.02x + 0.01$. (Note: non-uniform cross-sectional area.)
5. Derive the expression for all the entries for the expression for the element mass matrix, m_e, with the same properties as that in question 4 above.
6. Figure 4.9 shows a one-dimensional bar of length L and constant cross-sectional area A. The bar is made of functionally graded material (FGM). The Young's modulus of the FGM material $E(x)$ varies in the axial direction (x-direction). The bar is subjected to a uniformly distributed axial force F. The governing equation for the static problem of the bar is given by

$$\frac{\partial}{\partial x}\left(AE(x)\frac{\partial u}{\partial x}\right) + F = 0, \quad 0 \le x \le L$$

where u is the unknown axial displacement, and Young's modulus of the bar is given by

$$E(x) = E_0 + E_8\frac{x}{L}, \quad 0 \le x \le L$$

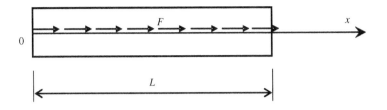

FIGURE 4.9

A one-dimensional bar of functionally graded material (FGM).

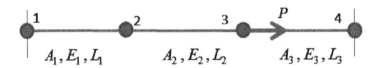

FIGURE 4.10

A truss structure made of 3 members along the horizontal direction.

where E_0 and E_g are given constants.
  a. Develop the finite element equations for a two-node linear element of length L.
  b. Discuss the results obtained.
7. Figure 4.10 shows a simple truss structure made of 3 members along the horizontal direction. The cross-sectional area, Young's modulus, and length are denoted as A, E, and L, respectively. Using FEM and 3 linear 1D elements:
  a. Give the global stiffness matrix **K**, in terms of A_1, A_2, A_3, E_1, E_2, E_3, L_1, L_2, and L_3, without considering boundary condition.
  b. Consider now nodes 1 and 4 are fixed, and let $A_1 = A_2 = A_3 = A$, $E_1 = E_2 = E_3 = E$, and $L_1 = L_2 = L_3 = L$. A force P acts at node 3 in the horizontal direction. Find expressions for the displacements at nodes 2 and 3 in terms of A, E, L, and P.
8. Work out the displacements and the stresses of the truss structure shown in Figure 4.11. All the truss members are of the same material ($E = 69.0 \, GPa$) and with the same cross-sectional area of $0.01 \, m^2$.
9. Figure 4.12 shows a truss structure with 5 uniform members made of the same material. The cross-sectional area of all the truss members is $0.01 \, m^2$, and the Young's modulus of the material is $2.1E11 \, N/m^2$. Using the finite element method:
  a. Calculate all the nodal displacements.
  b. Calculate the internal forces of all the truss members.
  c. Calculate the reaction forces at all the supports.

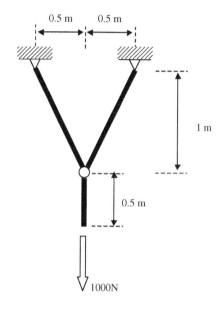

FIGURE 4.11

A 3 member planar truss structure.

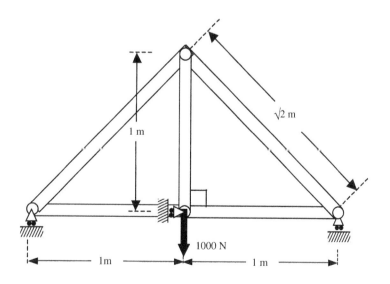

FIGURE 4.12

A 5 member planar truss structure.

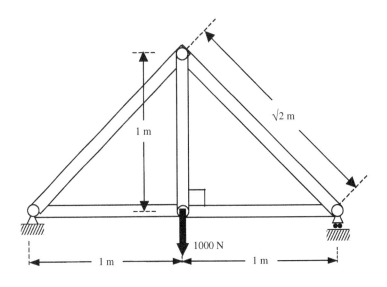

FIGURE 4.13

A 5 member planar truss structure.

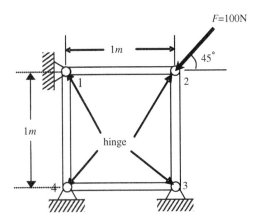

FIGURE 4.14

A 4 member planar truss structure.

10. Figure 4.13 indeed showed a simply supported structure with five uniform members made of the same material. The cross-sectional area of all the truss members is 0.01 m², and the Young's modulus of the material is 2.1E11 N/m². Using the finite element method:
   a. Calculate all the nodal displacements.
   b. Calculate the internal forces of all the truss members.
   c. Calculate the reaction forces at the supports.

11. Figure 4.14 shows a truss structure consisting of 4 truss members of the same dimensions. It is made of the same material with the Young's modulus $E = 210.0\,\text{GPa}$. The cross-sectional area of the truss members is $0.001\,\text{m}^2$. Using the finite element method calculate:
  a. The nodal displacement at node 2.
  b. The reaction forces at all the supports.
  c. The internal forces in all the truss members.

12. Figure 4.15 shows a three-node truss element of length L and a constant cross-sectional area A. It is made of a material of Young's modulus E and density ρ. The truss is subjected to a uniformly distributed force b:
  a. Derive the stiffness matrix for the element.
  b. Write down the expression for the element mass matrix, and obtain m_{11} in terms of L, E, ρ, and A.
  c. Derive the external force vector.

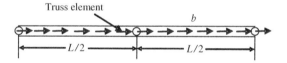

FIGURE 4.15

A truss element with 3 nodes.

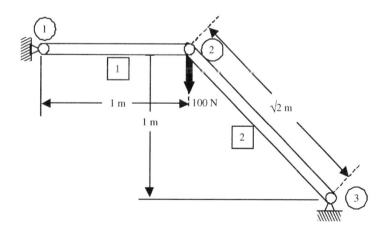

FIGURE 4.16

A truss element with 2 members.

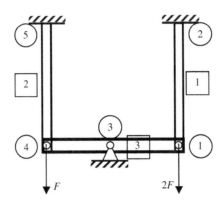

FIGURE 4.17

Two identical truss members connected by a ridged bar.

13. Figure 4.16 shows a truss structure with two uniform members made of the same material. The truss structure is constrained at two ends. The cross-sectional area of all the truss members is $0.01\,m^2$, and the Young's modulus of the material is $2.0E10\,N/m^2$. Using the finite element method, calculate:

    a. All the nodal displacements.

    b. The internal forces in all the truss members.

    c. The reaction forces at the supports.

14. Figure 4.17 shows two identical vertical truss members connected by a ridged horizontal bar. These three members are hinged together at nodes 1, 3, and 4. All the members are of length $L=2a$, cross-sectional area A, and Young's modulus E. Two forces are applied vertically at nodes 1 and 4. Using ONE truss element:

    a. Calculate the displacements at nodes 1 and 4, if any, in terms of F, L (or a), A, and E.

    b. Calculate the reaction forces at nodes 3, 2, and 5, if any, in terms of F, L (or a), A, and E.

FEM for Beams

5

5.1 Introduction

A *beam* is another simple but commonly used structural component. Geometrically, it is usually a slender straight *bar* of an arbitrary cross-section, but it deforms only in directions perpendicular to its axis. Note that the main difference between the beam and the truss is the type of load they carry and the resulting deformation directions. Beams are subjected to transverse loading, including transverse forces and moments that result in transverse deformation. Finite element equations for beams will be developed in this chapter, and the element developed is known as the *beam element*. The basic concepts, procedures, and formulations can also be found in many existing textbooks (see, for example, Petyt, 1990; Reddy, 1993; Rao, 1999; Zienkiewicz and Taylor, 2000).

In beam structures, the beams are joined together by welding (not by pins or hinges, as in the case of truss elements), so that both forces and moments can be transmitted between beams. In this book, the cross-section of the beam structure is assumed to be uniform. If a beam has a varying cross-section, one approach is to divide the beam into a series of shorter beams of constant cross-section to approximate the variation in cross-section. The finite element (FE) matrices for varying cross-sectional area can also be developed with ease using the same concepts that are introduced for a constant cross-section, by allowing the cross-sectional area changing as a function of the axis coordinate. The beam element developed in this chapter is based on the Euler–Bernoulli beam theory that is applicable for thin beams.

5.2 FEM equations

In planar beam elements there are two degrees of freedom (DOF) at a node in its local coordinate system. They are deflection in the y direction, v, and rotation in the $x-y$ plane, θ_z with respect to the z axis (see Section 2.5). Therefore, each two-noded beam element has a total of four DOFs.

5.2.1 Shape function construction

Consider a beam element of length $l=2a$ with nodes 1 and 2 at each end of the element, as shown in Figure 5.1. The local x-axis is taken in the axial direction of the element with its origin at the middle section of the beam. Similar to all other structures, in order to develop the FEM equations, shape functions for the interpolation of the variables from the nodal variables would have to be developed first. As there are four DOFs for a beam element, there should be four shape functions. It is often more convenient if the shape functions are derived from a special set of local coordinates, which is commonly known as the *natural coordinate system*. This natural coordinate system is dimensionless, has its origin at the center of the element, and the element is defined from -1 to $+1$, as shown in Figure 5.1.

The relationship between the natural coordinate system and the local coordinate system can simply be given as the following scaling function

$$\xi = \frac{x}{a} \tag{5.1}$$

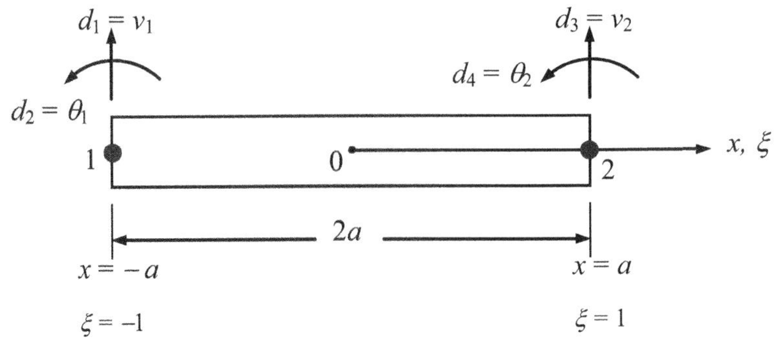

FIGURE 5.1

Beam element and its local coordinate systems: physical coordinates x, and natural coordinates ξ.

To derive the four shape functions in the natural coordinates, the displacement in an element is first assumed in the form of a third order polynomial of ξ that contains four unknown coefficients, a_0 to a_3:

$$v(\xi) = \alpha_0 + \alpha_1 \xi + \alpha_2 \xi^2 + \alpha_3 \xi^3 \tag{5.2}$$

The third order polynomial is chosen here because there are four unknowns in the polynomial, which can be related to the four nodal DOFs in the beam element. The above equation can have the following matrix form:

$$v(\xi) = \underbrace{[1 \ \xi \ \xi^2 \ \xi^3]}_{\mathbf{p}^T(\xi)} \underbrace{\begin{Bmatrix} \alpha_0 \\ \alpha_1 \\ \alpha_2 \\ \alpha_3 \end{Bmatrix}}_{\alpha} \tag{5.3}$$

or

$$v(\xi) = \mathbf{p}^T(\xi)\boldsymbol{\alpha} \tag{5.4}$$

where $\mathbf{p}$ is the vector of basis functions and $\boldsymbol{\alpha}$ is the vector of coefficients, as discussed in Chapters 3 and 4. The rotation θ can be obtained from the differential of Eq. (5.2) using the chain rule, bearing in mind the relationship between x and ξ in Eq. (5.1):

$$\theta = \frac{\partial v}{\partial x} = \frac{\partial v}{\partial \xi}\frac{\partial \xi}{\partial x} = \frac{1}{a}\frac{\partial v}{\partial \xi} = \frac{1}{a}(\alpha_1 + 2\alpha_2 \xi + 3\alpha_3 \xi^2) \tag{5.5}$$

The four unknown coefficients α_0 to α_3 can be determined by utilizing the following four conditions:

At $x = -a$ or $\xi = -1$:

$$
\begin{aligned}
&(1) \quad v(-1) = v_1 \\
&(2) \quad \left.\frac{dv}{dx}\right|_{\xi=-1} = \theta_1
\end{aligned} \tag{5.6}
$$

At $x = a$ or $\xi = 1$:

$$
\begin{aligned}
&(3) \quad v(1) = v_2 \\
&(4) \quad \left.\frac{dv}{dx}\right|_{\xi=1} = \theta_2
\end{aligned} \tag{5.7}
$$

Substituting Eqs. (5.2) and (5.5) into the above four conditions gives

$$
\begin{Bmatrix} v_1 \\ \theta_1 \\ v_2 \\ \theta_2 \end{Bmatrix} = \underbrace{\begin{bmatrix} 1 & -1 & 1 & -1 \\ 0 & 1/a & -2/a & 3/a \\ 1 & 1 & 1 & 1 \\ 0 & 1/a & 3/a & 3/a \end{bmatrix}}_{\mathbf{P}} \underbrace{\begin{Bmatrix} \alpha_0 \\ \alpha_1 \\ \alpha_2 \\ \alpha_3 \end{Bmatrix}}_{\alpha}
$$

$$\underbrace{\phantom{\begin{Bmatrix} v_1 \\ \theta_1 \\ v_2 \\ \theta_2 \end{Bmatrix}}}_{\mathbf{d}_e}$$

(5.8)

or

$$\mathbf{d}_e = \mathbf{P}\alpha \tag{5.9}$$

Since the moment matrix $\mathbf{P}$ is not singular, solving the above equation for α gives

$$\alpha = \mathbf{P}^{-1}\mathbf{d}_e \tag{5.10}$$

where

$$
\mathbf{P}^{-1} = \frac{1}{4}\begin{bmatrix} 2 & a & 2 & -a \\ -3 & -a & 3 & -a \\ 0 & -a & 0 & a \\ 1 & a & -1 & a \end{bmatrix} \tag{5.11}
$$

Substituting Eq. (5.10) back into Eq. (5.4) arrives at

$$v = \mathbf{N}(\xi)\mathbf{d}_e \tag{5.12}$$

where $\mathbf{N}$ is a matrix of *shape functions* given by

$$
\mathbf{N}(\xi) = \mathbf{p}^T(\xi)\mathbf{P}^{-1} = \underbrace{\begin{bmatrix} 1 & \xi & \xi^2 & \xi^3 \end{bmatrix}}_{\mathbf{p}^T(\xi)} \underbrace{\begin{bmatrix} 2 & a & 2 & -a \\ -3 & -a & 3 & -a \\ 0 & -a & 0 & a \\ 1 & a & -1 & a \end{bmatrix}\frac{1}{4}}_{\mathbf{P}^{-1}} \tag{5.13}
$$

$$= [N_1(\xi) \ N_2(\xi) \ N_3(\xi) \ N_4(\xi)]$$

in which the shape functions are found to be

$$N_1(\xi) = \frac{1}{4}(2 - 3\xi + \xi^3)$$
$$N_2(\xi) = \frac{a}{4}(1 - \xi - \xi^2 + \xi^3)$$
$$N_3(\xi) = \frac{1}{4}(2 + 3\xi + \xi^3)$$
$$N_4(\xi) = \frac{a}{4}(-1 - \xi + \xi^2 + \xi^3)$$

(5.14)

The first derivatives of this set of shape function becomes,

$$N_1'(\xi) = \frac{3}{4}(1 + \xi^2)$$
$$N_2'(\xi) = \frac{a}{4}(-1 - 2\xi + 3\xi^2)$$
$$N_3'(\xi) = \frac{-3}{4}(-1 + \xi^2)$$
$$N_4'(\xi) = \frac{a}{4}(-1 + 2\xi + 3\xi^2)$$

(5.15)

It can be easily confirmed that these shape functions given in Eq. (5.14) satisfy the Delta function property Eq. (3.34), and hence our interpolation using this set of shape functions will be passing node values. The two translational shape functions N_1 and N_3 also satisfy the conditions defined by Eq. (3.41), and hence the correct representation of translational rigid body movements is ensured. However, the two rotational shape functions N_2 and N_4 do not satisfy the conditions of Eq. (3.41). This is because these two shape functions are associated with the rotational DOFs, which are derived from the deflection functions. To ensure the correct representation of the rotational rigid body movement, the condition should be

$$N_2'(\xi = -1) = N_4'(\xi = 1)$$

(5.16)

which is also satisfied by Eq. (5.15). Therefore, all rigid body movements are properly represented in the beam element.

5.2.2 Strain matrix

Having now obtained the shape functions, the next step would be to obtain the element strain matrix. Substituting Eq. (5.12) into Eq. (2.47), which gives the relationship between the strain and the deflection, we obtain the normal stress,

$$\varepsilon_{xx} = \mathbf{B} \, \mathbf{d}_e$$

(5.17)

where the strain matrix **B** is given by

$$\mathbf{B} = -yL\mathbf{N} = -y\frac{\partial^2}{\partial x^2}\mathbf{N} = -\frac{y}{a^2}\frac{\partial^2}{\partial \xi^2}\mathbf{N} = -\frac{y}{a^2}\mathbf{N}'' \tag{5.18}$$

Note that Eqs. (2.48) and (5.1) have been used for the derivation of the above expression for **B**. Since the expressions for N_i have been derived in Eq. (5.14), we have

$$N'' = [N_1'' \quad N_2'' \quad N_3'' \quad N_4''] \tag{5.19}$$

where

$$N_1'' = \frac{3}{2}\xi, \; N_2'' = \frac{a}{2}(-1+3\xi)$$
$$N_3'' = \frac{3}{2}\xi, \; N_4'' = \frac{a}{2}(1+3\xi) \tag{5.20}$$

5.2.3 Element matrices

With the strain matrix **B** evaluated, we are now ready to obtain the element stiffness and mass matrices. By substituting Eq. (5.18) into Eq. (3.71), the stiffness matrix can be obtained as

$$\mathbf{k}_e = \int_V \mathbf{B}^T c\mathbf{B} \, dV = E \underbrace{\int_A y^2 dA}_{I_z} \int_{-a}^{a}\left(\frac{\partial^2}{\partial x^2}\mathbf{N}\right)^T\left(\frac{\partial^2}{\partial x^2}\mathbf{N}\right)dx$$

$$= EI_z \int_{-1}^{1}\frac{1}{a^4}\left[\frac{\partial^2}{\partial \xi^2}\mathbf{N}\right]^T\left[\frac{\partial^2}{\partial \xi^2}\mathbf{N}\right]a \, d\xi = \frac{EI_z}{a^3}\int_{-1}^{1}\mathbf{N}''^T\mathbf{N}'' d\xi \tag{5.21}$$

where $I_z = \int_A y^2 \, dA$ is the second moment of area (or moment of inertia) of the cross-section of the beam with respect to the z axis. Substituting Eq. (5.19) into (5.21), we obtain

$$\mathbf{k}_e = \frac{EI_z}{a^3}\int_{-1}^{1}\begin{bmatrix} N_1''N_1'' & N_1''N_2'' & N_1''N_3'' & N_1''N_4'' \\ N_2''N_1'' & N_2''N_2'' & N_2''N_3'' & N_2''N_4'' \\ N_3''N_1'' & N_3''N_2'' & N_3''N_3'' & N_3''N_4'' \\ N_4''N_1'' & N_4''N_2'' & N_4''N_3'' & N_4''N_4'' \end{bmatrix} d\xi \tag{5.22}$$

Evaluating the integrals in the above equation leads to

$$
\mathbf{k}_e = \frac{EI_z}{2a^3}
\begin{bmatrix}
3 & 3a & -3 & 3a \\
 & 4a^2 & -3a & 2a^2 \\
 & & 3 & -3a \\
sy. & & & 4a^2
\end{bmatrix}
\tag{5.23}
$$

For the mass matrix, we substitute Eq. (5.13) into Eq. (3.75):

$$
\mathbf{m}_e = \int_V \rho \mathbf{N}^T \mathbf{N}\, dV = \rho \int_A \mathbf{N}^T \mathbf{N}\, dx = \rho A \int_{-1}^{1} \mathbf{N}^T \mathbf{N} a\, d\xi
$$

$$
= \rho A a \int_{-1}^{1}
\begin{bmatrix}
N_1 N_1 & N_1 N_2 & N_1 N_3 & N_1 N_4 \\
N_2 N_1 & N_2 N_2 & N_2 N_3 & N_2 N_4 \\
N_3 N_1 & N_3 N_2 & N_3 N_3 & N_3 N_4 \\
N_4 N_1 & N_4 N_2 & N_4 N_3 & N_4 N_4
\end{bmatrix} d\xi
\tag{5.24}
$$

where A is the area of the cross-section of the beam. Evaluating the integral in the above equation leads to

$$
\mathbf{m}_e = \frac{\rho A a}{105}
\begin{bmatrix}
78 & 22a & 27 & -13a \\
 & 8a^2 & 13a & -6a^2 \\
 & & 78 & -22a \\
sy. & & & 8a^2
\end{bmatrix}
\tag{5.25}
$$

The other necessary element matrix would be the force vector. The nodal force vector for beam elements can again be obtained using the general expressions given in Eqs. (3.78), (3.79), and (3.81). We now assume that the element is loaded by an external, uniformly distributed force f_y along the x-axis, two concentrated forces F_{s1} and F_{s2}, and concentrated moments M_{s1} and M_{s2}, respectively, at nodes 1 and 2; the total nodal force vector can be obtained as

$$
\begin{aligned}
\mathbf{f}_e &= \int_V \mathbf{N}^T f_b\, dV + \int_{S_f} \mathbf{N}^T f_s\, dS_f \\
&= \int_{-a}^{a} A f_b \mathbf{N}^T\, dx + \int_{S_f} \mathbf{N}^T f_s\, dS_f \\
&= f_y\, a \int_{-1}^{1}
\begin{bmatrix}
N_1 \\ N_2 \\ N_3 \\ N_4
\end{bmatrix} d\xi +
\begin{Bmatrix}
F_{s1} \\ M_{s1} \\ F_{s2} \\ M_{s2}
\end{Bmatrix} =
\begin{Bmatrix}
f_y a + F_{s1} \\
f_y a^2/3 + M_{s1} \\
f_y a + F_{s2} \\
-f_y a^2/3 + M_{s2}
\end{Bmatrix}
\end{aligned}
\tag{5.26}
$$

Note that $f_y = A f_b$ and $dx = a\,d\xi$ when deriving the above expression. The final FEM equation for a beam element has the form of Eq. (3.89), and the element matrices are defined by Eqs. (5.21), (5.24), and (5.26).

5.3 Remarks

Theoretically, coordinate transformation can also be used to transform the beam element matrices from the local coordinate system into a global coordinate system. However, the transformation is necessary only if there is more than one beam element in the beam structure, and of these there are at least two beam elements of different orientations. A beam structure with at least two beam elements of different orientations is commonly termed a *frame* or *framework*. To analyze frames, frame elements, which carry both axial and bending forces, have to be used, and coordinate transformation is generally required. The formulation for frames is discussed in the next chapter.

5.4 Worked examples

EXAMPLE 5.1

A uniform cantilever beam subjected to a downward force
Consider the cantilever beam as shown in Figure 5.2. The beam is fixed at one end, and it has a uniform cross-sectional area as shown. The beam undergoes static deflection by a downward load of $P = 1000\,\text{N}$ applied at the free end. The dimensions of the beam are shown in the figure, and the beam is made of aluminum whose properties are shown in Table 5.1.

To emphasize the steps involved in solving this simple example, we just use one beam element to solve the deflection. The beam element would have a DOF as shown in Figure 5.1.

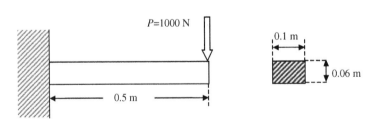

FIGURE 5.2

Cantilever beam of rectangular cross-section, subject to a static load.

Table 5.1 Material properties of aluminum.	
Young's Modulus, E (GPa)	**Poisson's Ratio, v**
69.0	0.33

Step 1: Obtaining the element matrices

The first step in formulating the FE equations is to form the element matrices and, in this case, being the only element used, the element matrices are actually the global FE matrices, i.e., no assembly is required. The shape functions for the four DOFs are given in Eq. (5.14).

The element stiffness matrix can be obtained using Eq. (5.21). Note that since this is a static problem, the mass matrix is not required here. The second moment of area of the cross-sectional area about the z axis can be given as

$$I_z = \frac{1}{12}bh^3 = \frac{1}{12}(0.1)\,(0.06)^3 = 1.8 \times 10^{-6}\ \mathrm{m}^4 \tag{5.27}$$

Since only one element is used, the final stiffness matrix of the structure is thus the same as the element stiffness matrix:

$$
\mathbf{K} = \mathbf{k}_e = \frac{(69 \times 10^9)\,(1.8 \times 10^{-6})}{2 \times 0.25^3}
\begin{bmatrix}
3 & 0.75 & -3 & 0.75 \\
0.75 & 0.25 & -0.75 & 0.125 \\
-3 & -0.75 & 3 & -0.75 \\
0.75 & 0.125 & -0.75 & 0.25
\end{bmatrix}
$$

$$
= 3.974 \times 10^6
\begin{bmatrix}
3 & 0.75 & -3 & 0.75 \\
0.75 & 0.25 & -0.75 & 0.125 \\
-3 & -0.75 & 3 & -0.75 \\
0.75 & 0.125 & -0.75 & 0.25
\end{bmatrix} \tag{5.28}
$$

The FE equation becomes

$$
3.974 \times 10^6
\begin{bmatrix}
3 & 0.75 & -3 & 0.75 \\
0.75 & 0.25 & -0.75 & 0.125 \\
-3 & -0.75 & 3 & -0.75 \\
0.75 & 0.125 & -0.75 & 0.25
\end{bmatrix}
\begin{Bmatrix}
v_1 \\ \theta_1 \\ v_2 \\ \theta_2
\end{Bmatrix}
=
\begin{Bmatrix}
Q_1 \\ M_1 \\ Q_2 = P \\ M_2 = 0
\end{Bmatrix} \tag{5.29}
$$

Note that, at node 1, the beam is clamped. Therefore, the shear force and moment at this node should be the reaction force and moment, which are unknowns before the FEM equation is solved for the displacements. To solve Eq. (5.29), we need to impose the displacement boundary condition at the clamped node.

Step 2: Applying boundary conditions

The beam is fixed or clamped at one end. This implies that at that end, the deflection, v_1, and the slope, θ_1, are both equal to zero:

$$v_1 = \theta_1 = 0 \tag{5.30}$$

The imposition of the above displacement boundary condition leads to the removal of the first and second rows and columns of the stiffness matrix:

$$3.974 \times 10^6 \begin{bmatrix} -3 & 0.75 & -3 & 0.75 \\ 0.75 & 0.25 & -0.75 & 0.125 \\ -3 & -0.75 & 3 & -0.75 \\ 0.75 & 0.125 & -0.75 & 0.25 \end{bmatrix} \begin{Bmatrix} v_1 = 0 \\ \theta_1 = 0 \\ v_2 \\ \theta_2 \end{Bmatrix} = \begin{Bmatrix} Q_1 \\ M_1 \\ Q_2 = P \\ M_2 = 0 \end{Bmatrix} \tag{5.31}$$

The reduced stiffness matrix becomes a 2×2 matrix of

$$\mathbf{K} = 3.974 \times 10^6 \begin{bmatrix} 3 & -0.75 \\ -0.75 & 0.25 \end{bmatrix} \tag{5.32}$$

The FE equation, after the imposition of the displacement condition, becomes

$$\mathbf{Kd} = \mathbf{F} \tag{5.33}$$

where

$$\mathbf{d}^T = [v_2 \; \theta_2] \tag{5.34}$$

and the force vector $\mathbf{F}$ is given as

$$\mathbf{F} = \begin{Bmatrix} -1000 \\ 0 \end{Bmatrix} \text{ N} \tag{5.35}$$

Note that, although we do not know the reaction shear force Q_1 and the moment M_1, it does not prevent it from solving the FEM equation because we know v_1 and θ_1 instead. This allows us to remove the unknowns of Q_1 and M_1 from the original FEM equation. We will come back to calculate the unknowns of Q_1 and M_1 after we have solved the FEM equations for all the displacements (deflections and rotations).

Step 3: Solving the FE matrix equation

The last step in this simple example would be to solve Eq. (5.33) to obtain v_2 and θ_2. In this case, Eq. (5.33) is actually two simultaneous equations involving two unknowns, and can be easily solved manually. Of course, when we have more unknowns or DOFs, some numerical methods of solving the matrix equation might be required. The solution to Eq. (5.33) is

$$v_2 = -3.3548 \times 10^{-4} \text{ m}$$
$$\theta_2 = -1.0064 \times 10^{-3} \text{ rad} \tag{5.36}$$

After v_2 and θ_2 have been obtained, they can be substituted back into the first two equations of Eq. (5.29) to obtain the reaction shear force at node 1:

$$Q_1 = 3.974 \times 10^6 \ (-3v_2 + 0.75\theta_2)$$
$$= 3.974 \times 10^6[-3 \times (-3.335 \times 10^{-4}) + 0.75 \times (-1.007 \times 10^{-3})]$$
$$= 1000.12 \ \text{N} \tag{5.37}$$

and the reaction moment at node 1:

$$M_1 = 3.974 \times 10^6 \ (-0.75v_2 + 0.125\theta_2)$$
$$= 3.974 \times 10^6[-0.75 \times (-3.335 \times 10^{-4}) + 0.125 \times (-1.007 \times 10^{-3})]$$
$$= 500.02 \ \text{Nm}$$
$$\tag{5.38}$$

This completes the solution process of this problem.

Note that this solution is exactly the same as the analytical solution in terms of the deflection. We again observe the reproduction feature of the FEM that was revealed in Example 4.1. In this case, it is because the exact solution of the deflection for the cantilever thin beam is a third order polynomial, which can be obtained easily by solving the strong form of the system equation of beam given by Eq. (2.59) with $f_y=0$. On the other hand, the shape functions used in our FEM analysis are also third order polynomials (see Eq. (5.14) or Eq. (5.2)). Therefore, the exact solution of the problem is included in the set of assumed deflections. The FEM based on Hamilton's principle has indeed reproduced the exact solution. This is, of course, also true if we were to calculate the deflection at anywhere else other than the nodes. For example, to compute the deflection at the center of the beam, we can use Eq. (5.12) with $x=0$, or in the natural coordinate system, $\xi=0$, and substituting the values calculated at the nodes:

$$v_{\xi=0} = \mathbf{N}_{\xi=0}\mathbf{d}_e = \begin{bmatrix} \frac{1}{2} & \frac{1}{16} & \frac{1}{2} & -\frac{1}{16} \end{bmatrix} \begin{Bmatrix} 0 \\ 0 \\ -3.355 \times 10^{-4} \\ -0.007 \times 10^{-3} \end{Bmatrix} = -1.048 \times 10^{-4} \text{m} \tag{5.39}$$

To calculate the rotation at the center of the beam, the derivatives of the shape functions are used as follows:

$$\theta_{\xi=0} = \left(\frac{dv}{dx}\right)_{\xi=0} = \left(\frac{d\mathbf{N}}{dx}\right)_{\xi=0}\mathbf{d}_e = \begin{bmatrix} -3 & -\frac{1}{4} & 3 & -\frac{1}{4} \end{bmatrix} \begin{Bmatrix} 0 \\ 0 \\ -3.355 \times 10^{-4} \\ -0.007 \times 10^{-3} \end{Bmatrix} \tag{5.40}$$

$$= -7.548 \times 10^{-4} \ \text{rad}$$

Note that in obtaining $d\mathbf{N}/dx$ above, the chain rule of differentiation is used, together with the relationship between x and ξ, as depicted in Eq. (5.1).

5.5 **Case study: resonant frequencies of micro-resonant transducer**

Making machines as small as insects, or even smaller, has been a dream of scientists for many years. Made possible by present lithographic techniques, such micro-systems are now being produced and applied in our daily lives. Such machines are called micro-electro-mechanical systems (MEMS), usually composed of mechanical and electrical devices. There are many MEMS devices being designed and manufactured today, from micro-actuators and sensors to micro-fluidic devices. The technology has very wide applications in areas such as communication, medical, aerospace, and robotics.

One of the most common MEMS devices is the resonant transducer. Resonant transducers convert externally induced beam strain into a beam resonant frequency change. This change in resonant frequency is then typically detected by implanted piezoresistors or optical techniques. Such resonant transducers are used for the measurement of pressure, acceleration, strain, and vibration. Figure 5.3 shows a micrograph of a micro-polysilicon resonant micro-beam transducer.

Figure 5.3 shows an overall view of the transducer, but the principle of the resonant transducer actually lies in the clamped–clamped bridge on top of a membrane. This bridge is actually located at the center of the micrograph. Figure 5.4 shows a schematic side view of the bridge structure. The resonant frequency of the bridge is related to the force applied to it (between anchor points), its material properties, cross-sectional area, and length. When the membrane deforms, for example, due to a change in pressure, the force applied to the bridge also changes, resulting in a change in the resonant frequency of the bridge.

It is thus important to analyze the resonant frequency of this bridge structure in the design of the resonant transducer. This case study first demonstrates the use of the beam element in the software ABAQUS to solve for the first three resonant frequencies of the bridge. Subsequently, results obtained using ANSYS for the same problem will also be shown for a comparison. More detailed discussions on using commercial software package are given in Chapter 13. The dimensions of the clamped–clamped bridge structure

FIGURE 5.3

Resonant micro-beam strain transducer.

Courtesy of Professor Henry Guckel and the University of Wisconsin-Madison.

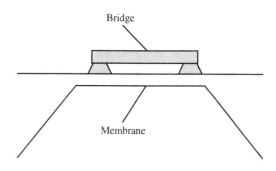

FIGURE 5.4

Bridge in a micro-resonant transducer.

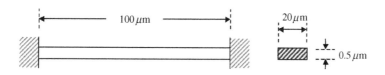

FIGURE 5.5

Geometrical dimensions of clamped–clamped bridge.

Table 5.2 Elastic properties of polysilicon.	
Elastic Properties of Polysilicon	
Young's modulus, E	169 GPa
Poisson's ratio, υ	0.262
Density, ρ	2300 kgm^{-3}

shown in Figure 5.5 are used to model a bridge in a micro resonant transducer. The material properties of polysilicon, of which the resonant transducer is normally made, are shown in Table 5.2.

5.5.1 Modeling

The modeling of the bridge is done using one-dimensional beam elements developed in this chapter. The beam is assumed to be clamped at two ends of the beam. The meshing of the structure should not pose any difficulty, but what is important here is the choice of how many elements to use to give sufficient accuracy. Because the exact solution of free vibration modes of the beam is no longer of a polynomial type, the FEM will not be able to produce the exact solution, but an approximated solution. One naturally becomes concerned with whether the results converge and whether they are accurate.

To start, the beam is meshed uniformly into ten two-nodal beam elements. This simple mesh will serve to show clearly the steps used in ABAQUS. Refined uniform meshes of 20, 40, and 60 elements will then be used to check the accuracy of the results obtained. This is a simplified way of performing what is commonly known as a convergence test. Remember that usually the greater the number of elements, the greater the accuracy. However, we cannot simply use as many elements as possible all the time, since there is usually a limit to the computer resources available. Hence, convergence tests are carried out to determine the optimum number of elements or nodes to be used for a certain problem. By "optimum," we mean the least number of elements or nodes to yield a desired accuracy within the acceptable tolerance.

5.5.2 ABAQUS input file

When using ABAQUS (or any other FEM software package), one often uses a graphic user interface, and almost everything can be done interactively on the screen. The analyst may not even need to know the input file for ABAQUS. This "black-box" setting is useful for beginners. However, understanding the data structure of the input file helps a lot in the appreciation of how the analysis is prepared for the computer to get the job done properly. Once getting used to the input file, the analyst can directly edit the file (in an "old" fashion) and have things (changing settings, modifying models, creating more running cases, etc.) done much more efficiently. Most of all, the analyst can have an important feeling of reassurance on what is being done by the computer. For this reason, we provide this section.

For ABAQUS, the pre-processor generates an input file in the form of a text file that includes information and parameters about the model. The ABAQUS input file for the above described FE model is shown below. In the early days, the analyst had to write these cards manually, but now it is generated by the pre-processors of FEM packages. Understanding the input file aids is important to the modeler in understanding the FEM and to effectively use the FEM packages. The ABAQUS input file for this problem is shown below. Note that the text boxes to the right of the input file are not part of the input file, but explain what the sections of the file mean.

```
* HEADING, SPARSE
ABAQUS job to calculate eigenvalues of beam
**
* NODE
1, 0., 0.
2, 10., 0.
3, 20., 0.
4, 30., 0.
5, 40., 0.
6, 50., 0.
7, 60., 0.
8, 70., 0.
9, 80., 0.
10, 90., 0.
11, 100., 0.
**
**
```

> *Nodal cards*
> These define the coordinates of the nodes in the model. The first entry is the node ID, while the second and third entries are the x and y coordinates of the position of the node, respectively.

```
* ELEMENT, TYPE=B23, ELSET=BEAM
1, 1, 2
2, 2, 3
3, 3, 4
4, 4, 5
5, 5, 6
6, 6, 7
7, 7, 8
8, 8, 9
9, 9, 10
10, 10, 11
**
** beam
**
```

Element (connectivity) cards
These define the element type and what nodes make up the element. B23 represents that it is a planar, cubic, Euler–Bernoulli beam element. There are many other beam element types in the ABAQUS element library. The "ELSET = BEAM" statement is simply for naming this set of elements so that it can be referenced when defining the material properties. In the subsequent data entry, the first entry is the element ID, and the following two entries are the nodes making up the element.

```
* BEAM SECTION, ELSET=BEAM, SECTION=RECT, MATERIAL=POLYSILI
20., 0.5,
0.0, 0.0, –1.0
**
**
**
**
**
**
**
```

Property cards
These define properties to the elements of set "BEAM." "SECT = RECT" describes the cross-section as a rectangle. ABAQUS provides a choice of other cross-sections. The first data line under "BEAM SECTION" defines the geometry of the cross-section. The second data line defines the normal, which in this case is for a planar beam. It will have the material properties defined under "POLYSILI."

```
** polysilicon
**
* MATERIAL, NAME=POLYSILI
**
* DENSITY
2.3E-15,
**
* ELASTIC, TYPE=ISO
169000., 0.262
**
**
```

Material cards
These define material properties under the name "POLYSILI." Density and elastic properties are defined. "TYPE=ISO" represents isotropic properties.

```
* BOUNDARY, OP=NEW
1, 1,, 0.
1, 2,, 0.
2, 1,, 0.
3, 1,, 0.
4, 1,, 0.
5, 1,, 0.
6, 1,, 0.
```

Boundary (BC) cards
These define boundary conditions. For nodes 1 and 11, the DOFs 1, 2, and 6 are constrained. For the rest, DOF 1 is constrained. Note that in ABAQUS, a planar beam has x and y translational displacements, as well as rotation about the z axis.

```
7, 1,, 0.
8, 1,, 0.
9, 1,, 0.
10, 1,, 0.
11, 1,, 0.
11, 2,, 0.
**
*BOUNDARY, OP=NEW
I, 6,, 0.
II, 6,, 0.
**
** Step 1, eigen
** LoadCase, Default
**
* STEP, NLGEOM
This load case is the default load case that always appears
*FREQUENCY
3, 0.,,, 30
**
**
**
* NODE PRINT, FREQ=1
U,
* NODE FILE, FREQ=1
U,
**
**
**
* END STEP
```

Control cards
These indicate the analysis step. In this case it is a "FREQUENCY" analysis or an eigenvalue analysis.

Output control cards
These define the output required. For example, in this case, we require the nodal output, displacement "U."

The input file above shows how a basic ABAQUS input file is set up. Note that all the input file does is provide the information necessary so that the program can utilize them to formulate and solve the FE equations. It may also be noticed that in the input file, there is no mention of the units of measurement used. This implies that the units must be consistent throughout the input file in all the information provided. For example, if the coordinate values of the nodes are in micrometers, the units for other values like the Young's modulus, density, forces, and so on must also undergo the necessary conversions in order to be consistent, before they are keyed into the pre-processor of ABAQUS and hence the input file. It is noted that in this case study, all the units are converted into micrometers to be consistent with the geometrical dimensions, as can be seen from the values of Young's modulus and density. This is the case for most FE software, and many times, errors in analysis occur due to negligence in ensuring the units' consistency. More details regarding ABAQUS input files will be provided in Chapter 13.

5.5.3 Solution process

Let's now try to relate the information provided in the input file with what is formulated in this chapter. The first part of the ABAQUS input normally describes the nodes and their coordinates (position). These lines are often called "nodal cards."[1] The second part of the input file are the so-called "element cards." Information regarding the definition of the elements using nodes is provided. For example, element 1 is formed by nodes 1 and 2. The element cards give the connectivity of the element or the order of the nodal number that forms the element. The connectivity is very important because a change in the order of the nodal numbers may lead to a breakdown of the computation. The connectivity is also used as the index for the direct assembly of the global matrices (see Example 4.2). This element and nodal information is required for determining the stiffness matrix (Eq. (5.21)) and the mass matrix (Eq. (5.24)).

The property cards define the properties (type of element, cross-sectional property, etc.) of the elements, as well as the material the element is made of. The cross-section of the element is defined here as it is required for computation of the moment of area about the z axis, which is in turn used in the stiffness matrix. The material properties defined are also a necessity for the computation of both the stiffness (elastic properties) and mass matrices (density).

The boundary cards (BC cards) define the boundary conditions for the model. In ABAQUS, a node of a *general beam element* (equivalent to the frame element, to be discussed in the next chapter) in the XY plane has three DOFs: Translational displacements in the x and y directions (1, 2), and the rotation about the z axis (6). To model just the transverse displacements and rotation as depicted in the formulation in this chapter, the x-displacement DOFs are constrained here. Hence it can be seen from the input file that the DOF "1" is constrained for all nodes. In addition to this, the two nodes at the ends, nodes 1 and 11, also have their "2" and "6" DOFs constrained to simulate clamped ends. Just as in the worked example previously, constraining these DOFs would effectively reduce the dimension of the matrix.

We should usually also have *load cards*. Because this case study is an eigenvalue analysis, there are no external loadings, and hence there is no need to define any loadings in the input file.

The control cards are used to control the analysis, such as defining the type of analysis required. ABAQUS uses the subspace iteration scheme by default to evaluate the eigenvalues of the equation of motion. This method is a very effective method of determining a number of lowest eigenvalues and corresponding eigenvectors for a very large system of several thousand DOFs. The procedure is as follows:

i. To determine the n lowest eigenvalues and eigenvectors, select a starting matrix $\mathbf{X}_1$ having m ($> n$) columns.

ii. Solve the equation $\mathbf{K}\overline{\mathbf{X}}_{k+1} = \mathbf{M}\mathbf{X}_k$ for $\overline{\mathbf{X}}_{k+1}$

iii. Calculate $\mathbf{K}_{k+1} = \overline{\mathbf{X}}_{k+1}^T \mathbf{K}\overline{\mathbf{X}}_{k+1}$ and $\mathbf{M}_{k+1} = \overline{\mathbf{X}}_{k+1}^T \mathbf{M}\overline{\mathbf{X}}_{k+1}$, where the dimension of $\mathbf{K}_{k+1}$ and $\mathbf{M}_{k+1}$ are of m by m.

iv. Solve the reduced eigenvalue problem $\mathbf{K}_{k+1}\psi_{k+1} - \mathbf{M}\psi_{k+1}\Lambda_{k+1} = 0$ for m eigenvalues, which are the diagonal terms of the diagonal matrix Λ_{k+1}, and for eigenvectors ψ_{k+1}.

---

[1] In the early 1980s, the input files were recorded on paper cards with punched holes, where each card recorded one line of letters or numbers represented by the arrangement of the holes. Optical devices were then used to read-in these cards for the computer. We do not use such "cards" anymore, but the name is still widely used.

v. Calculate the improved approximation to the eigenvectors of the original system using $\mathbf{X}_{k+1} = \overline{\mathbf{X}}_{k+1}\ \Psi_{k+1}$.

vi. Repeat the process until the eigenvalues and eigenvectors converge to the lowest eigenvectors to desired accuracy.

By specifying the line "*FREQUENCY" in the analysis step, ABAQUS will carry out a similar algorithm to that briefly explained above. The line after the "*FREQUENCY" contains some data which ABAQUS uses to aid the procedure. The first entry refers to the number of eigenvalues required (in this case, 3). The second refers to the maximum frequency of interest. This will limit the frequency range, and therefore anything beyond this frequency will not be calculated. In this case, no maximum frequency range will be specified. The third is to specify shift points, which are used to ensure that the stiffness matrix is not singular. Here, again, it is left blank since it is not necessary. The fourth is the number of columns of the starting matrix $\mathbf{X}_1$ to be used. It is left blank again, and thus ABAQUS will use its default setting. The last entry is the number of iterations, which in this case is 30.

Output control cards are used for selecting the data that needs to be output. This is very useful for large scale computation that produces huge data files; one needs to limit the output to what is really needed.

Once the input file is created, one can then invoke ABAQUS to execute the analysis, and the results will be written into an output file that can be read by the post-processor.

5.5.4 Results and discussion

Using the above input file, an analysis to calculate the eigenvalues, and hence the natural resonant frequencies of the bridge structure, is carried out using ABAQUS. Other than the 10-element mesh as shown in Figure 5.6, which is also depicted in the input file, a simple convergence test is carried out. Hence, there are similar uniform meshes using 20, 40, and 60 elements. All the frequencies obtained are given in Table 5.3. Because the clamped–clamped beam structure is a simple problem, it is possible to evaluate the natural frequencies analytically. The results obtained from analytical calculations are also shown in Table 5.3 for comparison.

From the table, it can be seen that the finite element results give very good approximations compared to the analytical results. Even with just 10 elements, the error of mode 1 frequency is about 0.016% from the analytical calculations. It can also be seen that as the number of elements increases, the finite element results get closer and closer to the analytical calculations, and converge such that the results obtained for 40 and 60 elements show no difference up to the fourth decimal place. What this implies is that in finite element analyses, the finer the mesh or the greater the number of elements used, the more accurate the results. However, using more elements will use up more computer resources and

FIGURE 5.6

Ten element mesh of clamped-clamped bridge.

Table 5.3   Resonant frequencies of bridge using FEA (ABAQUS) and analytical calculations.

Number of 2-Node Beam Elements	Natural Frequency (Hz)		
	Mode 1	Mode 2	Mode 3
10	4.4058×10^5	1.2148×10^6	2.3832×10^6
20	4.4057×10^5	1.2145×10^6	2.3809×10^6
40	4.4056×10^5	1.2144×10^6	2.3808×10^6
60	4.4056×10^5	1.2144×10^6	2.3808×10^6
Analytical calculations	4.4051×10^5	1.2143×10^6	2.3805×10^6

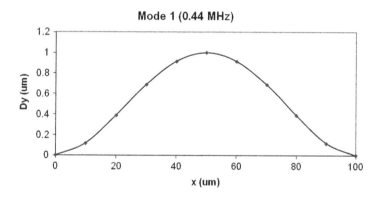

FIGURE 5.7

Mode 1 using 10 elements at 4.4285×10^5 Hz.

it will take a longer time to execute. Hence, it is advised to use the minimum number of elements which give the results of desired accuracy.

Other than the resonant frequencies, the mode shapes can also be obtained. Mode shapes can be considered to be the way in which the structure vibrates at a particular natural frequency. They correspond to the eigenvector of the finite element equation, just like the resonant frequencies correspond to the eigenvalues of the finite element equation. Mode shapes can be important in some applications, where the points of zero displacements, like the center of the beam in Figure 5.8, need to be identified for the installation of devices which should not undergo huge vibration.

The data for constructing the mode shape for each eigenvalue or natural frequency can be obtained from the displacement output for that natural frequency. Figures 5.7–5.9 show the mode shapes obtained by plotting the displacement components using 10 elements. The figures show how the clamped–clamped beam will vibrate at natural frequencies. Note that the output data usually consists of only the output at the nodes, and these are then used by the post-processor or any graph plotting applications to form a smooth curve. Most post-processors thus contain curve fitting functions to properly plot the curves using the data values.

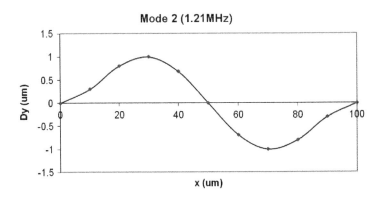

FIGURE 5.8

Mode 2 using 10 elements at 1.2284×10^6 Hz.

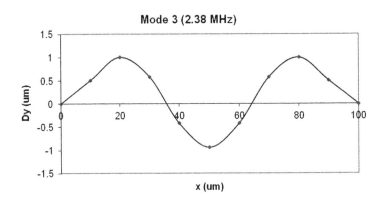

FIGURE 5.9

Mode 3 using 10 elements at 2.4276×10^6 Hz.

This simple case study points out some of the basic requirements needed in a finite element analysis. Like ABAQUS, most finite element software works on the same finite element principles. All that is needed is just to provide the necessary information for the software to use the necessary type of elements, and hence the shape functions; to build up the necessary element matrices; followed by the assembly of all the elements to form the global matrices; and finally, to solve the finite element equations. The next section in this case study will show a comparison with another popular FEM software package, ANSYS.

5.5.5 Comparison with ANSYS

In general, once one understands an FEM package well, it is generally much easier to learn than other packages. ANSYS is another popular FEM software package used widely by engineers. Although the user interface of ANSYS is different from that of ABAQUS, the FEM computation is essentially the same. Other differences between FEM software packages may include the use of different solvers. Because of these possible differences, the solutions can deviate from one another.

Most modelers using ANSYS run the simulation directly through its own software package via the graphical user interface (GUI). Nevertheless, the same information regarding the mesh (hence nodes and elements definition), material properties, beam section properties, analysis type, load, and boundary conditions are required by the software and they are keyed in through the GUI. Chapter 13 (Section 13.7) shows an example of setting up a simple cantilever beam model via the GUI in ANSYS. The procedures to set up this example model in ANSYS are similar to what is described in Chapter 13, except for the selection of the appropriate analysis type ("Modal" analysis is selected here, *cf.* Figures 13.12). Readers are referred to Chapter 13 to get an idea of how modeling can be done in ANSYS.

Figure 5.10 shows a screenshot of ANSYS with the deformation of the first mode plotted. The mode shape (eigenvector) of the first three modes obtained using ANSYS with 60 beam elements is the same as that obtained above using ABAQUS. From the inset window in Figure 5.10, the calculated natural frequencies for the first four modes are listed and they are found to be 4.4021×10^5 Hz, 1.2134×10^6 Hz, and 2.3786×10^6 Hz for modes 1, 2, and 3, respectively. While these values are close

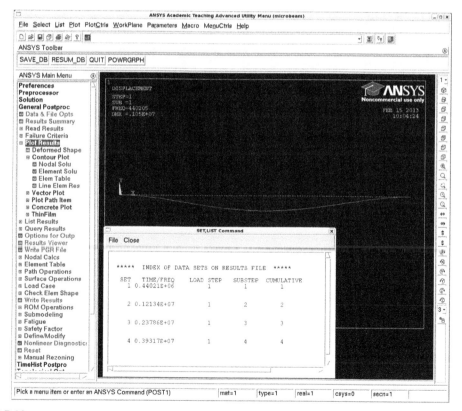

FIGURE 5.10

Screenshot of ANSYS showing the first eigenmode of the beam. The inset window shows the frequencies obtained using ANSYS.

to the analytical calculation in Table 5.3, some slight discrepancies in the frequencies can be observed. This can be explained by the different method chosen in ANSYS (amongst several options), to extract the eigenvalues for this problem. In this example, a Block Lanczos method is used to solve the eigenvalue problem, which can be as accurate as the subspace method, but is more efficient.

5.6 **Review questions**

1. How would you like to formulate a beam element that also carries axial forces?
2. Calculate the force vector for a simply supported beam subjected to a vertical force at the middle of the span, when only one beam element is used to model the beam, as shown in Figure 5.11.
3. Calculate the force vector for a simply supported beam subjected to a vertical force at the middle of the span, when two beam elements are used to model the beam, as shown in Figure 5.12.
4. Calculate the force vector for a simply supported beam subjected to a moment at the middle of the span, when one beam element is used to model the beam, as shown in Figure 5.13.
5. Figure 5.14 shows a uniform cantilever beam subjected to two vertical forces:
   a. Using one two-node beam element to model the entire beam, derive the external nodal force vector for the beam.
   b. Using two two-node elements of equal length, derive the external force vectors for these two elements.
   c. Discuss the accuracy of these two models.

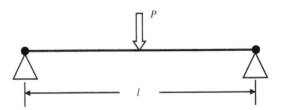

FIGURE 5.11

Simply supported beam modeled using one beam element.

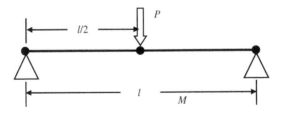

FIGURE 5.12

Simply supported beam modeled using two beam elements.

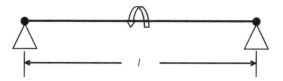

FIGURE 5.13

Simply supported beam modeled using one beam element.

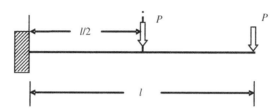

FIGURE 5.14

A cantilever beam subjected to two concentrated forces.

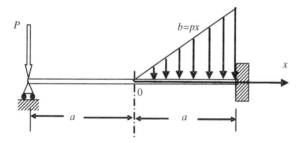

FIGURE 5.15

A beam subjected to concentrated and distributed forces.

6. Figure 5.15 shows a uniform beam subjected to a vertical concentrate force P and a distributed force $b = px$ on the right half span of the beam where p is a given constant.
   a. Using one two-node beam element to model the entire beam, derive the external nodal force vector for the beam.
   b. Using two two-node elements of equal length, give the formula to calculate the force vectors for these two elements (you are not required to carry out the integration for the right-hand side element, but full details of the integrand must be given).

7. For a cantilever beam subjected a vertical force at its free end, how many elements should be used to obtain the exact solution for the deflection of the beam?

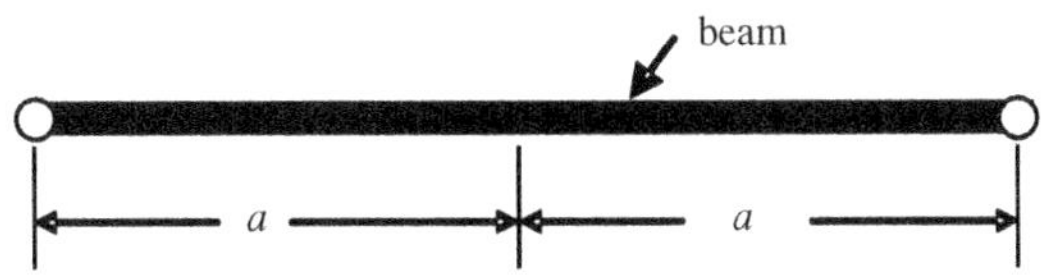

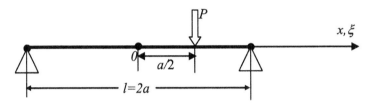

FIGURE 5.16

A beam with varying circular cross-section.

FIGURE 5.17

A simply supported beam.

8. Figure 5.16 shows a two-node beam of length $2a$. The beam is made of material of Young's modulus E and density ρ. The cross-section of the beam is circular, and the radius varies in the x-direction in a manner so that the area of the cross-section is a linear function of

$$A(x) = A_0 + \frac{A_1}{a}x.$$

   a. Write down the expression for the element stiffness matrix.
   b. Obtain k_{11} in terms of a, E, A_0, and A_1.
   c. Write down the expression for the element mass matrix.
   d. Obtain m_{11} in terms of a, ρ, A_0, and A_1.

9. Consider a beam with uniform rectangular cross-section, as shown in Figure 5.17. It is simply supported and subjected to a vertical load P at $x = a/2$. Using two elements for the entire beam, make use of the symmetry of the structure, and let $l = 2$ m, $E = 69$ GPa and $\upsilon = 0.3$, $b = 0.1$ m, $h = 0.06$ m, and

$$I_z = \tfrac{1}{12}bh^3$$

   a. Calculate the deflection v_c and rotation θ_c at the central point $(x = 0)$.
   b. Calculate the rotations at two ends.

FEM for Frames

6

6.1 Introduction

A frame element is formulated to model a straight *bar* of an arbitrary cross-section, which can deform not only in the axial direction but also in the directions perpendicular to the axis of the bar. The bar here is capable of carrying both axial and transverse forces, as well as moments. Therefore, a frame element is seen to possess the properties and capabilities of both truss and beam elements. In fact, the frame structure can be found in many real world structures. The developments of FEM equations for bar and beam elements facilitate the development of FEM equations for frame structures in this chapter.

The frame element developed in this chapter is also known in many commercial software packages as the *general beam element*, or even simply the *beam element*. Commercial software packages usually offer both pure beam and frame elements, but frame structures are more often used in actual engineering applications. A three-dimensional spatial frame structure can practically take forces and moments of all directions. Hence, it can be considered to be the most general form of element with a one-dimensional geometry.

Frame elements are applicable for the analysis of skeletal type systems of both *planar frames* (two-dimensional frames) and *space frames* (three-dimensional frames). A typical three-dimensional frame structure is shown in Figure 6.1. Structural members in a frame structure are joined together by welding so that both forces and moments can be transmitted between members. In the following formulation, we assume that the frame element has a uniform cross-sectional area. If a structure of varying cross-section is to be modeled using the formulation in this chapter, then one approach is to divide the structure into smaller elements each with a different but constant cross-sectional area so as to simulate the varying cross-section. If the variation in cross-section is too severe, then the equations for a varying cross-sectional area can also be formulated easily using the same concepts and procedures given in this chapter. The basic concepts, procedures, and formulations can also be found in many existing textbooks (see, for example, Petyt, 1990; Rao, 1999).

FIGURE 6.1

Example of a space frame structure.

6.2 **FEM equations for planar frames**

Consider a frame structure whereby the structure is divided into frame elements connected by nodes. Each element is of length $l_e = 2a$, and has two nodes at its two ends. The elements and nodes are numbered separately in a convenient manner. In a planar frame element, there are three degrees of freedom (DOFs) at one node in its local coordinate system, as shown in Figure 6.2. They are the axial deformation in the x direction, u; deflection in the y direction, v; and the rotation in the x–y plane and with respect to the z axis, θ_z. Therefore, each 2-node frame element has a total of six DOFs.

6.2.1 **The idea of superposition**

As mentioned, a frame element contains both the properties of the truss element and the beam element developed in the previous two chapters. Therefore, a simple way of formulating the element matrices is by superposition of the element matrices corresponding to the truss onto those of beam elements, without going through the detailed formulation process.

6.2.2 **Equations in the local coordinate system**

Considering the frame element shown in Figure 6.2 with nodes labeled 1 and 2 at each end of the element, it can be seen that the local x axis is taken as the axial direction of the element with its origin at the middle of the element. Recall that the truss element has only one DOF at each node (axial deformation), and the beam element has two DOFs at each node (transverse deformation and rotation).

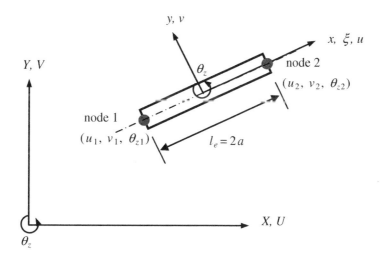

FIGURE 6.2

Planar frame element and the DOFs.

Combining these will give the three DOFs of a frame element, and the element displacement vector for a frame element can thus be written as

$$
\mathbf{d}_e = \begin{Bmatrix} d_1 \\ d_2 \\ d_3 \\ d_4 \\ d_5 \\ d_6 \end{Bmatrix} = \begin{Bmatrix} \left. \begin{matrix} u_1 \\ v_1 \\ \theta_{z1} \end{matrix} \right\} \text{displacement components at node } 1 \\ \left. \begin{matrix} u_2 \\ v_2 \\ \theta_{z2} \end{matrix} \right\} \text{displacement components at node } 2 \end{Bmatrix} \tag{6.1}
$$

To include the contribution of the stiffness matrix for the truss element, Eq. (4.15) is first extended to a 6×6 matrix corresponding to the order of the DOFs of the truss element in the element displacement vector in Eq. (6.1):

$$
\mathbf{k}_e^{\text{truss}} = \begin{matrix} d_1 = u_1 & & & d_4 = u_2 & & \\ \uparrow & & & \uparrow & & \\ \begin{bmatrix} AE/(2a) & 0 & 0 & -AE/(2a) & 0 & 0 \\ & 0 & 0 & 0 & 0 & 0 \\ & & 0 & 0 & 0 & 0 \\ & & & AE/(2a) & 0 & 0 \\ & & & & 0 & 0 \\ & s\,y. & & & & 0 \end{bmatrix} \begin{matrix} \rightarrow d_1 = u_1 \\ \\ \\ \rightarrow d_2 = u_2 \\ \\ \end{matrix} \end{matrix} \tag{6.2}
$$

Next, the stiffness matrix for the beam element, Eq. (5.23), is also extended to a 6×6 matrix corresponding to the order of the DOFs of the beam element in Eq. (6.1):

$$
\mathbf{k}_e^{\text{beam}} = \begin{matrix} & d_2(v_1) & d_3(\theta_{z1}) & & d_5(v_2) & d_6(\theta_{z2}) \\ & \uparrow & \uparrow & & \uparrow & \uparrow \\ \begin{bmatrix} 0 & 0 & 0 & 0 & 0 & 0 \\ & \dfrac{3EI_z}{2a^3} & \dfrac{3EI_z}{2a^2} & 0 & -\dfrac{3EI_z}{2a^3} & \dfrac{3EI_z}{2a^2} \\ & & \dfrac{2EI_z}{a} & 0 & -\dfrac{3EI_z}{2a^2} & \dfrac{EI_z}{a} \\ & & & 0 & 0 & 0 \\ & s\,y. & & & \dfrac{3EI_z}{2a^3} & -\dfrac{3EI_z}{2a^2} \\ & & & & & \dfrac{2EI_z}{a} \end{bmatrix} \begin{matrix} \\ \rightarrow d_2 = v_1 \\ \rightarrow d_3 = \theta_{z1} \\ \\ \rightarrow d_5 = v_2 \\ \rightarrow d_6 = \theta_{z2} \end{matrix} \end{matrix} \tag{6.3}
$$

The two matrices in Eqs. (6.2) and (6.3) are now superposed together to obtain the stiffness matrix for the frame element:

$$
\mathbf{k}_e = \begin{bmatrix}
\frac{AE}{2a} & 0 & 0 & -\frac{AE}{2a} & 0 & 0 \\
 & \frac{3EI_z}{2a^3} & \frac{3EI_z}{2a^2} & 0 & -\frac{3EI_z}{2a^3} & \frac{3EI_z}{2a^2} \\
 & & \frac{2EI_z}{a} & 0 & -\frac{3EI_z}{2a^2} & \frac{EI_z}{a} \\
 & & & \frac{AE}{2a} & 0 & 0 \\
 & s\,y. & & & \frac{3EI_z}{2a^3} & -\frac{3EI_z}{2a^2} \\
 & & & & & \frac{2EI_z}{a}
\end{bmatrix}
\tag{6.4}
$$

The element mass matrix of the frame element can also be obtained in the same way as the stiffness matrix. The element mass matrices for the truss element and the beam element, Eqs. (4.16) and (5.25), respectively, are extended into 6×6 matrices and added together to give the element mass matrix for the frame element:

$$
\mathbf{m}_e = \frac{\rho A a}{105} \begin{bmatrix}
70 & 0 & 0 & 35 & 0 & 0 \\
 & 78 & 22a & 0 & 27 & -13a \\
 & & 8a^2 & 0 & 13a & -6a^2 \\
 & & & 70 & 0 & 0 \\
 & s\,y. & & & 78 & -22a \\
 & & & & & 8a^2
\end{bmatrix}
\tag{6.5}
$$

The same simple procedure can be applied to the force vector as well. The element force vectors for the truss and beam elements, Eqs. (4.17) and (5.26), respectively, are extended into 6×1 vectors corresponding to their respective DOFs and added together. If the element is loaded by external distributed forces f_x and f_y along the x axis; concentrated forces F_{sx1}, F_{sx2}, F_{sy1}, and F_{sy2}; and concentrated moments M_{s1} and M_{s2}, respectively, at nodes 1 and 2, the total nodal force vector becomes

$$
\mathbf{f}_e = \begin{Bmatrix}
f_x a + F_{sx1} \\
f_y a + F_{sy1} \\
f_y a^2/3 + M_{s1} \\
f_x a + F_{sx2} \\
f_y a + F_{sy2} \\
-f_y a^2/3 + M_{s2}
\end{Bmatrix}
\tag{6.6}
$$

The final FEM equation for the frame element will thus have the form of Eq. (3.89) wherein the element matrices are given by Eqs. (6.4)–(6.6).

6.2.3 Equations in the global coordinate system

The matrices formulated in the previous section are in the local coordinate system of a frame element in a specific orientation. A full frame structure usually comprises numerous frame elements of different orientations joined together. As such, their local coordinate system would vary from one element to another. To assemble the element matrices together, all the matrices must first be expressed in a common coordinate system, which is the global coordinate system. The coordinate transformation process is the same as that discussed in Chapter 4 for truss structures.

Assume that local nodes 1 and 2 correspond to the global nodes i and j, respectively. The displacement at a local node should have two translational components in the x and y directions and one rotational deformation. They are numbered sequentially by u, v, and θ_z at each of the two nodes, as shown in Figure 6.3. The displacement at a global node should also have two translational components in the X and Y directions and one rotational deformation. They are numbered sequentially by D_{3i-2}, D_{3i-1}, and D_{3i} for the ith node, as shown in Figure 6.3. The same notation convention also applies to node j. The coordinate transformation gives the relationship between the displacement vector $\mathbf{d}_e$ based on the local coordinate system and the displacement vector $\mathbf{D}_e$ for the same element based on the global coordinate system:

$$\mathbf{d}_e = \mathbf{T}\mathbf{D}_e \tag{6.7}$$

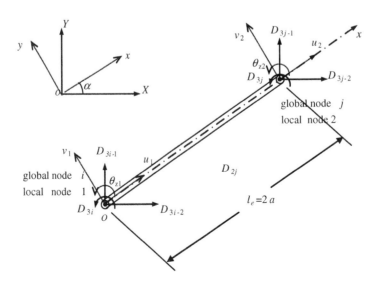

FIGURE 6.3

Coordinate transformation for 2D frame elements.

where

$$
\mathbf{D}_e =
\begin{Bmatrix}
D_{3i-2} \\
D_{3i-1} \\
D_{3i} \\
D_{3j-2} \\
D_{3j-1} \\
D_{3j}
\end{Bmatrix}
\tag{6.8}
$$

and $\mathbf{T}$ is the transformation matrix for the frame element given by

$$
\mathbf{T} =
\begin{bmatrix}
l_x & m_x & 0 & 0 & 0 & 0 \\
l_y & m_y & 0 & 0 & 0 & 0 \\
0 & 0 & 1 & 0 & 0 & 0 \\
0 & 0 & 0 & l_x & m_x & 0 \\
0 & 0 & 0 & l_y & m_y & 0 \\
0 & 0 & 0 & 0 & 0 & 1
\end{bmatrix}
\tag{6.9}
$$

in which

$$
\begin{aligned}
l_x &= \cos(x, X) = \cos \alpha = \frac{X_j - X_i}{l_e} \\
m_x &= \cos(x, Y) = \sin \alpha = \frac{Y_j - Y_i}{l_e}
\end{aligned}
\tag{6.10}
$$

where α is the angle between the local x axis and the global X axis, as shown in Figure 6.3, and

$$
\begin{aligned}
l_y &= \cos(y, X) = \cos(90° + \alpha) = -\sin \alpha = -\frac{Y_j - Y_i}{l_e} \\
m_y &= \cos(y, Y) = \cos \alpha = \frac{X_j - X_i}{l_e}
\end{aligned}
\tag{6.11}
$$

Note that the coordinate transformation in the X–Y plane does not affect the rotational DOF, as its direction is in the z direction (normal to the x–y plane), which still remains the same as the Z direction in the global coordinate system. The length of the element, l_e, can be calculated by

$$
l_e = \sqrt{(X_j - X_i)^2 + (Y_j - Y_i)^2}
\tag{6.12}
$$

Equation (6.7) can be easily verified, as it simply says that at node i, u_1 equals the summation of all the projections of D_{3i-2} and D_{3i-1} onto the local x axis, and v_1 equals the summation of all the projections of D_{3i-2} and D_{3i-1} onto the local y axis. The same can be said at node j. The matrix $\mathbf{T}$ for a frame element transforms a 6×6 matrix into another 6×6 matrix. Using the transformation matrix, $\mathbf{T}$, the matrices for the frame element in the global coordinate system become

$$\mathbf{K}_e = \mathbf{T}^T \mathbf{k}_e \mathbf{T} \tag{6.13}$$

$$\mathbf{M}_e = \mathbf{T}^T \mathbf{m}_e \mathbf{T} \tag{6.14}$$

$$\mathbf{F}_e = \mathbf{T}^T \mathbf{f}_e \tag{6.15}$$

Note that there is no change in dimension between the matrices and vectors in the local and global coordinate systems. Note also that the symmetry of the stiffness and mass matrices are all preserved in the coordinate transformation.

6.3 **FEM equations for space frames**
6.3.1 **Equations in the local coordinate system**

The approach used to develop the two-dimensional planar, frame elements can be used to develop the three-dimensional frame elements as well. The only difference is that there are more DOFs at a node in a 3D frame element than there are in a 2D frame element. There are altogether six DOFs at a node in a 3D frame element: three translational displacements in the x, y, and z directions, and three rotations with respect to the x, y, and z axes. Therefore, for an element with two nodes, there are altogether twelve DOFs, as shown in Figure 6.4.

The element displacement vector for a frame element in space can be written as

$$\mathbf{d}_e = \begin{Bmatrix} d_1 \\ d_2 \\ d_3 \\ d_4 \\ d_5 \\ d_6 \\ d_7 \\ d_8 \\ d_9 \\ d_{10} \\ d_{11} \\ d_{12} \end{Bmatrix} = \begin{Bmatrix} u_1 \\ v_1 \\ w_1 \\ \theta_{x1} \\ \theta_{y1} \\ \theta_{z1} \\ u_2 \\ v_2 \\ w_2 \\ \theta_{x2} \\ \theta_{y2} \\ \theta_{z2} \end{Bmatrix} \begin{matrix} \left. \vphantom{\begin{matrix} u_1 \\ v_1 \\ w_1 \\ \theta_{x1} \\ \theta_{y1} \\ \theta_{z1} \end{matrix}} \right\} \text{ displacement components at node 1} \\ \\ \left. \vphantom{\begin{matrix} u_2 \\ v_2 \\ w_2 \\ \theta_{x2} \\ \theta_{y2} \\ \theta_{z2} \end{matrix}} \right\} \text{ displacement components at node 2} \end{matrix} \tag{6.16}$$

The element matrices can be obtained by a similar process of obtaining the matrices of the truss element in space and that of beam elements, and adding them together. Because of the huge matrices involved, the details will not be shown in this book, but the stiffness matrix is listed here as follows, and can be easily confirmed simply by inspection:

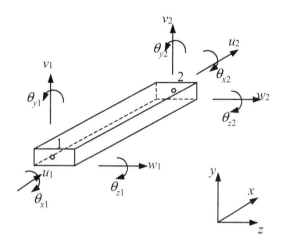

FIGURE 6.4

Frame element in space with twelve DOFs.

$$
\mathbf{k}_e =
\begin{array}{c}
\begin{array}{cccccccccccc}
u_1 & v_1 & w_1 & \theta_{x1} & \theta_{y1} & \theta_{z1} & u_2 & v_2 & w_2 & \theta_{x2} & \theta_{y2} & \theta_{z2} \\
\uparrow & \uparrow & \uparrow & \uparrow & \uparrow & \uparrow & \uparrow & \uparrow & \uparrow & \uparrow & \uparrow & \uparrow
\end{array} \\
\left[
\begin{array}{cccccccccccc}
\dfrac{AE}{2a} & 0 & 0 & 0 & 0 & 0 & -\dfrac{AE}{2a} & 0 & 0 & 0 & 0 & 0 \\[2mm]
 & \dfrac{3EI_z}{2a^3} & 0 & 0 & 0 & \dfrac{3EI_z}{2a^2} & 0 & -\dfrac{3EI_z}{2a^3} & 0 & 0 & 0 & \dfrac{3EI_z}{2a^2} \\[2mm]
 & & \dfrac{3EI_y}{2a^3} & 0 & \dfrac{3EI_y}{2a^2} & 0 & 0 & 0 & -\dfrac{3EI_y}{2a^3} & 0 & -\dfrac{3EI_y}{2a^2} & 0 \\[2mm]
 & & & \dfrac{GJ}{2a} & 0 & 0 & 0 & 0 & 0 & -\dfrac{GJ}{2a} & 0 & 0 \\[2mm]
 & & & & \dfrac{2EI_y}{a} & 0 & 0 & 0 & \dfrac{3EI_y}{2a^2} & 0 & \dfrac{EI_y}{a} & 0 \\[2mm]
 & & & & & \dfrac{2EI_z}{a} & 0 & -\dfrac{3EI_z}{2a^2} & 0 & 0 & 0 & \dfrac{EI_z}{a} \\[2mm]
 & & & & & & \dfrac{AE}{2a} & 0 & 0 & 0 & 0 & 0 \\[2mm]
 & & & & & & & \dfrac{3EI_z}{2a^3} & 0 & 0 & 0 & -\dfrac{3EI_z}{2a^2} \\[2mm]
 & \text{s } y. & & & & & & & \dfrac{3EI_z}{2a^3} & 0 & \dfrac{3EI_z}{2a^2} & 0 \\[2mm]
 & & & & & & & & & \dfrac{GJ}{2a} & 0 & 0 \\[2mm]
 & & & & & & & & & & \dfrac{2EI_y}{a} & 0 \\[2mm]
 & & & & & & & & & & & \dfrac{2EI_z}{a}
\end{array}
\right]
\end{array}
\quad (6.17)
$$

where I_y and I_z are the second moment of area (or moment of inertia) of the cross-section of the beam with respect to the y and z axes, respectively. Note that the fourth DOF is related to the *torsional* deformation. The formulation of a *torsional bar element* is very similar to that of a truss element. The only difference is that the axial deformation is replaced by the torsional angular deformation, and axial force is replaced by torque. Therefore, in the resultant stiffness matrix, the element tensile stiffness AE/l_e is replaced by the element torsional stiffness GJ/l_e, where G is the shear module and J is the polar moment of inertia of the cross-section of the bar.

The mass matrix can also be expressed by inspection as follows:

$$
\mathbf{m}_e = \frac{\rho A a}{105}
\begin{bmatrix}
70 & 0 & 0 & 0 & 0 & 0 & 35 & 0 & 0 & 0 & 0 & 0 \\
 & 78 & 0 & 0 & 0 & 22a & 0 & 27 & 0 & 0 & 0 & -13a \\
 & & 78 & 0 & -22a & 0 & 0 & 0 & 27 & 0 & 13a & 0 \\
 & & & 70r_x^2 & 0 & 0 & 0 & 0 & 0 & -35r_x^2 & 0 & 0 \\
 & & & & 8a^2 & 0 & 0 & 0 & -13a & 0 & -6a^2 & 0 \\
 & & & & & 8a^2 & 0 & 13a & 0 & 0 & 0 & -6a^2 \\
 & & & & & & 70 & 0 & 0 & 0 & 0 & 0 \\
 & & & & & & & 78 & 0 & 0 & 0 & -22a \\
 & & & & & & & & 78 & 0 & 22a & 0 \\
 & & & & & & & & & 70r_x^2 & 0 & 0 \\
 & & & & & & & & & & 8a^2 & 0 \\
 & & & & & & & & & & & 8a^2
\end{bmatrix}
\tag{6.18}
$$

where

$$
r_x^2 = \frac{I_x}{A}
\tag{6.19}
$$

in which I_x is the second moment of area (or moment of inertia) of the cross-section of the beam with respect to the x axis.

6.3.2 Equations in the global coordinate system

Having set up the element matrices in the local coordinate system, the next thing to do is to transform the element matrices into the global coordinate system to account for the differences in orientation of all the local coordinate systems (now in 3D) that are attached on individual frame members.

Assume that the local nodes 1 and 2 of the element correspond to global nodes i and j, respectively. The displacement at a local node should have three translational components in the x, y, and z directions, and three rotational components with respect to the x, y, and z axes. They are numbered sequentially by d_1–d_{12} corresponding to the physical deformations as defined by Eq. (6.16). The displacement at a global node should also have three translational components in the X, Y, and Z directions, and three rotational components with respect to the X, Y, and Z axes. They are numbered sequentially by $D_{6i-5}, D_{6i-4}, \ldots, D_{6i}$ for the ith node, as shown in Figure 6.5. The same sign convention applies to node j. The coordinate transformation gives the relationship between the displacement vector $\mathbf{d}_e$ based on the local coordinate system and the displacement vector $\mathbf{D}_e$ based on the global coordinate system:

$$\mathbf{d}_e = \mathbf{T}\mathbf{D}_e \tag{6.20}$$

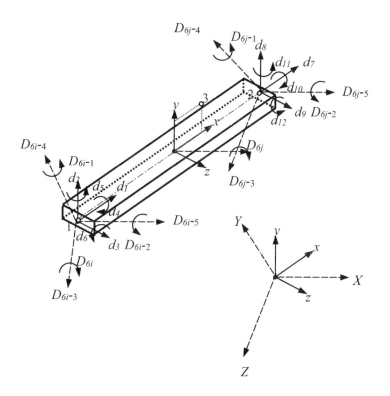

FIGURE 6.5

Coordinate transformation for a frame element in space.

where

$$\mathbf{D}_e = \begin{Bmatrix} D_{6i-5} \\ D_{6i-4} \\ D_{6i-3} \\ D_{6i-2} \\ D_{6i-1} \\ D_{6i} \\ D_{6j-5} \\ D_{6j-4} \\ D_{6j-3} \\ D_{6j-2} \\ D_{6j-1} \\ D_{6j} \end{Bmatrix} \tag{6.21}$$

and $\mathbf{T}$ is the transformation matrix for the truss element given by

$$\mathbf{T} = \begin{bmatrix} \mathbf{T}_3 & \mathbf{0} & \mathbf{0} & \mathbf{0} \\ \mathbf{0} & \mathbf{T}_3 & \mathbf{0} & \mathbf{0} \\ \mathbf{0} & \mathbf{0} & \mathbf{T}_3 & \mathbf{0} \\ \mathbf{0} & \mathbf{0} & \mathbf{0} & \mathbf{T}_3 \end{bmatrix} \tag{6.22}$$

in which

$$\mathbf{T}_3 = \begin{bmatrix} l_x & m_x & n_x \\ l_y & m_y & n_y \\ l_z & m_z & n_z \end{bmatrix} \tag{6.23}$$

where l_k, m_k, and n_k $(k=x,y,z)$ are direction cosines defined by

$$\begin{aligned}
l_x &= \cos(x,X), & m_x &= \cos(x,Y), & n_x &= \cos(x,Z) \\
l_y &= \cos(y,X), & m_y &= \cos(y,Y), & n_y &= \cos(y,Z) \\
l_z &= \cos(z,X), & m_z &= \cos(z,Y), & n_z &= \cos(z,Z)
\end{aligned} \tag{6.24}$$

To define these direction cosines, the position and the three-dimensional orientation of the frame element have to be defined first. With nodes 1 and 2, the location of the element is fixed on the local coordinate frame, and the orientation of the element has also been fixed in the x direction. However, the local coordinate frame can still rotate about the axis of the beam. One more additional point in the local coordinate has to be defined. This point can be chosen anywhere in the local x–y plane, but not on the x axis. Therefore, node 3 is chosen, as shown in Figure 6.6.

The position vectors $\vec{V}_1$, $\vec{V}_2$, and $\vec{V}_3$ can be expressed as

$$\vec{V}_1 = X_1\vec{X} + Y_1\vec{Y} + Z_1\vec{Z} \tag{6.25}$$

$$\vec{V}_2 = X_2\vec{X} + Y_2\vec{Y} + Z_2\vec{Z} \tag{6.26}$$

$$\vec{V}_3 = X_3\vec{X} + Y_3\vec{Y} + Z_3\vec{Z} \tag{6.27}$$

where X_k, Y_k, and Z_k ($k = 1, 2, 3$) are the coordinates for node k, and $\vec{X}$, $\vec{Y}$ and $\vec{Z}$ are unit vectors along the X, Y, and Z axes. We now define

$$\left.\begin{array}{l} X_{kl} = X_k - X_l \\ Y_{kl} = Y_k - Y_l \\ Z_{kl} = Z_k - Z_l \end{array}\right\} \quad k, l = 1, 2, 3 \tag{6.28}$$

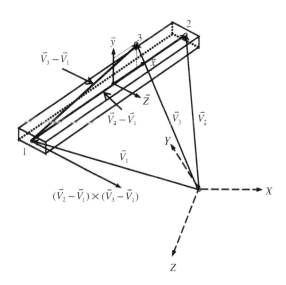

FIGURE 6.6

Vectors for defining the location and three-dimensional orientation of the frame element in space.

Vectors $(\vec{V}_2 - \vec{V}_1)$ and $(\vec{V}_3 - \vec{V}_1)$ can thus be obtained using Eqs. (6.25)–(6.28) as follows:

$$\vec{V}_2 - \vec{V}_1 = X_{21}\vec{X} + Y_{21}\vec{Y} + Z_{21}\vec{Z} \tag{6.29}$$

$$\vec{V}_3 - \vec{V}_1 = X_{31}\vec{X} + Y_{31}\vec{Y} + Z_{31}\vec{Z} \tag{6.30}$$

The length of the frame element can be obtained by

$$l_e = 2a = \left|\vec{V}_2 - \vec{V}_1\right| = \sqrt{X_{21}^2 + Y_{21}^2 + Z_{21}^2} \tag{6.31}$$

The unit vector along the x axis can thus be expressed as

$$\vec{x} = \frac{(\vec{V}_2 - \vec{V}_1)}{\left|\vec{V}_2 - \vec{V}_1\right|} = \frac{X_{21}}{2a}\vec{X} + \frac{Y_{21}}{2a}\vec{Y} + \frac{Z_{21}}{2a}\vec{Z} \tag{6.32}$$

Therefore, the direction cosines in the x direction are given as

$$l_x = \cos(x, X) = \vec{x} \cdot \vec{X} = \frac{X_{21}}{2a}$$
$$m_x = \cos(x, Y) = \vec{x} \cdot \vec{Y} = \frac{Y_{21}}{2a} \tag{6.33}$$
$$n_x = \cos(x, Z) = \vec{x} \cdot \vec{Z} = \frac{Z_{21}}{2a}$$

From Figure 6.6, it can be seen that the direction of the z axis can be defined by the cross product of vectors $(\vec{V}_2 - \vec{V}_1)$ and $(\vec{V}_3 - \vec{V}_1)$. Hence, the unit vector along the z axis can be expressed as

$$\vec{z} = \frac{(\vec{V}_2 - \vec{V}_1) \times (\vec{V}_3 - \vec{V}_1)}{\left|(\vec{V}_2 - \vec{V}_1) \times (\vec{V}_3 - \vec{V}_1)\right|} \tag{6.34}$$

Substituting Eqs. (6.29) and (6.30) into the above equation,

$$\vec{z} = \frac{1}{A_{123}}\{(Y_{21}Z_{31} - Y_{31}Z_{21})\vec{X} + (Z_{21}X_{31} - Z_{31}X_{21})\vec{Y} + (X_{21}Y_{31} - X_{31}Y_{21})\vec{Z}\} \tag{6.35}$$

where

$$A_{123} = \sqrt{(Y_{21}Z_{31} - Y_{31}Z_{21})^2 + (Z_{21}X_{31} - Z_{31}X_{21})^2 + (X_{21}Y_{31} - X_{31}Y_{21})^2} \tag{6.36}$$

Using Eq. (6.35), it is found that

$$l_z = \vec{z} \cdot \vec{X} = \frac{1}{A_{123}}(Y_{21}Z_{31} - Y_{31}Z_{21})$$

$$m_z = \vec{z} \cdot \vec{Y} = \frac{1}{A_{123}}(Z_{21}X_{31} - Z_{31}X_{21}) \qquad (6.37)$$

$$n_z = \vec{z} \cdot \vec{Z} = \frac{1}{A_{123}}(X_{21}Y_{31} - X_{31}Z_{21})$$

Since the y axis is perpendicular to both the x axis and the z axis, the unit vector along the y axis can be obtained by cross product,

$$\vec{y} = \vec{z} \times \vec{x} \qquad (6.38)$$

which gives

$$l_y = m_z n_x - n_z m_x$$

$$m_y = n_x l_x - l_z n_x \qquad (6.39)$$

$$n_y = l_z m_x - m_z l_x$$

in which l_x, m_x, n_x, l_z, m_z, and n_z have been obtained using Eqs. (6.33) and (6.37).

Using the transformation matrix, $\mathbf{T}$, the matrices for space frame elements in the global coordinate system can be obtained, as before, as

$$\mathbf{K}_e = \mathbf{T}^T \mathbf{k}_e \mathbf{T} \qquad (6.40)$$

$$\mathbf{M}_e = \mathbf{T}^T \mathbf{m}_e \mathbf{T} \qquad (6.41)$$

$$\mathbf{F}_e = \mathbf{T}^T \mathbf{f}_e \qquad (6.42)$$

6.4 Remarks

In the formulation of the matrices for the frame element in this chapter, the superposition of the truss element and the beam element have been used. This technique assumes that the axial effects are not coupled with the bending effects in the element. What this means is simply that the axial forces applied on the element will not result in any bending deformation, and the bending forces will not result in any axial deformation. Frame elements can also be used for frame structures with curved members. In such cases, the coupling effects can exist even in the elemental level. Therefore, depending on the curvature of the member, the meshing of the structure can be very important. For example, if the curvature is very large, resulting in a significant coupling effect, a finer mesh is required to provide the necessary accuracy. In practical structures, it is very rare to have beam structures subjected to purely transverse loading. Most skeletal structures are either trusses or frames that carry both axial and transverse loads.

It can now be seen that the beam element, developed in Chapter 5, as well as the truss element, developed in Chapter 4, are simply specific cases of the frame element. Therefore, in most commercial software packages, including ABAQUS, the frame element is just known generally as the beam element.

The beam element, formulated in Chapter 5, or general beam element, formulated in this chapter, is based on the so-called Euler–Bernoulli beam theory that is suitable for thin beams with a small thickness-to-span ratio ($<1/20$). For *thick* or *deep* beams of a large thickness-to-span ratio, corresponding beam theories should be used to develop thick beam elements. The procedure of developing thick beams is very similar to that of developing thick plates (discussed in Chapter 8) and most commercial software packages offer thick beam elements as well.

6.5 Case study: finite element analysis of a bicycle frame

In the design of many modern devices and pieces of equipment, the finite element method has become an indispensable tool for many of the successful products that we have come to use daily. In this case study, the analysis of a bicycle frame is carried out. Historically, intuition and trial-and-error in physical prototype testing have played important roles in coming out with the evolution of the diamond-shaped bicycle frame, as shown in Figure 6.7. However, using such a physical trial-and-error design procedure is costly and time consuming, and has its limitations, especially when new materials are introduced and when new applications or demands are placed on the structure. Hence, there is the need to use the finite element tool to help the designer come up with reliable properties in the design to meet the demands expected by consumers. Using the FEM to perform a virtual prototyping instead of a physical prototyping can save lots of cost, time, and effort. There can be numerous factors to consider when it

FIGURE 6.7

Diamond-shaped bicycle frame.

comes to designing a bicycle frame. For example, factors like the weight of the frame, the maximum load the frame can carry, the impact toughness of the frame, and so on. In an FE analysis, one would require information on the material properties, the boundary conditions, and the loading conditions. The actual loading conditions on a bicycle can be extremely complex, especially when a rider is riding a bicycle and going through different terrain. In this case study we simply consider a loading condition of a horizontal impact applied to the bicycle, simulating the effect of a low speed, head-on collision into a wall or curb. This case is one of the physical tests that manufacturers have to comply with before a bicycle design is approved.

6.5.1 Modeling

The bicycle frame to be modeled is made of aluminum, whose properties are shown in Table 6.1. The bicycle frame is meshed by two-nodal "beam elements" in ABAQUS. Note that in ABAQUS, as well as in many other software packages, a general beam element in space is basically the same as the frame element developed in this chapter. The "beam element" in space offered by ABAQUS has the same DOFs as the frame element in this chapter, that is, three translational DOFs and three rotational DOFs.

Altogether, 74 elements (71 nodes) are used in the skeletal model of the bicycle frame, as shown in Figure 6.8. Note that like all finite element meshes, connectivity at the nodes is very important.

Table 6.1 Material properties of aluminum.	
Young's Modulus, E (GPa)	**Poisson's Ratio, v**
69.0	0.33

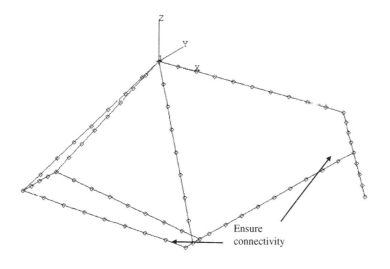

FIGURE 6.8

Finite element mesh of a bicycle frame.

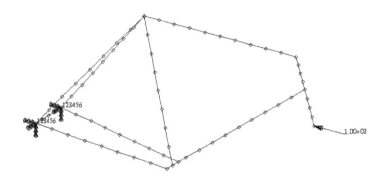

FIGURE 6.9

Loadings and boundary conditions on the bicycle frame.

It is important to make sure that at the joints, there is connectivity between the elements. No node should be left unattached to another element unless the node happens to be at the end of a structure.

To model the horizontal impact, the two nodes at the rear dropouts are constrained from any displacements and a horizontal force of 1000 N is applied to the front dropout. The forces and constraints are shown in Figure 6.9.

6.5.2 ABAQUS input file

The ABAQUS input file for the above described finite element model is shown below. The text boxes to the right of the input file are not part of the file, but explain what the sections of the file mean.

```
* HEADING, SPARSE
ABAQUS job to calculate low speed impact on bicycle frame
**
* NODE
1, –0.347, 0., –0.39
2, –0.303625, –0.06125, –0.34125
3, –0.347, –0.035, –0.39
⋮
68, 0.57829, 0., –0.12125
69, 0.586579, 0., –0.1475
70, 0.594868, 0., –0.17375
71, 0.603158, 0., –0.2
**
**
```

> *Nodal cards*
> These define the coordinates of the nodes in the model. The first entry being the node ID, while the second, third, and fourth entries are the x, y, and z coordinates of the position of the node, respectively.

```
* ELEMENT, TYPE=B33   ELSET=FRAME
1, 34, 38
2, 38, 42
3, 42, 46
4, 46, 48
5, 48, 50
6, 50, 52
:
```

Element (connectivity) cards
These define the element type and what nodes make up the element. B33 represents a spatial, 2-nodal, beam element following the Euler–Bernoulli theory with cubic interpolation. Note that this element has six DOFs per node—three translational and three rotational. There are many other element types in the ABAQUS element library. The "ELSET = FRAME" statement is simply for naming this set of elements so that it can be referenced when defining the material properties. In the subsequent data entry, the first entry is the element ID, and the following two entries are the nodes making up the element.

```
72, 68, 6
73, 69, 70
74, 70, 71
**
* BEAM SECTION, ELSET=FRAME, MATERIAL=ALU, SECTION=CIRC
0.012
**
* MATERIAL, NAME=ALU
**
* DENSITY
2710.,
**
* ELASTIC, TYPE=ISO
6.9E+10, 0.33
**
```

Material cards
These define material properties under the name "ALU." Density and elastic properties are defined. TYPE = ISO represents isotropic properties.

Property cards
These define properties to the elements of set "FRAME." "SECT = CIRC" describes the cross-section as a circle. ABAQUS provides a choice of other cross-sections. The first data line under "BEAM SECTION" defines the geometry of the cross-section. The material is defined by "ALU."

```
* BOUNDARY, OP=NEW
FIXED, 1,, 0.
FIXED, 2,, 0.
FIXED, 3,, 0.
FIXED, 4,, 0.
FIXED, 5,, 0.
FIXED, 6,, 0.
**
```

Boundary Conditions cards
These define boundary conditions. Nodes belonging to the node set, "FIXED" have all their DOFs constrained (= 0).

```
** Step 1, Default Static Step
** LoadCase, Default
**
* STEP, AMPLITUDE=RAMP, PERTURB
Linear Static Analysis
**
```

```
*STATIC
**
**
**
**
* NSET, NSET=FIXED
8, 9
* NSET, NSET=FORCE
71,
**
**
* CLOAD, OP=NEW
FORCE, 1, –1000.
**
**
* NODE PRINT, FREQ=1


U,
* NODE FILE, FREQ=1
U,
**
* EL PRINT, POS=INTEG, FREQ=1
1
S,
E,
* EL FILE, POS=INTEG, FREQ=1
1
S,
E,
**
* END STEP
```

Control cards
These indicate the analysis step. In this case it is a "STATIC" analysis.

Node sets
These group nodes into node sets "FIXED" and "FORCE."

Loading cards
These define the loads on the nodes according to the respective groups the nodes belong to. For example, in this case, node 71 belonging to "NSET = FORCE" is given a force of –1000 N in the 1 (x) direction.

Output control cards
These define the output required. For example, in this case, we require the nodal output, displacement "U," the stress, "S," and the strain, "E."

The above input file actually looks similar to that in Chapter 5. In fact, most ABAQUS input files have similar formats, only the information provided may be different depending on the problem required to solve.

6.5.3 Solution processes

The nodal and element cards provide information on the dimensions, position of nodes, and the connectivity of the elements. As discussed, these parameters play important roles in the formation of the element stiffness and mass matrices. Note that in the element cards, the element type specified is B33 which represents a general "beam element" in space with two nodes following the Euler–Bernoulli beam theory for its transverse displacement field. In ABAQUS, this would be exactly the same as the frame element developed in this chapter, since the DOFs also include the axial displacement component. ABAQUS, like most other software, does not have a pure beam element consisting of only the transverse displacement field. This is because not many real structures are a true beam structure without the axial displacement components. Hence, readers should not be confused by the use of beam elements in this case.

Next, the material properties and the cross-sectional geometry and dimensions are provided in the material cards and properties cards, respectively. As can be seen in Eqs. (6.4) and (6.5), the material properties, as well as the moment of area, are required to compute the stiffness and mass matrices. It is interesting to note that most of the information provided goes into forming the finite element matrices. In fact, that is the basic idea when it comes to applying the FEM: To form the finite element matrix equation and solve the equation to obtain the required field variables.

The next card would be the definition of the boundary conditions. In this case, the rear dropouts that are represented by nodes 8 and 9 and grouped as a node set, "FIXED," are constrained in all DOFs. Recall that in this case, since each node has six DOFs, we can actually reduce the size of the matrix here by eliminating 12 rows and columns each, just as we have done in worked examples in Chapters 4 and 5. For the loads, it is given as a concentrated force of $-1000\,\text{N}$ at node 71 in the X direction. This would be reflected in the force vector, Eq. (6.42).

The control cards control the type of analysis to be performed, which in this case is a static analysis. In static analyses, the static equation $\mathbf{KD}=\mathbf{F}$ is solved for $\mathbf{D}$, which is a vector of the displacements and rotations of the nodes in the model. The method used is usually efficient algorithms that solve the linear system of equations via the Gauss elimination method, which is mentioned in Section 3.5.

The last part of the input file would consist of the output control cards, which specify the type of output requested. In this case, the nodal displacement and rotation components (U) and the elemental stress (S) and strain (E) components are specified. With the input file written, one can then invoke ABAQUS to execute the analysis.

6.5.4 Results and discussion

Using the above input file, the finite element equation is solved and a deformation plot showing how the frame actually deforms under the specified loading is shown in Figure 6.10. The magnitude of the deformation is actually magnified 20 times as the true deformation is too small for viewing purposes. Such deformation plots, or the results containing the displacements of the nodes, are useful to a designer since he or she will then be able to visualize the way in which the frame will deform under specific conditions. Furthermore, knowing the magnitude of deformation also helps to gauge aspects of the frame in the design, as consumers would not want a bicycle which could not be ridden once one accidentally hits a wall or curb at low speed.

The other more important result that could be obtained from this case study would be the stresses that are incurred with this particular loading condition. Figure 6.11 shows the average axial stresses along the centroid of the aluminum beam members. It should be noted that it is possible to obtain stresses on different sections of the cross-section, for example, the upper or lower surface along the circumference of a circular cross-section. These stresses are useful for testing whether the bicycle frame

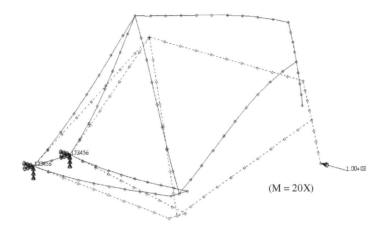

(M = 20X)

FIGURE 6.10

Deformation plot of the bicycle frame.

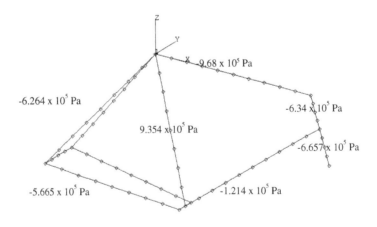

FIGURE 6.11

Stresses in the bicycle frame.

will fail under the applied loads. If these stresses are smaller than the yield stress or the design stress of the material, then it can be safely concluded that there is a high chance that the material will not fail or undergo plastic deformation. It is important that the deformations remain elastic, that is, it will return to the original shape upon removal of the applied load.

Hence, it can be seen how the FEM can aid the design engineer when it comes to designing a product. It would be disastrous if a new design of an engineering system was mass-produced without going through any sort of analysis to check its reliability.

6.6 Review questions

1. Explain why the superposition technique can be used to formulate the frame elements simply by using the formulations of the truss and beam elements. Under what conditions will this superposition technique fail?
2. In the transformation from the local to the global coordinate system in a planar truss, does the rotational degree of freedom undergo any transformation? Why?
3. Work out the displacements of the planar frame structure shown in Figure 6.12. All the members are of the same material ($E=69.0$ GPa, $v=0.33$) and with circular cross-sections. The areas of the cross-sections are $0.01 \, m^2$. Compare the results with those obtained for the Review Question 8 in Chapter 4.
4. Figure 6.13 shows a supporting system consisting of a beam and a truss member, which carries a vertical load of 2000N at point A that is at the middle span of the beam. Both the truss and beam members are of uniform cross-section and made of the same material with the Young's modulus $E=200.0$ GPa. The cross-section of the beam is circular, and the area of the cross-section is $0.01 \, m^2$. The area of the cross-section of the truss member is $0.0002 \, m^2$. Using the finite element method:
   a. Calculate the nodal displacement at point A.
   b. Calculate the reaction forces at the supports for the beam and truss member.
   c. Calculate the internal forces in the truss member.

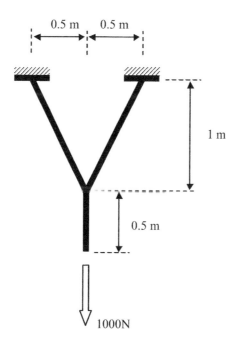

FIGURE 6.12

A three member planar truss structure.

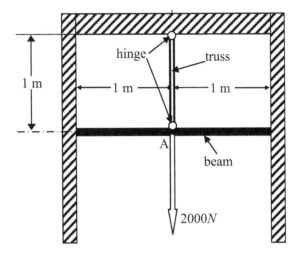

FIGURE 6.13

A supporting system consisting of a beam and a truss member.

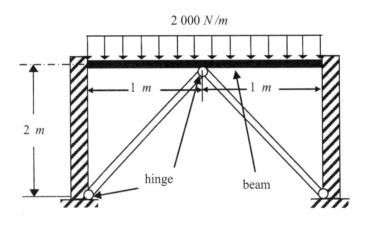

FIGURE 6.14

A supporting system consisting of one beam and two truss members.

5. Figure 6.14 shows a supporting system consisting of one beam and two truss members, which carries a vertical uniformly distributed load of 2000 N/m on the top surface of the beam. The beam is clamped at the two ends into the rigid walls. Both the truss and beam members are of uniform cross-section and made of the same material with the Young's modulus $E = 200.0$ GPa. The cross-section of the beam is circular, and the area of the cross-section is $0.01 \, \text{m}^2$. The area of the cross-section of the truss member is $0.0002 \, \text{m}^2$. Using the finite element method:

  a. Calculate the nodal displacement at the middle span of the beam.

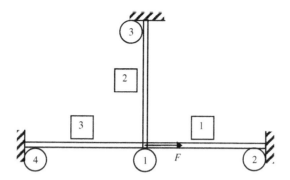

FIGURE 6.15

A frame made of three members.

    b. Calculate the reaction forces at the supports for the beam and truss member.

    c. Calculate the internal forces in the truss member.

6. Figure 6.15 shows a simple frame consisting of three identical members: two horizontal and one vertical. These three members are joined firmly together at node 1. All the members are of length $L = 2a$, cross-sectional area A, Young's modulus E and second moment of area I. A force F is applied horizontally at node 1. Using TWO frame elements, calculate the displacements and rotation at node 1, if any, in terms of F, L (or a), A, E, and I.

FEM for Two-Dimensional Solids

CHAPTER OUTLINE HEAD

7.1 Introduction

In this chapter, we develop finite element (FE) equations for the stress analysis of two-dimensional (2D) solids subjected to external loads. The basic concepts, procedures, and formulations can also be found in many existing textbooks (see Zienkiewicz and Taylor, 2000). The element developed here is called a 2D solid element that is used for solid mechanics problems where the loading (and hence the deformation) occur only within a plane. Though no real life structure can be truly 2D, experienced analysts can often idealize many 3D practical problems to 2D problems with satisfactory results by carrying out analyses using 2D models, which are usually much more efficient and cost effective compared to conducting full 3D analyses. As discussed in Chapter 2, there are 2D plane stress and plane strain problems, whereby correspondingly, plane stress and plane strain elements need to be used to solve them. For example, if we have a plate structure with loading acting in the plane of the plate as in Figure 7.1, we need to use 2D plane stress elements. When we want to model the effects of water pressure on a dam, as shown in Figure 7.2, we have to use 2D plane strain elements to model just the cross-section.

Note that in Figure 7.1, plane stress conditions are usually applied to structures that have relatively small thickness as compared to their other dimensions. Due to the absence of any off-plane external force, the normal stresses are negligible, which leads to a plane stress situation. In cases where plane strain conditions are applied, as in Figure 7.2, the thickness of the structure (in the z direction) is relatively large as compared to its other dimensions, and the loading (pressure) is uniform along the elongated direction. The deformation is, therefore, approximated to be the same throughout its thickness. In this case, the off-plane strain (strain components in the z direction) is negligible, which leads to a plane strain situation. In either a plane stress or plane strain situation, the governing system equation can be drastically simplified, as shown in Chapter 2. The formulations for plane stress and plane strain problems are very much the

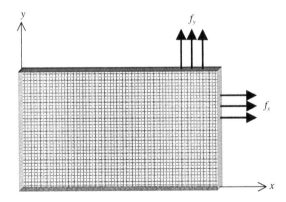

FIGURE 7.1

A typical 2D plane stress problem.

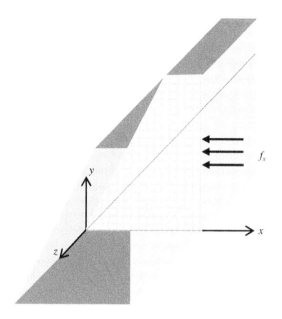

FIGURE 7.2

A typical 2D plane strain problem.

same, except for the difference in the material constant matrix. Another often encountered 2D problem comes from simplification of axisymmetric structures, which will also be discussed in this chapter.

A 2D solid element, be it plane strain or plane stress, can be triangular, rectangular or quadrilateral in shape with straight or curved edges. The most often used elements in engineering practice are linear elements (and thus with straight edges). Quadratic elements are also used for situations that require higher accuracy in stress, but they are less often used for practical problems. Higher order elements have also been developed, but they are generally less often used except for certain specific problems. The order of the 2D element is determined by the order of the shape functions used. A linear element uses linear shape functions, and therefore the edges of the element are straight. A quadratic element uses quadratic shape functions, and their edges can be curved. The same can be said for elements of the third order or higher.

In a 2D model, the elements can only deform in the plane where the model is defined, and in most situations, this is taken to be the $x–y$ plane. At any point, the variable, that is the displacement, has two components in the x and y directions, and so do the external forces. For plane strain problems, the thickness of the true structure is usually not important, and is normally treated as a unit quantity uniformly throughout the 2D model. However, for plane stress problems, the thickness is an important parameter for computing the stiffness matrix and stresses. Throughout this chapter, it is assumed that the elements have a uniform thickness of h. If the structure to be modeled has a varying thickness, the structure needs to be divided into small elements, where, in each element, a uniform thickness can be used. On the other hand, formulation of 2D elements with varying thicknesses can also be done easily, as the procedure is similar to that of a uniform element. However, very few commercially available software packages provide elements of varying thickness.

The equation system for a 2D element will be more complex as compared with the 1D element because of the higher dimension. The procedure for developing these equations is, however, very similar to that for the 1D truss elements, which is detailed in Chapter 4. These steps can be summarized in the following three-step procedure:

1. Construction of *shape functions* matrix **N** that satisfies Eqs. (3.34) and (3.41).
2. Formulation of the *strain matrix* **B**.
3. Calculation of $\mathbf{k}_e$, $\mathbf{m}_e$, and $\mathbf{f}_e$ using **N** and **B** and Eqs. (3.71), (3.75), and (3.81).

We shall be focusing on the formulation of three types of simple but very important elements: linear triangular, bilinear rectangular, and isoparametric linear quadrilateral elements. Once the formulation of these three types of element is understood, the development of other types of elements of higher orders is straightforward, because the same techniques can be utilized. Development of higher order elements will be discussed at the end of this chapter.

7.2 Linear triangular elements

The linear triangular element was the first type of element developed for 2D solids. The formulation is also the simplest among all the 2D solid elements. It has been found that the linear triangular element is less accurate compared to bilinear quadrilateral elements. For this reason, it is often thought to be ideal to use quadrilateral elements, but the reality is that the triangular element is still a very useful element for its adaptation to complex geometry. Triangular elements are normally used when we want to mesh a 2D model involving complex geometry with acute corners. Most importantly, the triangular configuration with the simplest topological feature makes it easier to develop automated meshing processors. Nowadays, analysts are hoping to use a fully automated mesh generator to perform the complex task of analysis that needs repeated or even adaptive re-meshing. Most automated mesh generators can only create triangular elements. There are automated mesh generators that can generate a quadrilateral mesh, but they still use triangular elements as patches for complex geometry, and end up with a mesh of mixed elements. Hence, despite being less accurate, we still have to use triangular elements for many practical engineering problems.

Consider a 2D model in the x–y plane, shown schematically in Figure 7.3. The 2D domain is divided in a *proper* manner into a number of *triangular elements*. The "proper" meshing of a domain will be outlined in Chapter 11, where a list of guidelines is provided. In a mesh of linear triangular elements, each triangular element has three nodes and three straight edges.

7.2.1 Field variable interpolation

Consider now a triangular element of uniform thickness h. The nodes of the element are numbered 1, 2, and 3 counter-clockwise, as shown in Figure 7.4. For 2D solid elements, the field variable is the displacement, which has two components (u and v), and hence each node has two degrees of freedom (DOFs). Since a linear triangular element has three nodes, the total number of DOFs of a linear triangular element is six. For the triangular element, the local coordinate of each element can be taken as the same as the global coordinate, since there is no advantage in specifying a different local coordinate system for each element.

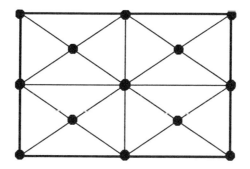

FIGURE 7.3

Rectangular domain meshed with triangular elements.

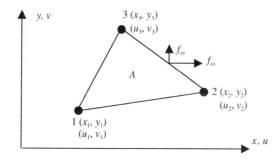

FIGURE 7.4

Linear triangular element.

Now, let's examine how a triangular element can be formulated. The displacement **U** is generally a function of the coordinates x and y, and we express the displacement at any point in the element using the displacements at the nodes and shape functions. It is therefore assumed that (see Section 3.4.2)

$$\mathbf{U}^h(x, y) = \mathbf{N}(x, y)\mathbf{d}_e$$

(7.1)

where the superscript h indicates that the displacement is approximated, and $\mathbf{d}_e$ is a vector of the *nodal displacements* arranged in the order of

$$\mathbf{d}_e = \begin{Bmatrix} u_1 \\ v_1 \\ u_2 \\ v_2 \\ u_3 \\ v_3 \end{Bmatrix} \begin{matrix} \Big\} \text{ displacements at node 1} \\ \Big\} \text{ displacements at node 2} \\ \Big\} \text{ displacements at node 3} \end{matrix}$$

(7.2)

and the matrix of shape functions $\mathbf{N}$ is arranged as

$$
\mathbf{N} = \begin{bmatrix} \underbrace{N_1 \quad 0}_{\text{Node 1}} & \underbrace{N_2 \quad 0}_{\text{Node 2}} & \underbrace{N_3 \quad 0}_{\text{Node 3}} \\ 0 \quad N_1 & 0 \quad N_2 & 0 \quad N_3 \end{bmatrix} \tag{7.3}
$$

in which N_i ($i = 1, 2, 3$) are three shape functions corresponding to the three nodes of the triangular element. Equation (7.1) was written in a compact matrix form, which can be explicitly expressed as

$$
\begin{aligned}
u^h(x, y) &= N_1(x, y)u_1 + N_2(x, y)u_2 + N_3(x, y)u_3 \\
v^h(x, y) &= N_1(x, y)v_1 + N_2(x, y)v_2 + N_3(x, y)v_3
\end{aligned} \tag{7.4}
$$

which implies that each of the displacement components at any point in the element is approximated by an interpolation from the nodal displacements using the shape functions. This is because the two displacement components (u and v) are basically independent from each other. The question now is how we can construct the shape functions, N_i, for our triangular element that satisfies the sufficient requirements: delta function property; partitions of unity; and linear field reproduction.

7.2.2 Shape function construction

Development of the shape functions is normally the first, and most important, step in developing FE equations for any type of element. In determining the shape functions N_i ($i = 1, 2, 3$) for the triangular element, we can of course follow exactly the standard procedure described in Sections 3.4.3 and 4.2.1, by starting with an assumption of the displacements using polynomial basis functions with unknown constants. These unknown constants are then determined using the nodal displacements at the nodes of the element. This standard procedure works in principle for the development of any type of element, but may not be the most convenient method. We demonstrate here another slightly different approach for constructing shape functions. We start with an assumption of shape functions directly using polynomial basis functions with unknown constants. These unknown constants are then determined using the property of the shape functions. The only difference here is that we assume the shape function in a polynomial form instead of assuming the displacements. For a linear triangular element, we assume that the shape functions are linear functions of x and y. They should, therefore, have the form of

$$
N_1 = a_1 + b_1 x + c_1 y \tag{7.5}
$$

$$
N_2 = a_2 + b_2 x + c_2 y \tag{7.6}
$$

$$
N_3 = a_3 + b_3 x + c_3 y \tag{7.7}
$$

where a_i, b_i, and c_i ($i = 1, 2, 3$) are coefficients to be determined. Equations (7.5)–(7.7) can be written in a concise form,

$$N_i = a_i + b_i x + c_i y, \quad i = 1, 2, 3 \tag{7.8}$$

We write the shape functions in the following matrix form:

$$N_i = \underbrace{\{1 \; x \; y\}}_{\mathbf{p}^T} \underbrace{\begin{Bmatrix} a_i \\ b_i \\ c_i \end{Bmatrix}}_{\alpha} = \mathbf{p}^T \alpha \tag{7.9}$$

where α is the vector of the three unknown constants, and $\mathbf{p}$ is the vector of polynomial basis functions (or monomials). Using Eq. (3.21), the moment matrix $\mathbf{P}$ corresponding to basis $\mathbf{p}$ can be given by

$$\mathbf{P} = \begin{bmatrix} 1 & x_1 & y_1 \\ 1 & x_2 & y_2 \\ 1 & x_3 & y_3 \end{bmatrix} \tag{7.10}$$

Note that the above equation is written for the shape functions, and not for the displacements. For this particular problem, we use up to the first order of polynomial basis. Depending upon the problem, we can use a higher order of polynomial basis functions. The *complete order* of polynomial basis functions in two-dimensional space up to the nth order can be given by using the so-called Pascal triangle, shown in Figure 3.2. The number of terms used in $\mathbf{p}$ depends upon the number of nodes the 2D element has. We usually try to use terms of lowest orders to make the basis as complete as possible in order. It is also possible to choose specific terms of higher orders for different types of elements. For our triangular element there are three nodes, and therefore the lowest terms with complete first order are used, as shown in Eq. (7.9). The assumption of Eqs. (7.5)–(7.7) implies that the displacement is assumed to vary linearly in the element. In these equations or Eq. (7.8) there are a total of nine constants to be determined. Our task now is to determine these constants.

If the shape functions constructed possess the delta function property, based on Lemmas 2 and 3 given in Chapter 3, the shape functions constructed will possess the partition of unity and linear field reproduction properties, as long as the moment matrix given in Eq. (7.10) is of full rank. Therefore, we can expect that the complete linear basis functions used in Eq. (7.9) guarantee that the shape functions to be constructed satisfy the sufficient requirements for FEM shape functions. What we need to do now is simply impose the delta function property on the assumed shape functions to determine the unknown coefficients a_i, b_i, and c_i.

The delta functions property states that the shape function must be a unit at its home node, and zero at all the remote nodes. For a two-dimensional problem, it can be expressed as

$$N_i(x_j, y_j) = \begin{cases} 1 \text{ for } i = j \\ 0 \text{ for } i \neq j \end{cases} \tag{7.11}$$

For a triangular element, this condition can be expressed explicitly for all three shape functions in the following equations. For shape function N_1, we have

$$
\begin{aligned}
N_1(x_1, y_1) &= 1 \\
N_1(x_2, y_2) &= 0 \\
N_1(x_3, y_3) &= 0
\end{aligned}
\tag{7.12}
$$

This is because node 1 at (x_1, y_1) is the home node of N_1, and nodes 2 at (x_2, y_2) and 3 at (x_3, y_3) are the remote nodes of N_1. Using Eqs. (7.5) and (7.12), we have

$$
\begin{aligned}
N_1(x_1, y_1) &= a_1 + b_1 x_1 + c_1 y_1 = 1 \\
N_1(x_2, y_2) &= a_1 + b_1 x_2 + c_1 y_2 = 0 \\
N_1(x_3, y_3) &= a_1 + b_1 x_3 + c_1 y_3 = 0
\end{aligned}
\tag{7.13}
$$

Solving the simultaneous Eq. (7.13) for a_1, b_1, and c_1, we obtain

$$
a_1 = \frac{x_2 y_3 - x_3 y_2}{2A_e}, \quad b_1 = \frac{y_2 - y_3}{2A_e}, \quad c_1 = \frac{x_3 - x_2}{2A_e},
\tag{7.14}
$$

where A_e is the area of the triangular element that can be calculated using the determinant of the moment matrix:

$$
A_e = \frac{1}{2} |\mathbf{P}| = \frac{1}{2} \begin{vmatrix} 1 & x_1 & y_1 \\ 1 & x_2 & y_2 \\ 1 & x_3 & y_3 \end{vmatrix} = \frac{1}{2}[(x_2 y_3 - x_3 y_2) + (y_2 - y_3)x_1 + (x_3 - x_2)y_1]
\tag{7.15}
$$

Note here that as long as the area of the triangular element is non-zero, or as long as the three nodes are not on the same line, the moment matrix $\mathbf{P}$ will be of full rank.

Substituting Eq. (7.14) into Eq. (7.5), we obtain

$$
N_1 = \frac{1}{2A_e}[(x_2 y_3 - x_3 y_2) + (y_2 - y_3)x + (x_3 - x_2)y]
\tag{7.16}
$$

which can be re-written as

$$
N_1 = \frac{1}{2A_e}[(y_2 - y_3) + (x - x_2) + (x_3 - x_2)(y - y_2)]
\tag{7.17}
$$

This equation clearly shows that N_1 is a plane in the space of (x, y) that passes through the line of 2–3, and vanishes at nodes 2 at (x_2, y_2) and 3 at (x_3, y_3). This plane also passes the point of $(x_1, y_1, 1)$,

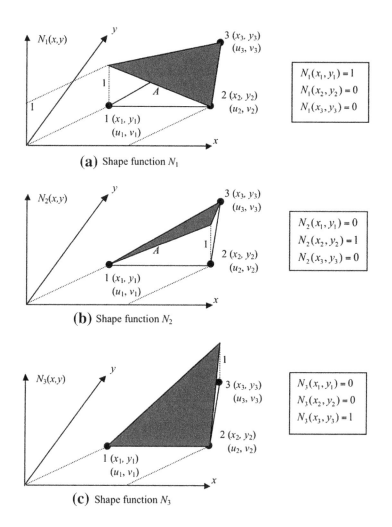

FIGURE 7.5

Linear triangular element and its shape functions.

which guarantees the unity of the shape function at the home node. Since the shape function varies linearly within the element, N_1 can then be easily plotted as in Figure 7.5a.

Similarly, making use of these features of N_2, we can immediately write out the other two shape functions for nodes 2 and 3. For node 2, the conditions are

$$N_2(x_1, y_1) = 0$$
$$N_2(x_2, y_2) = 1$$
$$N_2(x_3, y_3) = 0$$

(7.18)

and the shape function N_2 should pass through the line 3–1, which gives

$$N_2 = \frac{1}{2A_e}[(x_3y_1 - x_1y_3) + (y_3 - y_1)x + (x_1 - x_3)y]$$

$$= \frac{1}{2A_e}[(y_3 - y_1) + (x - x_3) + (x_1 - x_3)(y - y_3)]$$

$$(7.19)$$

which is plotted in Figure 7.5b. For node 3, the conditions are

$$N_3(x_1, y_1) = 0$$
$$N_3(x_2, y_2) = 0 \qquad\qquad (7.20)$$
$$N_3(x_3, y_3) = 1$$

and the shape function N_3 should pass through the line 1–2, and is given by

$$N_3 = \frac{1}{2A_e}[(x_1y_1 - x_1y_1) + (y_1 - y_2)x + (x_2 - x_1)y]$$

$$= \frac{1}{2A_e}[(y_1 - y_2) + (x - x_1) + (x_2 - x_1)(y - y_1)]$$

$$(7.21)$$

which is plotted in Figure 7.5c. The process of determining these constants is basically simple, algebraic manipulation.

Finally, the shape functions are summarized in the following concise form:

$$N_i = a_i + b_ix + c_iy \qquad\qquad (7.22)$$

with

$$a_i = \frac{1}{2A_e}(x_jy_k - x_ky_j)$$

$$b_i = \frac{1}{2A_e}(y_j - y_k) \qquad\qquad (7.23)$$

$$c_i = \frac{1}{2A_e}(x_k - x_j)$$

where the subscript i varies from 1 to 3, and j and k are determined by the cyclic permutation in the order of i, j, k. For example, if $i = 1$, then $j = 2$, $k = 3$. When $i = 2$, then $j = 3$, $k = 1$.

7.2.3 Area coordinates

Another alternative and effective method for creating shape functions for triangular elements is to use what is called *area coordinates* L_1, L_2, and L_3. The use of the area coordinates will immediately lead to the shape functions for triangular elements. However, we first need to define the area coordinates.

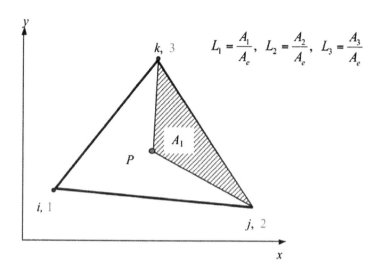

FIGURE 7.6

Definition of area coordinates.

In defining L_1, we consider a point P at (x, y) inside the triangular element, as shown in Figure 7.6, and form a sub-triangle of 2–3–P. The area of this sub-triangle is noted as A_1, and it can be calculated using the formula

$$A_1 = \frac{1}{2} \begin{vmatrix} 1 & x & y \\ 1 & x_2 & y_2 \\ 1 & x_3 & y_3 \end{vmatrix} = \frac{1}{2}[(x_2 y_3 - x_3 y_2) + (y_2 - y_3)x + (x_3 - x_2)y] \tag{7.24}$$

The area coordinate L_1 is then defined as

$$L_1 = \frac{A_1}{A_e} \tag{7.25}$$

Similarly, for L_2 we form sub-triangle 3–1–P with an area of A_2 given by

$$A_2 = \frac{1}{2} \begin{vmatrix} 1 & x & y \\ 1 & x_3 & y_3 \\ 1 & x_1 & y_1 \end{vmatrix} = \frac{1}{2}[(x_3 y_1 - x_1 y_3) + (y_3 - y_1)x + (x_1 - x_3)y] \tag{7.26}$$

The area coordinate L_2 is then defined as

$$L_2 = \frac{A_2}{A_e} \tag{7.27}$$

For L_3, we naturally write

$$L_3 = \frac{A_3}{A_e} \tag{7.28}$$

where A_3 is the area of the sub-triangle 1–2–P and is calculated using

$$A_3 = \frac{1}{2} \begin{vmatrix} 1 & x & y \\ 1 & x_1 & y_1 \\ 1 & x_2 & y_2 \end{vmatrix} = \frac{1}{2}[(x_1 y_2 - x_2 y_1) + (y_1 - y_2)x + (x_2 - x_1)y] \tag{7.29}$$

It is very easy to confirm the unity property of the area coordinates L_1, L_2, and L_3. First, they are partitions of unity, i.e.,

$$L_1 + L_2 + L_3 = 1 \tag{7.30}$$

that can be proven using the definition of the area coordinates:

$$L_1 + L_2 + L_3 = \frac{A_1}{A_e} + \frac{A_2}{A_e} + \frac{A_3}{A_e} = \frac{A_1 + A_2 + A_3}{A_e} = 1 \tag{7.31}$$

Secondly, these area coordinates are of delta function properties. For example, L_1 will definitely be zero if P is at the remote nodes 2 and 3, and it will be a unit if P is at its home node 1. The same arguments are also valid for L_2 and L_3.

These two properties are exactly those defined for shape functions. Therefore, we immediately have

$$N_1 = L_1, \quad N_2 = L_2, \quad N_3 = L_3 \tag{7.32}$$

The previous equation can also be easily confirmed by comparing Eqs. (7.16) with (7.25), (7.19) with (7.27) and (7.21) with (7.28). The area coordinates are very convenient for constructing higher order shape functions for triangular elements.

Once the shape function matrix has been developed, one can write the displacement at any point in the element in terms of nodal displacements in the form of Eq. (7.1). The next step is to develop the strain matrix so that we can write the strain, and hence the stress, at any point in the element in terms of the nodal displacements. This will further lead to the element matrices.

7.2.4 Strain matrix

Let's now move to the second step, which is to derive the strain matrix required for computing the stiffness matrix of the element. According to the discussion in Chapter 2, there are only three major stress components, $\boldsymbol{\sigma}^T = \{\sigma_{xx}\ \sigma_{yy}\ \sigma_{xy}\}$ in a 2D solid, and the corresponding strains, $\boldsymbol{\varepsilon}^T = \{\varepsilon_{xx}\ \varepsilon_{yy}\ \gamma_{xy}\}$ can be expressed as

$$\begin{aligned}
\varepsilon_{xx} &= \frac{\partial u}{\partial x} \\
\varepsilon_{yy} &= \frac{\partial v}{\partial y} \\
\gamma_{xy} &= \frac{\partial u}{\partial y} + \frac{\partial v}{\partial x}
\end{aligned} \tag{7.33}$$

or in a concise matrix form,

$$\boldsymbol{\varepsilon} = \mathbf{L}\mathbf{U} \tag{7.34}$$

where $\mathbf{L}$ is called a differential operation matrix, and can be obtained simply by inspection of Eq. (7.33):

$$\mathbf{L} = \begin{bmatrix} \partial/\partial x & 0 \\ 0 & \partial/\partial y \\ \partial/\partial y & \partial/\partial x \end{bmatrix} \tag{7.35}$$

Substituting Eq. (7.1) into Eq. (7.34), we have

$$\boldsymbol{\varepsilon} = \mathbf{L}\mathbf{U} = \underbrace{\mathbf{L}\mathbf{N}}_{\mathbf{B}}\,\mathbf{d}_e = \mathbf{B}\mathbf{d}_e \tag{7.36}$$

where $\mathbf{B}$ is termed the *strain matrix*, which can be obtained by the following equation once the shape function is known:

$$\mathbf{B} = \mathbf{L}\mathbf{N} = \begin{bmatrix} \partial/\partial x & 0 \\ 0 & \partial/\partial y \\ \partial/\partial y & \partial/\partial x \end{bmatrix} \mathbf{N} \tag{7.37}$$

Equation (7.36) implies that the strain is now expressed by the nodal displacement of the element using the strain matrix. Equations (7.36) and (7.37) are applicable for all types of 2D elements.

Using Eqs. (7.3), (7.22), (7.23), (7.37), and evaluating the partial derivatives for the shape functions, the strain matrix $\mathbf{B}$ for the linear triangular element can be easily obtained, to have the following simple form:

$$\mathbf{B} = \begin{bmatrix} b_1 & 0 & b_2 & 0 & b_3 & 0 \\ 0 & c_1 & 0 & c_2 & 0 & c_3 \\ c_1 & b_1 & c_2 & b_2 & c_3 & b_3 \end{bmatrix} \tag{7.38}$$

It can be clearly seen that the strain matrix $\mathbf{B}$ for a linear triangular element is a constant matrix. This implies that the strain within a linear triangular element is constant, and thus so is the stress.

Therefore, the linear triangular elements are also referred to as *constant strain elements* or *constant stress elements*. In reality, stress or strain varies across the structure. Using linear triangular elements with a coarse mesh will therefore result in inaccurate stress or strain distribution. We would need to have a fine mesh of linear triangular elements in order to show an appropriate variation of stress or strain across the structure.

7.2.5 Element matrices

Having obtained the shape function and the strain matrix, the displacement and strain (hence the stress) can all be expressed in terms of the nodal displacements of the element. The element matrices, like the stiffness matrix $\mathbf{k}_e$, mass matrix $\mathbf{m}_e$, and the nodal force vector $\mathbf{f}_e$, can then be found using the equations developed in Chapter 3.

The element stiffness matrix $\mathbf{k}_e$ for 2D solid elements can be obtained using Eq. (3.71):

$$\mathbf{k}_e = \int_{V_e} \mathbf{B}^T \mathbf{c} \mathbf{B} dV = \int_{A_e} \underbrace{\left(\int_0^h dz \right)}_{h} \mathbf{B}^T \mathbf{c} \mathbf{B} dA = \int_{A_e} h\mathbf{B}^T \mathbf{c} \mathbf{B} dA \tag{7.39}$$

Note that the material constant matrix $\mathbf{c}$ has been given by Eqs. (2.31) and (2.32) for plane stress and plane strain problems, respectively. Since the strain matrix $\mathbf{B}$ is a constant matrix, as shown in Eq. (7.38), and the thickness of the element is assumed to be uniform, the integration in Eq. (7.39) can be carried out very easily, which leads to

$$\mathbf{k}_e = hA_e \mathbf{B}^T \mathbf{c} \mathbf{B} \tag{7.40}$$

The element mass matrix $\mathbf{m}_e$ can also be easily obtained by substituting the shape function matrix into Eq. (3.75):

$$\mathbf{m}_e = \int_{V_e} \rho \mathbf{N}^T \mathbf{N} dV = \int_{A_e} \underbrace{\left(\int_0^h dz \right)}_{h} \rho \mathbf{N}^T \mathbf{N} dA = \int_{A_e} h\rho \mathbf{N}^T \mathbf{N} dA \tag{7.41}$$

For elements with uniform thickness and density, they can be taken out of the integral and we can rewrite Eq. (7.41) as

$$\mathbf{m}_e = h\rho \int_{Ae} \begin{bmatrix} N_1N_1 & 0 & N_1N_2 & 0 & N_1N_3 & 0 \\ 0 & N_1N_1 & 0 & N_1N_2 & 0 & N_1N_3 \\ N_2N_1 & 0 & N_2N_2 & 0 & N_2N_3 & 0 \\ 0 & N_2N_1 & 0 & N_2N_2 & 0 & N_2N_3 \\ N_3N_1 & 0 & N_3N_2 & 0 & N_3N_3 & 0 \\ 0 & N_3N_1 & 0 & N_3N_2 & 0 & N_3N_3 \end{bmatrix} dA \tag{7.42}$$

The integration of all the terms in the mass matrix can be carried out simply by using a mathematical formula developed by Eisenberg and Malvern (1973):

$$\int_A L_1^m L_2^n L_3^p \, dA = \frac{m!n!p!}{(m+n+p+2)!} 2A \tag{7.43}$$

where $L_i = N_i$ is the *area coordinates* for triangular elements that is the same as the shape function, as we have seen in Section 7.2.2. The element mass matrix $\mathbf{m}_e$ is found to be

$$\mathbf{m}_e = \frac{\rho h A}{12} \begin{bmatrix} 2 & 0 & 1 & 0 & 1 & 0 \\ & 2 & 0 & 1 & 0 & 1 \\ & & 2 & 0 & 1 & 0 \\ & & & 2 & 0 & 1 \\ & sy. & & & 2 & 0 \\ & & & & & 2 \end{bmatrix} \tag{7.44}$$

The nodal force vector for 2D solid elements can be obtained using Eqs. (3.78), (3.79), and (3.81). For the case when the element is loaded by a distributed force $\mathbf{f}_s$ on the edge 2–3 of the element, as shown in Figure 7.4, the nodal force vector involves the surface integral, which in this case is the line integral around the perimeter of the triangular element. Since the external force is zero on all edges except along the edge 2–3, the line integral can be written as

$$\mathbf{f}_e = \int_l \mathbf{N}^T \bigg|_{2-3} \begin{Bmatrix} f_{sx} \\ f_{sy} \end{Bmatrix} dl \tag{7.45}$$

If the load is uniformly distributed, f_{sx} and f_{sy} are constants within the element, so the above equation becomes

$$\mathbf{f}_e = \frac{1}{2} l_{2-3} \begin{Bmatrix} 0 \\ 0 \\ f_{sx} \\ f_{sy} \\ f_{sx} \\ f_{sy} \end{Bmatrix} \tag{7.46}$$

where l_{2-3} is the length of the edge 2–3 of the element. The force vector above is obtained by the integral of the linear shape functions, N_2 and N_3, along the edge 2–3, which can be easily shown to be $l_{2-3}/2$. Note that for a uniformly distributed load, the equivalent force components acting on the edge 2–3 ($f_{sx}l_{2-3}$ and $f_{sy}l_{2-3}$) are shared equally by the two nodes 2 and 3.

Once the element stiffness matrix $\mathbf{k}_e$, mass matrix $\mathbf{m}_e$ and nodal force vector $\mathbf{f}_e$ have been obtained, the global FE equation can be obtained by assembling the element matrices by summing up the contribution from all the adjacent elements at the shared nodes.

7.3 Linear rectangular elements

Triangular elements are usually not preferred by many analysts nowadays, unless there are difficulties with the meshing and re-meshing of models of complex geometry. The main reason is that the triangular elements are usually much less accurate than rectangular or quadrilateral (four-sided) elements. As shown in the previous section, the strain matrix of the linear triangular elements is constant, accounting partially for the inaccuracy. For the rectangular element, the strain matrix is not a constant, as will be shown in this section. This will provide a more realistic presentation in the strain, and hence the stress distribution, across the solid. The formulation of the equations for the rectangular elements is simpler compared to the more general quadrilateral elements, because the shape functions can be formed very easily due to the regularity in the shape of the rectangular element. The simple three-step procedure is applicable, and will be shown in the following sections.

7.3.1 Shape function construction

Consider a 2D domain. The domain is discretized into a number of *rectangular elements* with four nodes and four straight edges, as shown in Figure 7.7. As always, we number the nodes in each element 1, 2, 3, and 4 in a counter-clockwise direction. Note also that, since each node has two DOFs, the total DOFs for a linear rectangular element would be eight. The dimension of the element is defined here as $2a \times 2b \times h$. A dimensionless local *natural coordinate system* (ξ, η) with its origin located at the center of the rectangular element is defined. The relationship between the physical coordinate (x, y) and the local natural coordinate system (ξ, η) is given by

$$\xi = \frac{1}{a}\left(x - \frac{x_1 + x_2}{2}\right), \quad \eta = \frac{1}{b}\left(y - \frac{y_1 + y_2}{2}\right) \tag{7.47}$$

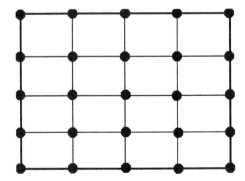

FIGURE 7.7

Rectangular domain meshed by rectangular elements.

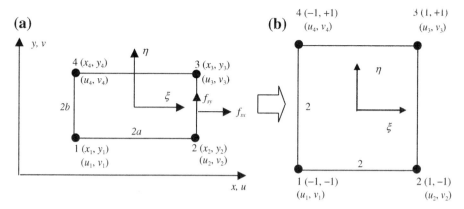

FIGURE 7.8

Rectangular element and the coordinate systems. (a) Rectangular element in physical system; (b) Square element in natural coordinate system.

Equation (7.47) defines a very simple coordinate *mapping* between physical and natural coordinate systems for rectangular elements as shown in Figure 7.8. Our formulation can now be based on the natural coordinate system. The benefits of the use of natural coordinates include the ease of constructing shape functions and of evaluating matrix integrations over a dimensionless domain. This kind of coordinate mapping technique is one of the most frequently used techniques in the FEM. It is extremely powerful when used for developing elements of complex shapes.

We perform the field variable interpolation and express the displacement within the element as an interpolation of the nodal displacements using shape functions. The displacement vector $\mathbf{U}$ is assumed to have the form

$$\mathbf{U}^h(x, y) = \mathbf{N}(x, y)\mathbf{d}_e \tag{7.48}$$

where the nodal displacement vector $\mathbf{d}_e$ is arranged in the form

$$\mathbf{d}_e = \begin{Bmatrix} u_1 \\ v_1 \\ u_2 \\ v_2 \\ u_3 \\ v_3 \\ u_4 \\ v_4 \end{Bmatrix} \begin{matrix} \left. \vphantom{\begin{matrix} u_1 \\ v_1 \end{matrix}} \right\} \text{displacements at node 1} \\ \left. \vphantom{\begin{matrix} u_2 \\ v_2 \end{matrix}} \right\} \text{displacements at node 2} \\ \left. \vphantom{\begin{matrix} u_3 \\ v_3 \end{matrix}} \right\} \text{displacements at node 3} \\ \left. \vphantom{\begin{matrix} u_4 \\ v_4 \end{matrix}} \right\} \text{displacements at node 4} \end{matrix} \tag{7.49}$$

and the matrix of shape functions has the form

$$\mathbf{N} = \begin{bmatrix} \underbrace{N_1 \quad 0}_{\text{Node 1}} & \underbrace{N_2 \quad 0}_{\text{Node 2}} & \underbrace{N_3 \quad 0}_{\text{Node 3}} & \underbrace{N_4 \quad 0}_{\text{Node 4}} \\ 0 \quad N_1 & 0 \quad N_2 & 0 \quad N_3 & 0 \quad N_4 \end{bmatrix} \tag{7.50}$$

where the shape functions N_i $(i=1,2,3,4)$ are the shape functions corresponding to the four nodes of the rectangular element.

In determining these shape functions N_i $(i=1,2,3,4)$, we can follow exactly the same steps used in Sections 4.2.1 or 7.2.2, by starting with an assumption of the displacement or shape functions using polynomial basis functions with unknown coefficients. These unknown coefficients are then determined using the displacements at the nodes of the element or the property of the shape functions. The only difference here is that we need to use four terms of monomials of basis functions. As we have seen in Section 7.2.2, the process can be troublesome and lengthy. In many cases one often uses "shortcuts" to construct shape functions. One of these "shortcuts" is by inspection, and involves utilizing the properties of shape functions.

Due to the regularity of the square element in the natural coordinates, the shape functions in Eq. (7.50) can be written out directly as follows, without going through the detailed process that we described in the previous section for triangular elements:

$$\begin{aligned} N_1 &= \frac{1}{4}(1-\xi)(1-\eta) \\ N_2 &= \frac{1}{4}(1+\xi)(1-\eta) \\ N_3 &= \frac{1}{4}(1+\xi)(1+\eta) \\ N_4 &= \frac{1}{4}(1-\xi)(1+\eta) \end{aligned} \tag{7.51}$$

While constructing these shape functions given in Eq. (7.51) by inspection, we make sure that they satisfy the delta function property of Eq. (3.34). For example, for N_3 we have

$$\begin{aligned} N_3\big|_{\text{at node 1}} &= \frac{1}{4}(1+\xi)(1+\eta)\Big|_{\substack{\xi=-1 \\ \eta=-1}} = 0 \\ N_3\big|_{\text{at node 2}} &= \frac{1}{4}(1+\xi)(1+\eta)\Big|_{\substack{\xi=-1 \\ \eta=-1}} = 0 \\ N_3\big|_{\text{at node 3}} &= \frac{1}{4}(1+\xi)(1+\eta)\Big|_{\substack{\xi=-1 \\ \eta=-1}} = 1 \\ N_3\big|_{\text{at node 4}} &= \frac{1}{4}(1+\xi)(1+\eta)\Big|_{\substack{\xi=-1 \\ \eta=-1}} = 0 \end{aligned} \tag{7.52}$$

The same examination of N_1, N_2, and N_4 will confirm the same property.

It is also very easy to show that all the shape functions given in Eq. (7.51) satisfy the partition of unity property of Eq. (3.41):

$$\sum_{I=1}^{4} N_i = N_1 + N_2 + N_3 + N_4$$

$$= \frac{1}{4}[(1 - \xi)(1 - \eta) + (1 + \xi)(1 - \eta) + (1 + \xi)(1 + \eta) + (1 - \xi)(1 + \eta)] \qquad (7.53)$$

$$= \frac{1}{4}[2(1 - \xi) + 2(1 + \xi)] = 1$$

The partitions of unity property can also be easily confirmed using Lemma 1 in Chapter 3.

Equation (7.51) should be called a bilinear shape function to be exact, because it varies linearly in both the ξ and η directions, but varies quadratically in any other direction. Denoting the natural coordinates of node j by (ξ_j, η_j), the bilinear shape function N_j can be re-written in the following concise form:

$$N_j = \frac{1}{4}(1 + \xi_j\xi)(1 + \eta_j\eta) \qquad (7.54)$$

7.3.2 Strain matrix

Using the same procedure as for the case of the triangular element, the strain matrix **B** would have the same form as in Eq. (7.37), that is

$$\mathbf{B} = \mathbf{LN}$$

$$= \frac{1}{4}\begin{bmatrix} -\frac{1-\eta}{a} & 0 & \frac{1-\eta}{a} & 0 & \frac{1+\eta}{a} & 0 & -\frac{1+\eta}{a} & 0 \\ 0 & -\frac{1-\xi}{b} & 0 & -\frac{1+\xi}{b} & 0 & \frac{1+\xi}{b} & 0 & \frac{1-\xi}{b} \\ -\frac{1-\xi}{b} & -\frac{1-\eta}{a} & -\frac{1+\xi}{b} & \frac{1-\eta}{a} & \frac{1+\xi}{b} & \frac{1+\eta}{a} & \frac{1-\xi}{b} & -\frac{1+\eta}{a} \end{bmatrix} \qquad (7.55)$$

It is now clear that the strain matrix for a bilinear rectangular element is no longer a constant matrix since each element in the matrix contains a linear function of either ξ or η. This implies that the strain, and hence the stress, within a bilinear rectangular element is not constant, in contrast to that of the triangular elements given in Eq. (7.38).

7.3.3 Element matrices

Having obtained the shape function and the strain matrix **B**, the element stiffness matrix $\mathbf{k}_e$, mass matrix $\mathbf{m}_e$, and the nodal force vector $\mathbf{f}_e$ can be obtained using the equations presented in Chapter 3. Using first the relationship given in Eq. (7.47), we have

$$dx\, dy = ab\, d\xi\, d\eta \qquad (7.56)$$

Substituting Eq. (7.56) into Eq. (7.39), we obtain

$$\mathbf{k}_e = \int_A h\mathbf{B}^{\mathrm{T}}\mathbf{c}\mathbf{B}\,\mathrm{d}A = \int_{-1}^{+1} abh\mathbf{B}^{\mathrm{T}}\mathbf{c}\mathbf{B}\,\mathrm{d}\xi\,\mathrm{d}\eta \tag{7.57}$$

The material constant matrix $\mathbf{c}$ has been given by Eqs. (2.31) and (2.32) for plane stress and plane strain problems, respectively. Evaluation of the integral in Eq. (7.57) would not be as straightforward, since the strain matrix $\mathbf{B}$ is a function of ξ and η. Even though it is still possible to obtain the closed form for the stiffness matrix by carrying out the integrals in Eq. (7.57) analytically, in practice, we often use a numerical integration scheme to evaluate the integral, and the commonly used *Gauss integration scheme* will be introduced here. The Gauss integration scheme is a very simple and efficient procedure for numerical integration, and it is briefly outlined here.

7.3.4 Gauss integration

Consider first a one-dimensional integral. Using the Gauss integration scheme, the integral is evaluated simply by a summation of the integrand evaluated at m *Gauss points* multiplied by corresponding *weight coefficients* as follows:

$$I = \int_{-1}^{+1} f(\xi)\mathrm{d}\xi = \sum_{j=1}^{m} w_j f(\xi_j) \tag{7.58}$$

The locations of the Gauss points and the weight coefficients have been found for different m, and are given in Table 7.1. So to perform the integration, all a computer code has to do is to call up the lookup table containing the locations of the Gauss points and the corresponding weight coefficients and perform the summation. In general, the use of more Gauss points will produce more accurate results for the integration. However, excessive use of Gauss points will increase the computational time and use up

Table 7.1   Gauss integration points and weight coefficients.

m	ξ_j	w_j	Accuracy n
1	0	2	1
2	$-1/\sqrt{3}, 1/\sqrt{3}$	1, 1	3
3	$-\sqrt{0.6}, 0, \sqrt{0.6}$	5/9, 8/9, 5/9	5
4	−0.861136, −0.339981, 0.339981, 0.861136	0.347855, 0.652145, 0.652145, 0.347855	7
5	−0.906180, −0.538469, 0, 0.538469, 0.906180	0.236927, 0.478629, 0.568889, 0.478629, 0.236927	9
6	−0.932470, −0.661209, −0.238619, 0.238619, 0.661209, 0.932470	0.171324, 0.360762, 0.467914, 0.467914, 0.360762, 0.171324	11

more computational resources, and it may not necessarily give better results. The appropriate number of Gauss points to be used depends upon the complexity of the integrand.

Considering polynomial integrands, it has been proven that the use of m Gauss points gives the exact results of a polynomial integrand of up to an order of $n=2m-1$. For example, if the integrand is a linear function (straight line), we have $2m-1=1$, which gives $m=1$. This means that for a linear integrand, one Gauss point will be sufficient to give the exact result of the integration. If the integrand is of a polynomial of a third order, we have $2m-1=3$, which gives $m=2$. This means that for an integrand of a third order polynomial, the use of two Gauss points will be sufficient to give the exact result. The use of more than two points will still give the same results, but takes more computation time. For two-dimensional integrations, the Gauss integration is sampled in two directions, as follows:

$$I = \int_{-1}^{+1} \int_{-1}^{+1} f(\xi, \eta)d\xi\,d\eta = \sum_{i=1}^{n_x}\sum_{j=1}^{n_y} w_1 w_j f(\xi_i, \eta_1) \tag{7.59}$$

Figure 7.9b shows the locations of four Gauss points used for integration in a square region.

The element stiffness matrix $\mathbf{k}_e$ can be obtained by numerically carrying out the integrals in Eq. (7.57) using the Gauss integration scheme shown in Eq. (7.59). 2×2 Gauss points shown in Figure 7.9b are sufficient to obtain the exact solution for the stiffness matrix given by Eq. (7.57). This is because the entry in the strain matrix, $\mathbf{B}$, is a linear function of ξ or η. The integrand in Eq. (7.57) consists of $\mathbf{B}^T \mathbf{cB}$, which implies multiplications of two linear functions, and hence this becomes a quadratic function. In Table 7.1, having two Gauss points sampled in each direction is sufficient to obtain the exact results for a polynomial function in that direction of an order up to 3. Figure 7.9a and c show some other different functions and possible number of integration points in a square region.

To obtain the element mass matrix $\mathbf{m}_e$, we substitute Eq. (7.56) into Eq. (3.75) to obtain

$$\mathbf{m}_e = \int_V \rho\mathbf{N}^T\mathbf{N}\,dV = \int_A \left(\underbrace{\int_0^h dz}_{h}\right)\rho\mathbf{N}^T\mathbf{N}dA = \int_A h\rho\mathbf{N}^T\mathbf{N}\,dA$$

$$= \int_{-1}^{+1}\int_{-1}^{+1} abh\rho\mathbf{N}^T\mathbf{N}\,d\xi\,d\eta \tag{7.60}$$

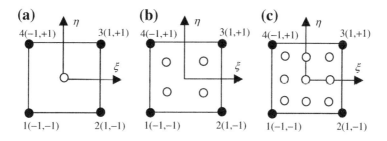

FIGURE 7.9

Integration points for $n_x=n_y=1$, 2, and 3 in a square region.

Upon evaluation of the integral, after substitution of Eq. (7.50) into Eq. (7.60), the element mass matrix $\mathbf{m}_e$ is obtained explicitly as

$$\mathbf{m}_e = \frac{\rho hab}{9} \begin{bmatrix} 4 & 0 & 2 & 0 & 1 & 0 & 2 & 0 \\ & 4 & 0 & 2 & 0 & 1 & 0 & 2 \\ & & 4 & 0 & 2 & 0 & 1 & 0 \\ & & & 4 & 0 & 2 & 0 & 1 \\ & & & & 4 & 0 & 2 & 0 \\ & & & & & 4 & 0 & 2 \\ & sy. & & & & & 4 & 0 \\ & & & & & & & 4 \end{bmatrix} \tag{7.61}$$

To evaluate integrals in the mass matrix, the following has been carried out and repeatedly used:

$$\begin{aligned} m_{ij} &= \rho hab \int_{-1}^{+1}\int_{-1}^{+1} N_i N_j \, d\xi \, d\eta \\ &= \frac{\rho hab}{16} \int_{-1}^{+1}(1+\xi_i\xi)(1+\xi_j\xi)\,d\xi \int_{-1}^{+1}(1+\eta_i\eta)(1+\eta_j\eta)d\eta \\ &= \frac{\rho hab}{4}\left(1+\frac{1}{3}\xi_i\xi_j\right)\left(1+\frac{1}{3}\eta_i\eta_j\right) \end{aligned} \tag{7.62}$$

where i and j are the usual indices for the nodes. For example, in calculating m_{33}, we use the above equation to obtain

$$m_{33} = \frac{\rho hab}{4}\left(1+\frac{1}{3}\times 1 \times 1\right)\left(1+\frac{1}{3}\times 1 \times 1\right) = 4 \times \frac{\rho hab}{9} \tag{7.63}$$

In practice, the integrals in Eq. (7.60) are often calculated numerically using the Gauss integration scheme.

The nodal force vector for a rectangular element can be obtained by using Eqs. (3.78), (3.79), and (3.81). Suppose the element is loaded by a distributed force $\mathbf{f}_s$ on edge 2–3 of the element, as shown in Figure 7.8; the nodal force vector becomes

$$\mathbf{f}_e = \int_l \mathbf{N}^T \Big|_{2-3} \begin{Bmatrix} f_{sx} \\ f_{sy} \end{Bmatrix} \, dl \tag{7.64}$$

If the load is uniformly distributed within the element edge, and f_{sx} and f_{sy} are constant, the above equation becomes

$$\mathbf{f}_e = b \begin{Bmatrix} 0 \\ 0 \\ f_{sx} \\ f_{sy} \\ f_{sx} \\ f_{sy} \\ 0 \\ 0 \end{Bmatrix} \tag{7.65}$$

where b is the half length of the side 2–3. Equation (7.65) suggests that the evenly distributed load is divided equally onto nodes 2 and 3.

The stiffness matrix $\mathbf{k}_e$, mass matrix $\mathbf{m}_e$, and nodal force vector $\mathbf{f}_e$ can be used directly to assemble the global FE equation, Eq. (3.96). Coordinate transformation is needed if the orientation of the local natural coordinate does not coincide with that of the global coordinate system. In such a case, *quadrilateral elements* are often used, which will be developed in the next section.

7.4 Linear quadrilateral elements

Though the rectangular element can be very useful, and is usually much more accurate than the triangular element, it is difficult to use for problems with any geometry other than rectangles. Hence, its practical application is very limited. A much more practical and useful element would be the general quadrilateral element, that can have unparalleled edges. However, there can be a problem for the integration of the mass and stiffness matrices for a quadrilateral element, because of the irregular shape of the integration domain. The Gauss integration scheme cannot be implemented directly with quadrilateral elements. Therefore, what is required first is to *map* the quadrilateral element into the natural coordinates system to become a square element, so that the shape functions and the integration method used for the rectangular element can be utilized. Hence, the key in the development of a quadrilateral element is the coordinate mapping. Once the mapping is established, the rest of the procedure is exactly the same as that used for formulating the rectangular element in the previous section.

7.4.1 Coordinate mapping

Figure 7.10 shows a 2D domain with the shape of an airplane wing. As you can imagine, dividing such a domain into rectangular elements of parallel edges is impossible. The job can be easily accomplished by the use of *quadrilateral elements* with four straight but unparallel edges, as shown in Figure 7.10. In developing the quadrilateral elements, we use the same idea of coordinate mapping that was used for the rectangular elements in the previous section. Due to the deviation from regularity of the element

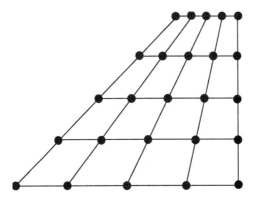

FIGURE 7.10

2D domain meshed by quadrilateral elements.

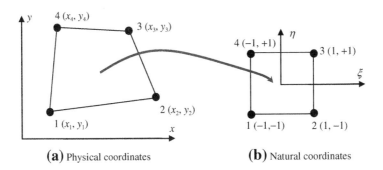

(**a**) Physical coordinates   (**b**) Natural coordinates

FIGURE 7.11

Coordinates mapping between coordinate systems.

shape, the mapping will become a little more involved, but the procedure is basically the same. In fact, the mapping carried out for the rectangular element is just a specific case of the more general procedure that will be described for the quadrilateral element.

Consider now a quadrilateral element with four nodes numbered 1, 2, 3, and 4 in a counter-clockwise direction, as shown in Figure 7.11. The coordinates for the four nodes are indicated in Figure 7.11a in the *physical coordinate system*. The physical coordinate system can be the same as the global coordinate system for the entire structure. As there are two DOFs at a node, a linear quadrilateral element has a total of eight DOFs, like the rectangular element. A local *natural coordinate system* (ξ, η), as shown in Figure 7.11b, with its origin at the center of the squared element mapped from the global coordinate system is used to construct the shape functions, and the displacement is interpolated using the equation

$$\mathbf{U}^h(\xi, \eta) = \mathbf{N}(\xi, \eta)\mathbf{d}_e \tag{7.66}$$

Equation (7.66) represents a field variable interpolation using the nodal displacements. Using a similar concept, we can also interpolate the coordinates x and y themselves. In other words, we let coordinates x and y be interpolated from the nodal coordinates using the shape functions which are expressed as functions of the natural coordinates. This coordinate interpolation is mathematically expressed as

$$\mathbf{X}(\xi, \eta) = \mathbf{N}(\xi, \eta)\mathbf{x}_e \qquad (7.67)$$

where $\mathbf{X}$ is the vector of the physical coordinates,

$$\mathbf{X} = \begin{Bmatrix} x \\ y \end{Bmatrix} \qquad (7.68)$$

and $\mathbf{N}$ is the matrix of shape functions given by Eq. (7.50). In Eq. (7.67), $\mathbf{x}_e$ represents the physical coordinates at the nodes of the element, given by

$$\mathbf{x}_e = \begin{Bmatrix} \begin{Bmatrix} x_1 \\ y_1 \end{Bmatrix} & \text{coordinates at node 1} \\ \begin{Bmatrix} x_2 \\ y_2 \end{Bmatrix} & \text{coordinates at node 2} \\ \begin{Bmatrix} x_3 \\ y_3 \end{Bmatrix} & \text{coordinates at node 3} \\ \begin{Bmatrix} x_4 \\ y_4 \end{Bmatrix} & \text{coordinates at node 4} \end{Bmatrix} \qquad (7.69)$$

Equation (7.67) can also be expressed explicitly as

$$x = \sum_{i=1}^{4} N_i(\xi, \eta)x_i$$

$$y = \sum_{i=1}^{4} N_i(\xi, \eta)y_i \qquad (7.70)$$

where N_i is the shape function defined for the rectangular element in Eqs. (7.53), or (7.54). Each of the physical coordinates (x and y) is now a function of the natural coordinates, ξ and η. Note that, due to the unique property of the shape functions, the interpolation at these nodes will be exact. For example, substituting $\xi = 1$ and $\eta = -1$ in Eq. (7.70) gives $x = x_2$ and $y = y_2$, as shown in Figure 7.11. Physically, this means that point 2 in the natural coordinate system is mapped to point 2 in the physical coordinate system, and vice versa. The same can also be easily observed for points 1, 3, and 4. The coordinate mapping that was carried out for the rectangular element is a specific case of this more general mapping using the shape functions. For the case of the rectangular element, the regularity results in only a scaling of the x and y axes such that x is a function of only ξ and y is a function of only η.

Let's now analyze this mapping more closely. Substituting $\xi = 1$ into Eq. (7.70) gives

$$
\begin{aligned}
x &= \frac{1}{2}(1 - \eta)x_2 + \frac{1}{2}(1 + \eta)x_3 \\
y &= \frac{1}{2}(1 - \eta)y_2 + \frac{1}{2}(1 + \eta)y_3
\end{aligned}
\tag{7.71}
$$

or

$$
\begin{aligned}
x &= \frac{1}{2}(x_2 + x_3) + \frac{1}{2}\eta(x_3 - x_2) \\
y &= \frac{1}{2}(y_2 + y_3) + \frac{1}{2}\eta(y_3 - y_2)
\end{aligned}
\tag{7.72}
$$

Eliminating η from the above two equations gives

$$
y = \frac{(y_3 - y_2)}{(x_3 - x_2)}\left\{ x - \frac{1}{2}(x_2 + x_3) \right\} + \frac{1}{2}(y_2 + y_3)
\tag{7.73}
$$

which represents a straight line connecting the points (x_2, y_2) and (x_3, y_3). This means that edge 2–3 in the physical coordinate system is mapped onto edge 2–3 in the natural coordinate system. The same can be observed for the other three edges. Hence, we can see that the four straight edges of the quadrilateral in the physical coordinate system correspond to the four straight edges of the square in the natural coordinate system. Therefore, the full domain of the quadrilateral element is mapped onto a squared natural coordinate system.

7.4.2 Strain matrix

After mapping is performed for the coordinates, we can evaluate the strain matrix **B**. To do so in this case, it is necessary to express the differentials in terms of the natural coordinates, since the relationship between the x and y coordinates and the natural coordinates is no longer a trivial case of scaling in the ξ and η, respectively, as in the case for rectangular elements. Utilizing the chain rule in application to partial differentiation, we have

$$
\begin{aligned}
\frac{\partial N_i}{\partial \xi} &= \frac{\partial N_i}{\partial x}\frac{\partial x}{\partial \xi} + \frac{\partial N_i}{\partial y}\frac{\partial y}{\partial \xi} \\
\frac{\partial N_i}{\partial \eta} &= \frac{\partial N_i}{\partial x}\frac{\partial x}{\partial \eta} + \frac{\partial N_i}{\partial y}\frac{\partial y}{\partial \eta}
\end{aligned}
\tag{7.74}
$$

The above equations can be written in the matrix form

$$
\begin{bmatrix} \partial N_i/\partial \xi \\ \partial N_i/\partial \eta \end{bmatrix} = \mathbf{J} \begin{bmatrix} \partial N_i/\partial x \\ \partial N_i/\partial y \end{bmatrix}
\tag{7.75}
$$

where **J** is the *Jacobian matrix* defined by

$$
\mathbf{J} = \begin{bmatrix} \partial x/\partial \xi & \partial y/\partial \xi \\ \partial x/\partial \eta & \partial y/\partial \eta \end{bmatrix}
\tag{7.76}
$$

We now substitute the interpolation of the coordinates defined by Eq. (7.70) into the above equation, and obtain

$$\mathbf{J} = \begin{bmatrix} \partial N_1/\partial \xi & \partial N_2/\partial \xi & \partial N_3/\partial \xi & \partial N_4/\partial \xi \\ \partial N_1/\partial \eta & \partial N_2/\partial \eta & \partial N_3/\partial \eta & \partial N_4/\partial \eta \end{bmatrix} \begin{bmatrix} x_1 & y_1 \\ x_2 & y_2 \\ x_3 & y_3 \\ x_4 & y_4 \end{bmatrix} \tag{7.77}$$

Rewriting Eq. (7.75) to obtain

$$\begin{bmatrix} \partial N_i/\partial x \\ \partial N_i/\partial y \end{bmatrix} = \mathbf{J}^{-1} \begin{bmatrix} \partial N_i/\partial \xi \\ \partial N_i/\partial \eta \end{bmatrix} \tag{7.78}$$

which gives the relationship between the differentials of the shape functions with respect to x and y with those with respect to ξ and η. We can now use the equation $\mathbf{B} = \mathbf{LN}$ to compute the strain matrix $\mathbf{B}$, by replacing all the differentials of the shape functions with respect to x and y with those with respect to ξ and η, obtained using Eq. (7.78). This process can be easily implemented by a computer code.

7.4.3 Element matrices

Once the strain matrix $\mathbf{B}$ has been obtained, we can proceed to evaluate the element matrices. The stiffness matrix can be obtained by Eq. (7.39). To evaluate the integration, the following formula, which has been proven by Murnaghan (1951), can be used:

$$dA = \det |\mathbf{J}| d\xi\, d\eta \tag{7.79}$$

where det $|\mathbf{J}|$ is the determinate of the Jacobian matrix. Hence, the element stiffness matrix can be written as

$$\mathbf{k}_e = \int_{-1}^{+1} \int_{-1}^{+1} h\mathbf{B}^T \mathbf{cB} \det |\mathbf{J}| d\xi\, d\eta \tag{7.80}$$

The above integrals can then be evaluated using the Gauss integration scheme discussed in the previous section. Notice how the coordinate mapping enables us to use the Gauss integration scheme over a simple squared area.

The shape function defined by Eq. (7.53) is a bilinear function of ξ and η. The elements in the strain matrix $\mathbf{B}$ are obtained by differentiating these bilinear functions with respect to ξ and η, and by multiplying the inverse of the Jacobian matrix whose elements are also bilinear functions. Therefore, the elements of $\mathbf{B}^T\mathbf{cB}$det$|\mathbf{J}|$ are quite complicated and may not be expressed by polynomials. This means that such a stiffness matrix may not, in general, be evaluated exactly using the Gauss integration scheme, unlike the case for the rectangular element.

The element mass matrix $\mathbf{m}_e$ can also be evaluated in the same way as for the rectangular element using Eq. (7.60):

$$\mathbf{m}_e = \int_V \rho \mathbf{N}^T \mathbf{N} dV = \int_A \underbrace{\left(\int_0^h dz \right)}_{h} \rho \mathbf{N}^T \mathbf{N} dA = \int_A h\rho \mathbf{N}^T \mathbf{N} dA$$

$$= \int_{-1}^{+1} \int_{-1}^{+1} h\rho \mathbf{N}^T \mathbf{N} \det |\mathbf{J}| d\xi \, d\eta$$

(7.81)

The element force vector is obtained in the same way as described for the rectangular element. For a distributed load acting on one of the edges, the integral only involves one-dimensional line integrals like before and there is thus no change in the way the integration is carried out. Having obtained the element matrices, the usual method of assembling the element matrices is carried out to obtain the global matrices.

7.4.4 Remarks

The shape functions used to interpolate the coordinates in Eq. (7.70) are the same as those used for interpolation of the displacements. Such an element is called an *isoparametric element*. However, the shape functions for coordinate and displacement interpolations do not necessarily have to be the same. Using different shape functions for coordinate and displacement interpolations will lead to the development of what is known as *subparametric* or *superparametric* elements. These elements have been studied in academic research, but are less often used in practical applications. Details of such elements will not be covered in this book.

7.5 Elements for axisymmetric structures

In engineering applications, there are structures made by revolving a plane of arbitrary shape with respect to an axis, forming so-called *axisymmetric structures*, as shown in Figure 7.12. Such axisymmetric structures can often be modeled as 2D plane problems. In such cases, we formulate the problem in the *cylindrical coordinate system* of r, z, and θ, where r is the radial direction from the axis of rotation, z is the direction along the axis of rotation, and θ is the circumferential direction (following counter-clockwise convention). If the geometry, boundary conditions, loadings, and material properties of the geometry are independent of the circumferential direction, θ, the problem is essentially a 2D one. We can then reduce the problem domain drastically to a 2D "cross-section" in the r–z plane or θ=constant plane as shown schematically in Figure 7.12.

In order to formulate 2D axisymmetric elements, we shall define the displacement field by two components in the r–z plane: u for displacement in the r or radial direction; and w for displacement in the z or axial direction. Because of the axisymmetric behavior, the asymmetric strain components shall vanish, and we have

$$\gamma_{r\theta} = \gamma_{z\theta} = 0$$

(7.82)

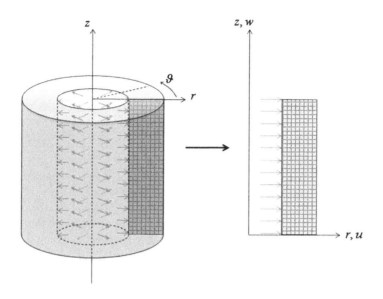

FIGURE 7.12

Reduction of a 3D axisymmetric problem into a 2D planar problem using axisymmetric elements.

Note that the so-called *hoop strain*, $\varepsilon_{\theta\theta} = u/r$, is not zero in general. The strain components are defined as:

$$\varepsilon_{rr} = \frac{\partial u}{\partial r}$$
$$\varepsilon_{zz} = \frac{\partial w}{\partial z}$$
$$\varepsilon_{\theta\theta} = \frac{u}{r} \tag{7.83}$$
$$\gamma_{rz} = \frac{\partial u}{\partial z} + \frac{\partial w}{\partial r}$$

Expressing this in a matrix form, we obtain

$$\underbrace{\begin{Bmatrix} \varepsilon_{rr} \\ \varepsilon_{zz} \\ \varepsilon_{\theta\theta} \\ \gamma_{rz} \end{Bmatrix}}_{\varepsilon} = \underbrace{\begin{bmatrix} \partial/\partial r & 0 \\ 0 & \partial/\partial z \\ 1/r & 0 \\ \partial/\partial z & \partial/\partial r \end{bmatrix}}_{\mathbf{L}} \underbrace{\begin{Bmatrix} u \\ w \end{Bmatrix}}_{\mathbf{U}} = \mathbf{LU} \tag{7.84}$$

where $\mathbf{L}$ is the differential operator matrix for 2D axisymmetric problems. Equation (7.84) shows that the strain definition is similar to plane stress or plane strain problems described earlier in this chapter except for the additional hoop strain, $\varepsilon_{\theta\theta}$, that does not change with θ. As a result, the third row in the differential operator matrix, $\mathbf{L}$, does not contain a differential operator but a multiplier with the radial displacement, u. Because of this extra hoop strain, a uniform displacement in the radial displacement, without additional boundary conditions in the radial direction, no longer produces a rigid body motion, but instead produces a circumferential strain. Correspondingly, the stress components associated with the 2D axisymmetric element are σ_{rr}, σ_{zz}, $\sigma_{\theta\theta}$, and σ_{rz}, where $\sigma_{\theta\theta}$ is the hoop stress. If we approximate the displacement field by $\mathbf{U} = \mathbf{N}\mathbf{d}_e$, like before, Eq. (7.84) can then be written as

$$\varepsilon = \mathbf{B}\mathbf{d}_e \tag{7.85}$$

where $\mathbf{B} = \mathbf{L}\mathbf{N}$ and the shape functions N_i in $\mathbf{N}$ will depend on the type of element (triangular, rectangular or quadrilateral) as described in earlier sections of this chapter.

The matrix of elastic constants, $\mathbf{c}$, for the 2D axisymmetric problem, can be obtained from that of a 3D solid by imposing the conditions of $\gamma_{r\theta} = \gamma_{z\theta} = 0$ and by assuming that the shear strain, γ_{rz}, is not coupled with the hoop stress, $\sigma_{\theta\theta}$. For a 2D, axisymmetric isotropic material, the matrix of elastic constants is given as

$$\mathbf{c} = \frac{E(1-\upsilon)}{(1+\upsilon)(1-2\upsilon)} \begin{bmatrix} 1 & \upsilon/(1-\upsilon) & \upsilon/(1-\upsilon) & 0 \\ \upsilon/(1-\upsilon) & 1 & \upsilon/(1-\upsilon) & 0 \\ \upsilon/(1-\upsilon) & \upsilon/(1-\upsilon) & 1 & 0 \\ 0 & 0 & 0 & (1-2\upsilon)/(2(1-\upsilon)) \end{bmatrix} \tag{7.86}$$

The element matrices can then be obtained in a similar way as for a 2D element, but with special care on the integration. By considering the small volume element, dV, as small volumetric "rings," it becomes

$$dV = 2\pi r dA \tag{7.87}$$

and the element stiffness matrix can be obtained using Eq. (3.71) as

$$\mathbf{k}_e = \int_{V_e} \mathbf{B}^T \mathbf{c} \mathbf{B} \, dV = 2\pi \int_{A_e} \mathbf{B}^T \mathbf{c} \mathbf{B} r \, dA \tag{7.88}$$

Similarly, the element mass matrix can be obtained from Eq. (3.75) as

$$\mathbf{m}_e = \int_{V_e} \rho \mathbf{N}^T \mathbf{N} \, dV = 2\pi \int_{A_e} \rho \mathbf{N}^T \mathbf{N} r \, dA \tag{7.89}$$

The element force vector can be obtained from Eqs. (3.78), (3.79), and (3.81) with the necessary substitution of Eq. (7.87) for the body force term (Eq. (3.78)). For the surface force term, Eq. (7.79), dS becomes

$$dS = 2\pi r \, dl \tag{7.90}$$

and, therefore, the element force vector can be expressed as

$$\mathbf{f}_e = 2\pi \left(\int_{A_e} \mathbf{N}^T \mathbf{f}_b r \, dA + \int_l \mathbf{N}^T \mathbf{f}_s r \, dl \right)$$

(7.91)

where $\mathbf{f}_b$ is the vector of body force per unit volume and $\mathbf{f}_s$ is the vector of traction force per unit area. With the element matrices obtained, the remaining procedures of assembling the element matrices into the global matrix equation, prescription of boundary conditions and loads, and solving the system of linear equations follow.

It should also be noted that the plane strain element is an extreme case of the axisymmetric element when r becomes very large. In this case, the contribution of the hoop strain, $\varepsilon_{\theta\theta}$, is negligible and, therefore, the third row and third column of Eq. (7.86) can be ignored and the $\mathbf{c}$ matrix becomes that for the case of plane strain problems.

7.6 Higher order elements—triangular element family

7.6.1 General formulation of shape functions

In developing higher order elements, we make use of the area coordinate system. Figure 7.13 shows a general triangular element of order p that has n_d nodes calculated by

$$n_d = (p + 1)(p + 2)/2$$

(7.92)

Node i (I, J, K) is located at the Ith node in the L_1 direction, at the Jth node in the L_2 direction, and at the Kth node in the L_3 direction. I, J, and K are indexed from zero as illustrated in Figure 7.13. From Figure 7.13, we have at any node that

$$I + J + K = p$$

(7.93)

The shape function can be written in the form (Argyris et al., 1968)

$$N_i = l_I^I(L_1) l_J^J(L_2) l_K^K(L_3)$$

(7.94)

where l_I^I, l_J^J, and l_K^K are defined by Eq. (4.73), but the coordinate ξ is replaced by the area coordinates, i.e.,

$$l_\beta^\beta(L_\alpha) = \frac{(L_\alpha - L_{\alpha 0})(L_\alpha - L_{\alpha 1}) \cdots (L_\alpha - L_{\alpha(\beta-1)})}{(L_{\alpha\beta} - L_{\alpha 0})(L_{\alpha\beta} - L_{\alpha 1}) \cdots (L_{\alpha\beta} - L_{\alpha(\beta-1)})}$$

(7.95)

where $\alpha = 1, 2, 3$; $\beta = I, J, K$. For example, when $\alpha = 1$ and $\beta = I$, we have

$$l_I^I(L_1) = \frac{(L_1 - L_{10})(L_1 - L_{11}) \cdots (L_1 - L_{1(I-1)})}{(L_{1I} - L_{10})(L_{1I} - L_{11}) \cdots (L_{1I} - L_{1(I-1)})}$$

(7.96)

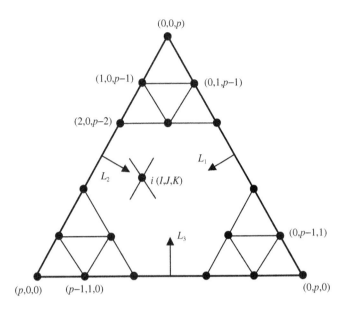

FIGURE 7.13

Triangular element of order p defined under the area coordinate system. Node i (I, J, K) is located at Ith node in the L_1 direction, at Jth node in the L_2 direction, and at Kth node in the L_3 direction. At any node, we have $I+J+K=p$.

Since

$$l_\beta^\beta(L_\alpha) = \begin{cases} 1 \text{ when } L_\alpha = L_{\alpha I} \\ 0 \text{ otherwise} \end{cases} \tag{7.97}$$

it is easy to verify that the delta function property is satisfied, i.e.,

$$N_i = \begin{cases} 1 \text{ at nodes } \quad L_1 = L_{1I}, \quad L_1 = L_{1J} \quad \text{and} \quad L_1 = L_{1K} \\ 0 \text{ other nodes} \end{cases} \tag{7.98}$$

From Eqs. (7.94) and (7.95), the order of the shape function can be found to be the same as

$$(L_1)^I \quad (L_2)^J \quad (L_3)^K \tag{7.99}$$

Since L_α is a linear function of x and y, the order of the shape function will be

$$I + J + K = p \tag{7.100}$$

7.6.2 Quadratic triangular elements

Consider a quadratic triangular element as shown in Figure 7.14. The element has six nodes: three corner nodes and three mid-side nodes. Using Eqs (7.94) and (7.95), the shape functions can be obtained very easily. Here we demonstrate the calculation of N_1. Note that for the element shown in Figure 7.14, the area coordinate L_1 has three coordinate values:

$$
\begin{aligned}
L_{10} &= 0 \quad \text{at nodes 2, 3 and 5} \\
L_{11} &= 0.5 \quad \text{at nodes 4 and 6} \\
L_{12} &= 1.0 \quad \text{at node 1}
\end{aligned}
\tag{7.101}
$$

Using Eq. (7.94), we have

$$
\begin{aligned}
N_1 &= l_2^2(L_1)l_0^0(L_2)l_0^0(L_3) = l_2^2(L_1) \times 1 \times 1 = l_2^2(L_1) \\
&= \frac{(L_1 - L_{10})(L_1 - L_{11})}{(L_{12} - L_{10})(L_{12} - L_{11})} = \frac{(L_1 - 0)(L_1 - 0.5)}{(1 - 0)(1 - 0.5)} \\
&= (2L_1 - 1)L_1
\end{aligned}
\tag{7.102}
$$

For the other two corner nodes, 2, 3, we should have exactly the same equation:

$$
N_2 = (2L_2 - 1)L_2, \quad N_3 = (2L_3 - 1)L_3
\tag{7.103}
$$

For the mid-side node 4, we have

$$
\begin{aligned}
N_4 &= l_1^1(L_1)l_1^1(L_2)l_0^0(L_3) = l_1^1(L_1)l_1^1(L_2) \times 1 = l_1^1(L_1)l_1^1(L_2) \\
&= \frac{(L_1 - L_{10})(L_2 - L_{20})}{(L_{11} - L_{10})(L_{21} - L_{20})} = \frac{(L_1 - 0)(L_2 - 0)}{(0.5 - 0)(0.5 - 0)} \\
&= 4L_1L_2
\end{aligned}
\tag{7.104}
$$

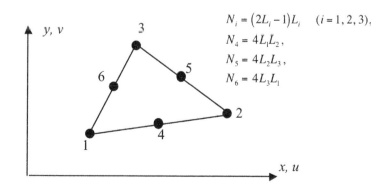

$$
\begin{aligned}
N_i &= (2L_i - 1)L_i \quad (i = 1, 2, 3), \\
N_4 &= 4L_1L_2, \\
N_5 &= 4L_2L_3, \\
N_6 &= 4L_3L_1
\end{aligned}
$$

FIGURE 7.14

Quadratic triangular element.

This equation is also valid for the other two mid-nodes, and therefore we have

$$N_5 = 4L_2L_3, \quad N_6 = 4L_1L_3 \tag{7.105}$$

7.6.3 Cubic triangular elements

For the cubic triangular element shown in Figure 7.15 that has nine nodes, the shape function can also be obtained using Eq. (7.94), as well as four area coordinate values of (taking L_1 as an example)

$$
\begin{aligned}
L_{10} &= 0 && \text{at nodes } 2, 6, 7 \text{ and } 3 \\
L_{11} &= \frac{1}{3} && \text{at nodes } 5, \ 10 \text{ and } 8 \\
L_{12} &= \frac{2}{3} && \text{at nodes } 4 \text{ and } 9 \\
L_{13} &= 1 && \text{at node } 1
\end{aligned}
\tag{7.106}
$$

We omit the details of the evaluation process and list the results below. The reader is encouraged to confirm the results. For corner nodes (1, 2, and 3):

$$N_1 = N_2 = N_2 = \frac{1}{2}(3L_1 - 1)(3L_1 - 2)L_1 \tag{7.107}$$

For side nodes (4–9):

$$N_4 \sim N_9 = \frac{9}{2}L_1L_2(3L_1 - 1) \tag{7.108}$$

For the interior node (10):

$$N_{10} = 27L_1L_2L_3 \tag{7.109}$$

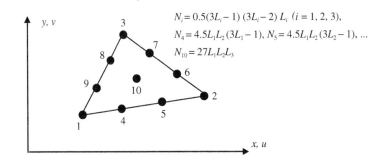

$N_i = 0.5(3L_i - 1)(3L_i - 2)L_i \ (i = 1, 2, 3)$,

$N_4 = 4.5L_1L_2(3L_1 - 1)$, $N_5 = 4.5L_1L_2(3L_2 - 1)$, ...

$N_{10} = 27L_1L_2L_3$

FIGURE 7.15

Cubic triangular element.

7.7 **Rectangular Elements**
7.7.1 **Lagrange type elements**

Considering a rectangular element with $n_d = (n+1)(m+1)$ nodes, shown in Figure 7.16. The element is defined in the domain of $(-1 \leq \xi \leq 1, -1 \leq \eta \leq 1)$ in the natural coordinates ξ and η. Due to the regularity of the nodal distribution along both the ξ and η directions, the shape function of the element can be simply obtained by multiplying one-dimensional shape functions with respect to the ξ and η directions using the Lagrange interpolants defined in Eq. (4.73) (Zienkiewicz and Taylor, 2000):

$$N_i = N_I^{1D} N_J^{1D} = l_I^n(\xi) l_J^m(\eta) \tag{7.110}$$

Due to the delta function property of the 1D shape functions given in Eq. (4.74), it is easy to confirm that the N_i given by Eq. (7.110) also has the delta function property.

Using Eqs. (7.110) and (4.73), the nine-node quadratic element shown in Figure 7.17 can be given by

$$N_1 = N_0^{1D}(\xi) N_0^{1D}(\eta) = \frac{1}{4}\xi(1-\xi)\eta(1-\eta)$$

$$N_2 = N_2^{1D}(\xi) N_0^{1D}(\eta) = -\frac{1}{4}\xi(1+\xi)\eta(1-\eta)$$

$$N_3 = N_2^{1D}(\xi) N_2^{1D}(\eta) = \frac{1}{4}\xi(1+\xi)(1+\eta)\eta$$

$$N_4 = N_0^{1D}(\xi) N_2^{1D}(\eta) = -\frac{1}{4}\xi(1-\xi)(1+\eta)\eta$$

$$N_5 = N_1^{1D}(\xi) N_0^{1D}(\eta) = -\frac{1}{2}(1+\xi)(1-\xi)(1-\eta)\eta \tag{7.111}$$

$$N_6 = N_2^{1D}(\xi) N_1^{1D}(\eta) = \frac{1}{2}\xi(1+\xi)(1+\eta)(1-\eta)$$

$$N_7 = N_1^{1D}(\xi) N_2^{1D}(\eta) = \frac{1}{2}(1+\xi)(1-\xi)(1+\eta)\eta$$

$$N_8 = N_0^{1D}(\xi) N_1^{1D}(\eta) = -\frac{1}{2}\xi(1-\xi)(1-\eta)(1+\eta)$$

$$N_9 = N_1^{1D}(\xi) N_1^{1D}(\eta) = (1-\xi^2)(1-\eta^2)$$

From Eq. (7.111), it can easily be seen that all the shape functions are formed using the same set of nine basis functions:

$$1, \xi, \eta, \xi\eta, \xi^2, \eta^2, \xi^2\eta, \eta^2\xi, \xi^2\eta^2 \tag{7.112}$$

which are linearly-independent. From Lemma 2 in Chapter 3, we can expect that the shape functions given in Eq. (7.111) to be partitions of unity. In addition, because the basis functions also contain the linear basis functions, these shape functions can also be expected to have linear field reproduction (Lemma 3), at least. Hence, they satisfy the sufficient requirements for FEM shape functions. Any other high order Lagrange type of rectangular element can be created in exactly the same way as for the nine-node element.

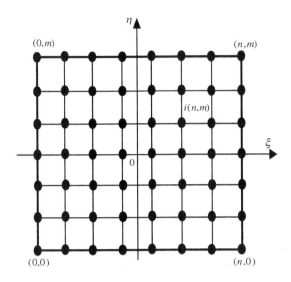

FIGURE 7.16

Rectangular element of arbitrary high orders.

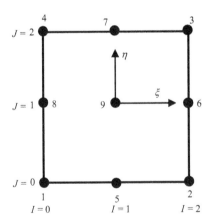

FIGURE 7.17

9-node rectangular element.

7.7.2 Serendipity type elements

The method used in constructing the Lagrange type elements is very systematic. However, the Lagrange type of elements is not very widely used, due to the presence of the interior nodes. Serendipity type elements are created by inspective construction methods. For these elements, we intentionally construct high order elements without interior nodes.

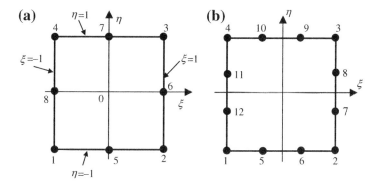

FIGURE 7.18

High order serendipity element. (a) 8-node element; (b) 12-node element.

Consider the eight-node element shown in Figure 7.18a. The element has four corner nodes and four mid-side nodes. The shape functions in the natural coordinates for the quadratic rectangular element are given as

$$N_j = \frac{1}{4}(1 + \xi_j\xi)(1 + \eta_j\eta)\ (\xi_j\xi + \eta_j\eta - 1) \text{ for corrner nodes } j = 1, 2, 3, 4$$

$$N_j = \frac{1}{2}(1 - \xi^2)(1 + \eta_j\eta) \text{ for mid-side nodes } j = 5, 7 \tag{7.113}$$

$$N_j = \frac{1}{2}(1 + \xi_j\xi)(1 - \eta^2) \text{ for mid-side nodes } j = 6, 8$$

where (ξ_j, η_j) are the natural coordinates of node j. It is very easy to observe that the shape functions possess the delta function property. The shape function is constructed by simple inspections making use of the shape function properties. For example, for the corner node 1 (where $\xi_1 = -1$, $\eta_1 = -1$), the shape function N_1 has to pass the following three lines as shown in Figure 7.19 to ensure its vanishing at remote nodes:

$$1 - \xi = 0 \Rightarrow \text{ vanishes at nodes } 2, 6, 3$$
$$1 - \eta = 0 \Rightarrow \text{ vanishes at nodes } 3, 4, 7 \tag{7.114}$$
$$-\xi - \eta - 1 = 0 \Rightarrow \text{ vanishes at nodes } 5, 8$$

The shape N_1 can then immediately be written as

$$N_1 = C(1 - \xi)(1 - \eta)\ (-\xi - \eta - 1) \tag{7.115}$$

where C is a constant to be determined using the condition that it has to be unity at node 1 at ($\xi_1 = -1$, $\eta_1 = -1$), which gives

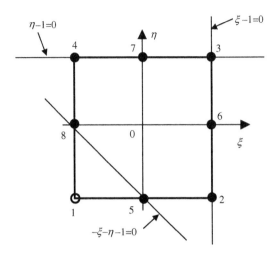

FIGURE 7.19

Construction of the 8-node element serendipity element. Three straight lines passing through the remote nodes of node 1 are used.

$$C = \frac{1}{(1 - (-1))(1 - (-1))(-(-1) - (-1) - 1)} = \frac{1}{4} \qquad (7.116)$$

We finally have

$$N_1 = \frac{1}{4}(1 + \xi_1 \xi)(1 + \eta_1 \eta)(\xi_1 \xi + \eta_1 \eta - 1) \qquad (7.117)$$

which is the first equation in Eq. (7.113) for $j = 1$.

Shape functions at all the other corner nodes can be constructed in exactly the same manner. As for the mid-side nodes, say node 5, we enforce the shape function to pass through the following three lines as shown in Figure 7.20:

$$
\begin{aligned}
1 - \xi &= 0 \Rightarrow \text{vanishes at nodes } 2, 6, 3 \\
1 + \xi &= 0 \Rightarrow \text{vanishes at nodes } 1, 8, 4 \\
1 - \eta &= 0 \Rightarrow \text{vanishes at nodes } 3, 4, 7
\end{aligned}
\qquad (7.118)
$$

The shape N_5 can then immediately be written as

$$N_5 = C(1 + \xi)(1 - \xi)(1 - \eta) \qquad (7.119)$$

where C is a constant to be determined using the condition that it has to be unity at node 5 at ($\xi_5 = 0$, $\eta_5 = -1$), which gives

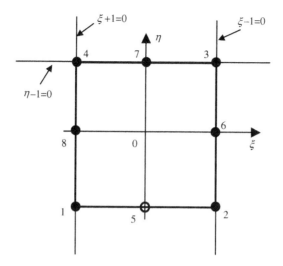

FIGURE 7.20

Construction of the 8-node element serendipity element. Three straight lines passing through the remote nodes of node 5 are used.

$$C = \frac{1}{(1 + \xi)(1 - \xi)(1 - \eta)} = \frac{1}{(1 + 0)(1 - 0)(1 - (-1))} = \frac{1}{2} \tag{7.120}$$

We finally have

$$N_5 = \frac{1}{2}(1 - \xi^2)(1 + \eta_5 \eta) \tag{7.121}$$

which is the second equation in Eq. (7.113) for $j = 5$.

Because the delta functions property is used for the shape functions given in Eq. (7.113), they will of course possess the delta function property. It can be easily seen that all the shape functions can be formed using the same set of basis functions

$$1, \xi, \eta, \xi\eta, \xi^2, \eta^2, \xi^2\eta, \eta^2\xi \tag{7.122}$$

that are linearly-independent. From Lemmas 2 and 3, we confirm that the shape functions are partitions of unity, and at least linear field reproductions. Hence, they satisfy the sufficient requirements for FEM shape functions.

Following a similar procedure, the shape functions for the 12-node cubic element shown in Figure 7.18b can be written as

$$N_j = \frac{1}{32}(1 + \xi_j\xi)(1 + \eta_j\eta)(9\xi^2 + 9\eta^2 - 10)$$

for corner nodes $j = 1, 2, 3, 4$

$$N_j = \frac{9}{32}(1 + \xi_j\xi)(1 + \eta^2)(1 + 9\eta_j\eta)$$

for side nodes $j = 7, 8, 11, 12$ where $\xi_j = \pm 1$ and $\eta_j = \pm \frac{1}{3}$ (7.123)

$$N_j = \frac{9}{32}(1 + \eta_j\eta)(1 + \xi^2)(1 + 9\xi_j\xi)$$

for side nodes $j = 5, 6, 9, 10$ where $\xi_j = \pm\frac{1}{3}$ and $\eta_j = \pm 1$

The reader is encouraged to figure out what lines or curves should be used to form the shape functions listed in Eq. (7.123). When $\eta = \eta_i = 1$, the above equations reduce to the one-dimensional cases of quadratic and cubic elements defined by Eqs. (4.75) and (4.76), respectively.

7.8 Elements with curved edges

Using high order elements, elements with curved edges can be used in the modeling. Two frequently used higher order elements of curved edges are shown in Figure 7.21a. In formulating these types of

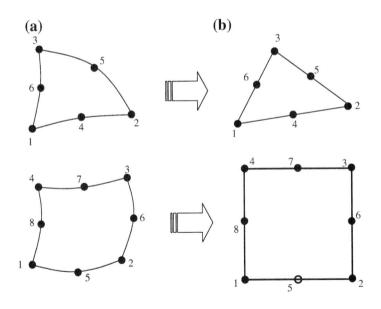

FIGURE 7.21

2D solid elements with curved edges. (a) Curved elements in the physical coordinate system; (b) Elements with straight edges obtained by mapping.

elements, the same mapping technique used for the linear quadrilateral elements (Section 7.4) can be used. In the physical coordinate system, elements with curved edges, as shown in Figure 7.21a, are first formed in the problem domain. These elements are then mapped into the natural coordinate system using Eq. (7.67). The elements mapped in the natural coordinate system will have straight edges, as shown in Figure 7.21b.

Higher order elements of curved edges are often used for modeling curved boundaries. Note that elements with excessively curved edges may cause problems in the numerical integration. Therefore, a finer mesh should be used where the curvature of the geometry is large. In addition, it is recommended that in the internal portion of the domain, elements with straight edges should be used whenever possible. More details on modeling issues will be discussed in detail in Chapter 11.

7.9 Comments on Gauss integration

When the Gauss integration scheme is used, one has to decide how many Gauss points should be used. Theoretically, for a one-dimensional integral, using m points can give the exact solution for the integral of a polynomial integrand of up to an order of $(2m - 1)$. As a general rule of thumb, more points should be used for a higher order of elements. It is also noted that using a smaller number of Gauss points tends to counteract the *over-stiff behavior* associated with the displacement-based FEM.

This over-stiff behavior of the displacement-based FEM comes about primarily because of the use of the shape function. As discussed, the displacement in an element is assumed using shape functions interpolated from the nodal displacements. This implies that the deformation of the element is actually prescribed in the fashion of the shape function. This gives a constraint to the element, and thus the element behaves more stiffly than it should. It is often observed that higher order elements are usually softer than lower order ones. This is because the use of more nodes decreases the constraint on the element.

Coming back to the Gauss integration issue, two Gauss points for bilinear elements, and about two or three Gauss points in each direction for quadratic elements, should be sufficient for most cases. Many of the explicit FEM codes based on explicit formulation tend to use one-point integration for bilinear elements to achieve the best performance in saving central processing unit (CPU) time.

7.10 Case study: Side drive micro-motor

In this case study, we analyze another microelectromechanical systems (MEMs) device: A common micro-actuator in the form of a side drive electrostatic micro-motor, as shown in Figure 7.22. Such micro-motors are usually made from polysilicon using lithographic techniques. Their diameters vary depending on the design, with the first designs having diameters of 60–120 μm. Of course, the actual working dynamics of the micro-motor will be rather complex to model, though it can still be readily done if required. Therefore, to illustrate certain points pertaining to the use of basic 2D solid elements, we use the geometrical and material information for this micro-motor and apply arbitrary loading and boundary conditions to it.

FIGURE 7.22

SEM image of an electrostatic micro-motor with 8 rotor and 12 stator poles.

Table 7.2 Elastic properties of polysilicon.	
Young's Modulus, E	169 GPa
Poisson's ratio, υ	0.262
Density, ρ	2300 kg m^{-3}

Isotropic material properties will be employed here to makes things less complicated. The material properties of polysilicon are shown in Table 7.2. We shall do a stress analysis on the rotor with some loading condition on the rotor blades. Examining the geometry, loading, and boundary conditions of the rotor in Figure 7.22, we can see that it is symmetrical, i.e., we need not model the full rotor, but rather we can just model say one quarter of the rotor and apply the necessary boundary conditions. We can do this since this one-quarter model will be repeated geometrically anyway. Of course, we can even model one eighth of the model and the results will be the same if the condition of repetitive symmetry is properly applied. Hence, this becomes a neat and efficient way of modeling repetitive or symmetrical geometry.

7.10.1 **Modeling**

Figure 7.23 shows one-quarter of the full model of the micro-motor rotor. We take the diameter of the whole rotor to be 100 μm and the depth or thickness to be 13 μm, to correspond with realistic values of micro-motor designs. The geometry can be easily drawn using pre-processors like PATRAN, ABAQUS CAE or using basic CAD software, after which it can be imported into appropriate pre-processors for meshing. Note that pre-processors are software packages used to aid us in visualizing the geometry, and to mesh up the geometry using finite elements, especially for complicated geometries. To illustrate the formulation of the FE equations clearly, we would initially mesh up the geometry in Figure 7.23 with a coarse mesh, as shown in Figure 7.24. Four-nodal, quadrilateral elements are used with a total of 24 elements and 41 nodes in the model. We shall increase the number of elements (and nodes) in later analyses to compare the results. Since the depth or thickness of the motor is much smaller than the other dimensions, and the external forces are assumed to be within the plane of the rotor, we can assume plane stress conditions.

In the above figure, it can also be seen that a distributed force of 10 N/m is applied compressively to the rotor blades. The center hole in the rotor, which is supposed to be the location for a "hub" to keep the rotor in place, is assumed to be constrained. The nodes along the edge $y=0$ are constrained in the y direction and the nodes along the edge $x=0$ are constrained in the x direction. These are to simulate the symmetrical boundary conditions of the model, since those nodes are not supposed to move in the direction normal to the plane of symmetry.

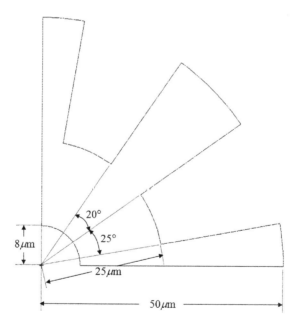

FIGURE 7.23

Plan view (2D) of a quarter of micro-motor rotor.

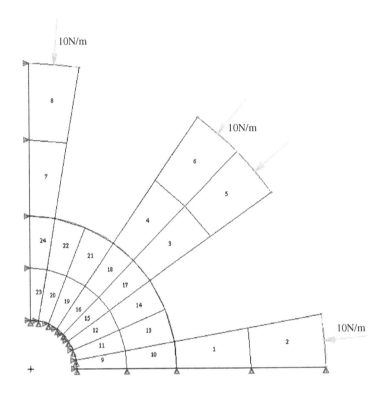

FIGURE 7.24

Finite element mesh with 24 2D quadrilateral, 4 nodal elements.

7.10.2 **ABAQUS input file**

The ABAQUS input file for the above described FE model is shown below. Note that some parts of the input file containing the data values are left out to limit the length of the file in this book. The text boxes to the right of the input file are not part of the file, but rather explain what the sections of the file mean.

*HEADING, SPARSE
Static analysis of micro-motor
**
*NODE
1, 8., 0.
2, 7.87846, 1.38919
3, 16.5, 0.
4, 16.2493, 2.8652
5, 25., 0.

Nodal cards
These define the coordinates of the nodes in the model. The first entry being the node ID while the second and third entries are the x and y coordinates of the position of the node, respectively.

.
.
.
38, 6.5118, 36.9303
39, −2.73294E−7, 37.5
40, 8.68241, 49.2404
41, 5.46197E−7, 50.
**
**

*ELEMENT, TYPE=CPS4,
ELSET=PLSTRESS
1, 1, 3, 4, 2
2, 3, 5, 6, 4
3, 2, 4, 11, 8

Element (connectivity) cards
These define the element type and what nodes make up
the element. CPS4 represents that it is a plane stress,
four-nodal, quadrilateral element. There are many
other element types in the ABAQUS element library.
For example, if we were to use a plane strain element,
the element type would be CPE4. The "ELSET=
PLSTRESS"statement is simply for naming this set of
elements so that it can be referenced when defining the
material properties. In the subsequent data entry, the
first entry is the element ID, and the following four
entries are the nodes making up the element. The order
of the nodes for all elements must be consistent and
counter-clockwise.

4, 8, 11, 12, 9
.
.
.
20, 23, 50, 51, 24
21, 49, 52, 53, 50
22, 50, 53, 54, 51
23, 33, 57, 58, 39
24, 57, 59, 60, 58
**
** plstress
**
*SOLID SECTION, ELSET=PLSTRESS, MATERIAL=POLYSILI
13 ,
**
** polysilicon
**
*MATERIAL, NAME=POLYSILI
**
*DENSITY

Material cards
These define material
properties under the
name "POLYSILI."
Density and elastic
properties are defined.
TYPE = ISO represents
isotropic properties.

Property cards
These define properties
to the elements of set
"PLSTRESS." It will
have the material
properties defined
under "POLYSILI."

2.3E−15,
**

```
*ELASTIC, TYPE=ISO
169000., 0.262
**
** Fixed
**
*BOUNDARY, OP=NEW
FIXED, 1,, 0.
FIXED, 2,, 0.
**
** Fixed-x
**
*BOUNDARY, OP=NEW
FIXED-X, 1,, 0.
**
** Fixed-y

**
*BOUNDARY, OP=NEW
FIXED-Y, 2,, 0.
**
**
*STEP, AMPLITUDE=RAMP,
PERTURBATION
Linear Static Analysis
**
**
*STATIC
**
*NSET, NSET=FIXED
1, 2, 8, 9, 17, 18, 26, 27,
35,
*NSET, NSET=FIXED-X
37, 39, 58, 60
*NSET, NSET=FIXED-Y
3, 5, 42, 44
*ELSET, ELSET=PRESS
18, 21, 22, 24
**
** press
**
*DLOAD, OP=NEW
PRESS, P2, 10.
**
**
```

> *BC cards*
> These define boundary conditions. For example,
> the first one labeled "FIXED" represents that nodes
> belonging to the set "FIXED" have the "1" and "2"
> directions constrained.

> *Control card*
> This indicates the analysis step. In this case it is a
> "STATIC" analysis.

> *Node sets*
> Sets of nodes are defined to be used for referencing
> when defining boundary and loading conditions. For
> example, the nodes 37, 39, 58 and 60 are grouped up as
> a set labeled "FIXED-X."

> *Load cards*
> "DLOAD" defines distributed loading on the element set
> "PRESS" defined earlier.

*NODE PRINT, FREQ=1
U,
*NODE FILE, FREQ=1
U,
**
*EL PRINT, POS=INTEG, FREQ=1
S,
E,
*EL FILE, POS=INTEG, FREQ=1
S,
E,
**
*END STEP

> *Output control cards*
> These define the output required. For example for nodal output, we require the displacement "U," while for element output, we require the stress, "S" and strain "E."

The input file above shows how a basic ABAQUS input file can be set up. It should be noted that the units used in this case study are micrometers, and all the conversions of the necessary inputs is done for consistency, as before.

7.10.3 **Solution process**

Let's now try to relate the information we provided in the input file with what is covered in this chapter. As before, the first sets of data usually defined are the nodes and their coordinates. Then, there are the element cards containing the connectivity information. The importance of this information has already been mentioned in previous case studies. Looking at Figure 7.24, it is not difficult to guess that the element used is an isoparametric quadrilateral element (CPS4–2D, quadrilateral, bilinear, plane stress elements), rather than that of a rectangular element.

Obviously, it can be visualized that using rectangular elements would pose a problem in meshing the geometry here. In fact, the use of purely rectangular elements is so rare that most software (including ABAQUS) only provides the more versatile quadrilateral element. This information from the nodal and element cards will be used for constructing the element matrices (see, Eqs. (7.80) and (7.81)).

Next, the property cards define the properties of the elements, and also specify the material the elements should possess. For the plane stress elements, the thickness of the elements must be specified ($13\,\mu$m in this case), since it is required in the stiffness and mass matrices (the mass matrix is actually not required in this case study, since this is a static analysis). Similarly, the elastic properties of the polysilicon material defined in the material card are also required in the element matrices. It should be noted that in ABAQUS, the integral in Eq. (7.80) is evaluated using the Gauss integration scheme, and the default number of Gauss points for the bilinear element is 4.

The boundary (BC) cards define the boundary conditions for the model. To model the symmetrical boundary conditions, at the lines of symmetry ($x=0$ and $y=0$), the nodal displacement component normal to the line is constrained to zero. The nodes (node set, FIXED) along the center hole where the hub should be is also fully clamped in. The load cards defined specify the distributed loading on the motor, as shown in Figure 7.24. These will be used to form the force vector, which is similar in form to that of Eq. (7.64).

The control cards are used to control the analysis, which in this case defines that this is a static analysis. Finally, the output cards define the necessary output requested, which here are the displacement components, the stress components, and the strain components.

Once the input file has been created, one can then invoke ABAQUS to execute the analysis, and the results will be written into an output file that can be read by the post-processor.

7.10.4 Results and discussion

Using the above ABAQUS input file that describes the problem, a static analysis is carried out. Figure 7.25 shows the Von Mises stress distribution obtained with 24 bilinear quadrilateral elements. It should be noted here that 24 elements (41 nodes) for such a problem may not be sufficient for accurate results. Analyses with a denser mesh (129 nodes and 185 nodes) using the same element type are also carried out. Their input files will be similar to that shown, but with more nodes and elements.

Figure 7.26 and Figure 7.27 show the Von Mises stress distribution obtained using 96 (129 nodes) and 144 elements (185 nodes), respectively. Figure 7.28 also shows the results obtained when 24 eight-nodal elements (105 nodes in total) used instead of four-nodal elements. The element type in ABAQUS for an eight-nodal, plane stress, quadratic element is "CPS8." Finally, linear, triangular elements (CPS3) are also used for comparison, and the stress distribution obtained is shown in Figure 7.29.

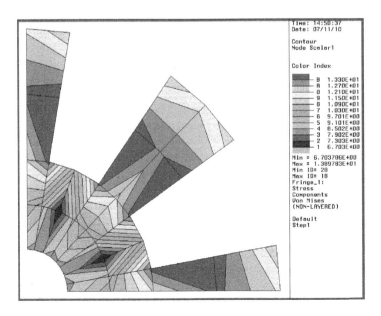

FIGURE 7.25

Analysis no. 1: Von Mises stress distribution using 24 bilinear quadrilateral elements.

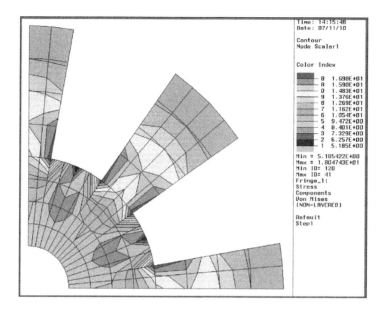

FIGURE 7.26

Analysis no. 2: Von Mises stress distribution using 96 bilinear quadrilateral elements.

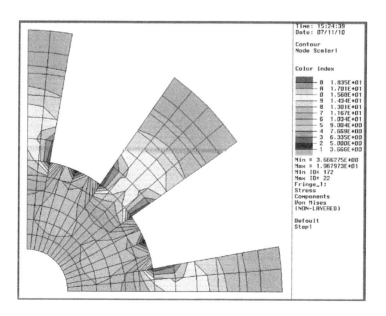

FIGURE 7.27

Analysis no. 3: Von Mises stress distribution using 144 bilinear quadrilateral elements.

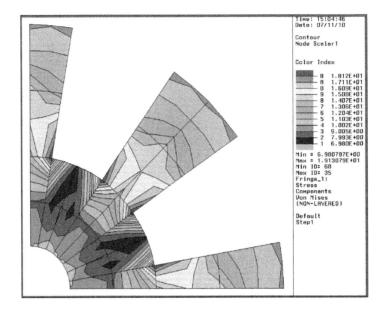

FIGURE 7.28

Analysis no. 4: Von Mises stress distribution using 24 eight-nodal, quadratic elements.

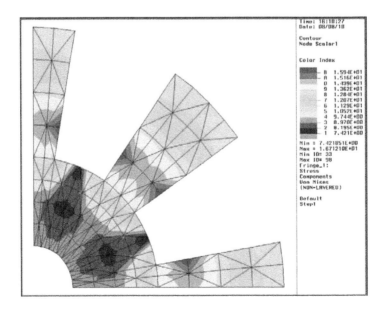

FIGURE 7.29

Analysis no. 5: Von Mises stress distribution using 192 three-nodal, triangular elements.

Table 7.3   Maximum Von Mises stress.

Analysis No.	Number/Type of Elements	Total Number of Nodes in Model	Maximum Von Mises Stress (GPa)
1	24 bilinear, quadrilateral	41	0.0139
2	96 bilinear, quadrilateral	129	0.0180
3	144 bilinear, quadrilateral	185	0.0197
4	24 quadratic, quadrilateral	105	0.0191
5	192 linear, triangular	129	0.0167

From the results obtained, it can be noted that analysis 1, which uses 24 bilinear elements, does not seem as accurate as the other three. Table 7.3 shows the maximum Von Mises stress for the five analyses. It can be seen that the maximum Von Mises stress using just 24 bilinear, quadrilateral elements (41 nodes) is just about 0.0139 GPa, which is a bit low when compared with the other analyses. The other analyses, especially from analyses 2 to 4 using quadrilateral elements, obtained results that are quite close to one another when we compare the maximum Von Mises stress. We can conclude that using just 24 bilinear, quadrilateral elements is definitely not sufficient in this case. The comparison also shows that using quadratic elements (eight-nodal) with a total of 105 nodes, yielded results that are close to analysis 3 with the bilinear elements and 185 nodes. In this case, the quadratic elements also have curved edges, instead of straight edges, and this would better define the curved geometry. Looking at the maximum Von Mises stress obtained using triangular elements in analysis 5, we can see that, despite having the same number of nodes as in analysis 2, the results obtained showed some deviation. This clearly shows that quadrilateral elements in general provide better accuracy than triangular elements. However, it is still convenient to use triangular elements to mesh complex geometry containing sharp corners.

From the stress distribution, it can generally be seen that there is stress concentration at the corners of the rotor structure, as expected. Therefore, if structural failure is to occur, it would be at these areas of stress concentration.

7.11 **Review questions**

1. Figure 7.30 shows a wall that is very thin in the z direction (out of paper), and can be modeled as a 2D problem. It is subjected to a uniform pressure in the x direction on the left edge. Use FEM to perform stress analysis for this problem.
   a. Is the problem a plane strain or plane stress problem? Justify your answer.
   b. In FEA procedure, what is the difference in modeling a plane strain and a plane stress problem?
   c. Explain your strategy to model this particular problem.
   d. Sketch a mesh for this problem domain of your model using about 50 elements. (You are not required to consider the stress concentration at the corners.)
   e. Provide all the boundary conditions at relevant nodes that are needed to solve this problem using your FEM model.

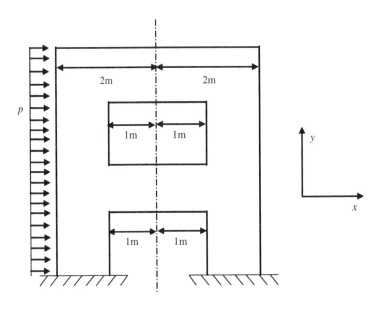

FIGURE 7.30

A thin wall subjected to a uniform pressure in the x direction on the left edge.

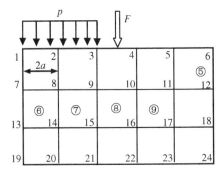

FIGURE 7.31

A simple mesh of square elements for a 2D domain.

2. F.igure 7.31 shows a mesh with 15 equal square elements of width $2a$ and 24 nodes for a 2D domain. A uniform pressure is applied on edges 1–2–3, and a concentrate load is applied at the middle of edge 3–4. In the input file of the FEM model, there are the following lines defining the connectivity for elements ⑤, ⑥, ⑦, ⑧, and ⑨.

5, 12, 11, 5, 6
6, 7, 8, 13, 14
7, 8, 9, 14, 15

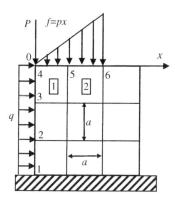

FIGURE 7.32

A 2D solid that is modeled using uniformly distributed square elements.

8, 15, 16, 10, 9

9, 10, 16, 11, 17

   a. Indicate the lines that wrongly define (unconventional) the connectivity of the element. Correct these lines and justify your answer.

   b. Give the external force vector. Propose a better way to re-number the nodes and justify your answer.

3. Figure 7.32 shows a square 2D solid that is modeled using uniformly distributed square elements. The solid is subjected to a vertical concentrated force P at the top-left corner, a distributed force $f = px$ over part of the top surface, and a uniformly distributed force q on the left edge.

   a. Using bilinear square elements for the entire solid, derive the *external* nodal force vector for nodes 1, 2, 3, 4, 5, and 6.

   b. If 8-node quadratic square elements are used instead of the bilinear elements, how is the external nodal force vector computed? You are required only to provide the formulae and explain the procedure.

4. Figure 7.33 shows two right-angle triangular elements of base b and height h. Derive the shape functions and the strain matrix **B** for these two elements.

5. Figure 7.34 shows two right-angle triangular elements of base b and height h. Derive the shape functions and the strain matrix **B** for these two elements.

6. Consider a plane strain element as shown in Figure 7.35.

   a. Calculate the shape functions for the element.

   b. The nodal displacements are given as

$$u_1 = 0.005 \text{ mm } u_2 = 0.0 \text{ mm } u_3 = 0.005 \text{ mm}$$

$$v_1 = 0.002 \text{ mm } v_2 = 0.0 \text{ mm } v_3 = 0.0 \text{ mm}$$

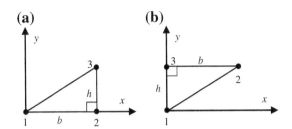

FIGURE 7.33

Two right-angle triangular elements for 2D solids.

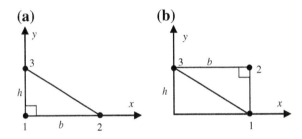

FIGURE 7.34

Two right-angle triangular elements for 2D solids.

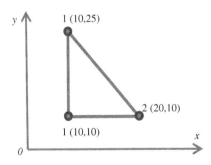

FIGURE 7.35

A plane strain element (unit: mm).

c. Calculate the stress components $\sigma_{xx}, \sigma_{yy}, \sigma_{xy}$ and the principle stresses σ_1 and σ_2 for the element, with Young's modulus $E=70\,\text{GPa}$ and Poisson ratio $\nu=0.3$.

d. Determine the principal stresses σ_1 and σ_2 and the principal angles θ_p, using the following equations:

$$\sigma_1, \sigma_2 = \frac{\sigma_{xx} + \sigma_{yy}}{2} \pm \sqrt{\left(\frac{\sigma_{xx} - \sigma_{yy}}{2}\right)^2 + \sigma_{xy}^2}$$

$$\theta_p = \tfrac{1}{2}\tan^{-1}\left(\frac{2\sigma_{xy}}{\sigma_{xx}-\sigma_{yy}}\right)$$

7. Figure 7.36 shows two triangular elements, and the unit for the coordinates is in cm. Let $E=2.1 \times 10^{11}\,\text{Pa}$, $\nu=0.25$, and thickness $t=5\,\text{mm}$. Assume plane stress conditions.
  a. Calculate shape functions for these triangular elements;
  b. Calculate the strain matrix for these triangular elements;
  c. Calculate the stiffness matrix for these triangular elements;
  d. Determine the stresses $\sigma_{xx}, \sigma_{yy}, \sigma_{xy}$ in the element for given nodal displacements (in m) of

$$u_1 = 0.0 \qquad u_2 = 0.0001 \quad u_3 = 0.0$$
$$v_1 = 0.0002 \quad v_2 = 0.0 \qquad v_3 = 0.0002$$

  e. Determine the principal stresses σ_1 and σ_2 and the principal angles θ_p, using the Equation given in the previous question.

8. Show that the stiffness matrix of an isotropic linear triangular element whose thickness varies linearly in the element is

$$\mathbf{k}_e = \bar{h} A_e \mathbf{B}^T \mathbf{c} \mathbf{B}$$

where $\mathbf{B}$ is the strain matrix, $\mathbf{c}$ is the matrix of material constants, A_e is the area of the triangle, and $\bar{h}$ is the average thickness $(h_1 + h_2 + h_3)/3$, where $h_1, h_2,$ and h_3 are the nodal thickness at the node.

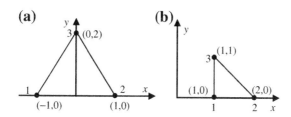

FIGURE 7.36

Two triangular elements for 2D solids (unit: cm).

9. Show that the mass matrix of a linear triangular element whose thickness varies linearly within the plane of the element is

$$
\mathbf{m}_e = \frac{\rho \bar{h} A_e}{60}
\begin{bmatrix}
6 + 4\alpha_1 & 0 & 6 - \alpha_3 & 0 & 6 - \alpha_2 & 0 \\
 & 6 + 4\alpha_1 & 0 & 6 - \alpha_3 & 0 & 6 - \alpha_2 \\
 & & 6 + 4\alpha_2 & 0 & 6 - \alpha_1 & 0 \\
 & & & 6 + 4\alpha_2 & 0 & 6 - \alpha_1 \\
 & sy. & & & 6 + 4\alpha_3 & 0 \\
 & & & & & 6 + 4\alpha_3
\end{bmatrix}
$$

where ρ is the density, A_e is the area, $\bar{h}$ is the mean thickness, and $\alpha_I = h_i/\bar{h}$ with $i = 1, 2, 3$ for the three nodes.

10. The thickness variation of a linear rectangular element is given by

$$
h(\xi, \eta) = \sum_{i=1}^{4} N_i(\xi, \eta) h_i
$$

where N_i is the bilinear shape function, and h_i is the thickness value at node i. How many Gauss points are required to evaluate exactly the mass and stiffness matrices.

11. If the thickness variation of a linear quadrilateral is the same as the rectangular element in Problem (3), how many Gauss points are required to evaluate the mass matrix exactly? How many Gauss points are required to integrate the volume of the element exactly? How many Gauss points are required to integrate the stiffness matrix exactly?

12. Construct the shape functions for the 12-node rectangular element for one corner node and one side node.

13. Figure 7.37 shows a plane strain problem to be solved using only two triangular elements.
    a. Determine the nodal displacements; (Hint: use results from Question 4 for matrix **B**).
    b. Derive expressions for displacements and the element stresses.

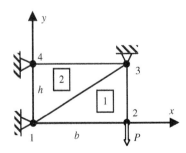

FIGURE 7.37

A rectangular 2D solid subjected to a vertical force at node 2.

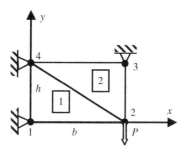

FIGURE 7.38

A rectangular 2D solid meshed by two triangular elements subjected to a vertical force at node 2.

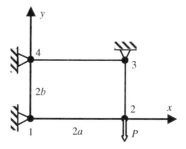

FIGURE 7.39

A rectangular element for 2D solid subjected to a vertical force at node 2.

14. Figure 7.38 shows a plane strain problem to be solved using only two triangular elements.
   a. Determine the nodal displacements; (Hint: use results from Question 4 for matrix **B**).
   b. Derive the expressions for the displacements and element stresses.
   c. Calculate the averaged element stresses.
15. Figure 7.39 shows a plane strain problem to be solved using only one rectangular element. Determine the nodal displacements and element stresses.

FEM for Plates and Shells

8

CHAPTER OUTLINE HEAD

8.1 Introduction

In this chapter, finite element (FE) equations for plates and shells are developed. The procedure is to first develop FE matrices for plate elements, and the FE matrices for shell elements are then obtained by superposing the matrices for plate elements and those for 2D solid plane stress elements developed in Chapter 7 (akin to superposing the truss and beam elements for the 1D frame elements). Unlike the 2D solid elements in the previous chapter, plate and shell elements are computationally more tedious as they involve more degrees of freedom (DOFs). The constitutive equations may seem daunting to one who may not have a strong background in the mechanics theory of plates and shells, and the integration may be quite involved if it is to be carried out analytically. However, the basic concepts of formulating the FE equation always remain the same. Readers are advised to pay more attention to the finite element concepts and the procedures outlined in developing plate and shell elements. After all, the computer can handle many of the tedious calculations/integrations that are required in the process of forming the elements. The basic concepts, procedures, and formulations can also be found in many existing textbooks (see, Petyt, 1990; Rao, 1999; Zienkiewicz and Taylor, 2000).

8.2 **Plate elements**

As discussed in Chapter 2, a plate structure is geometrically similar to the structure of the 2D plane stress problem, but it usually carries only transverse loads that lead to bending deformation in the plate. For example, consider the horizontal boards on a bookshelf that support the books. Those boards can be approximated as a plate structure, and the transverse loads are of course the weight of the books. Higher floors of a building are a typical plate structure that carries most of us every day, as are the wings of aircraft, which usually carry loads like the engines, as shown in Figure 2.13. The plate structure can be schematically represented by its middle plane laying on the x–y plane, as shown in Figure 8.1. The deformation caused by the transverse loading on a plate is represented by the deflection and rotation of the normals of the middle plane of the plate, and they will be independent of z and a function of only x and y. The element to be developed to model such plate structures is aptly known as the *plate element*. The formulation of a plate element is very much the same as for the 2D solid element, except for the process for deriving the strain matrix in which the theory of plates is used.

Plate elements are normally used to analyze the bending deformation of plate structures and the resulting forces such as shear forces and moments. In this respect, it is similar to the beam element developed in Chapter 5, except that the plate element is two dimensional whereas the beam element is one dimensional. Like the 2D solid element, a plate element can also be triangular, rectangular or quadrilateral in shape. In this book, we cover the development of the rectangular element only, as it is so widely used. Matrices for the triangular element can also be easily developed using similar procedures,

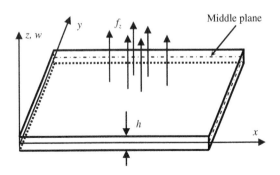

FIGURE 8.1

A plate and its coordinate system.

and those for the quadrilateral element can be developed using the idea of an isoparametric element discussed for 2D solid elements. In fact, the development of a quadrilateral element is very much the same as the rectangular element, except for an additional procedure of coordinate mapping, as shown for the case of 2D solid elements.

There are a number of theories that govern the deformation of plates. In this chapter, rectangular elements based on the Mindlin plate theory that works for thick plates will be developed. Most books go into great detail to first cover plate elements based on the thin plate theory. However, most finite element packages do not use plate elements based on thin plate theory. In fact, most analysis packages like ABAQUS do not even offer the choice of plate elements. Instead, one has to use the more general shell elements, also discussed in this chapter. Furthermore, using the thin plate theory to develop the finite element equations has a problem, in that the elements developed are usually incompatible or "nonconforming." This means that some components of the rotational displacements may not be continuous on the edges between elements. This is because the rotation depends only upon the deflection w in the thin plate theory, and hence the assumed function for w has to be used to calculate the rotation. Many texts go into even greater detail to explain the concept, and to prove the conformability of many kinds of thin plate elements. In our opinion, there is really no need, for most practical purposes, to understand such a concept and proof for readers who are interested in using the FEM to solve real-life problems. In addition, many structures may not be considered as a "thin plate," or rather their transverse shear strains cannot be ignored. Therefore, the Reissner–Mindlin plate theory is more suitable in general, and the elements developed based on the Reissner–Mindlin plate theory are more practical and useful. This book will only discuss the elements developed based on the Reissner–Mindlin plate theory.

There are a number of higher order plate theories that can be used for the development of finite elements. Since these higher order plate theories are extensions of the Reissner–Mindlin plate theory, there should not be any difficulty for readers who can formulate the Mindlin plate element to understand the formulation of higher order plate elements.

It is assumed that the element has a uniform thickness h. If the plate structure has a varying thickness, the structure can be divided into small elements, each of uniform thickness, to approximate the overall variation in thickness. However, the formulation of plate elements with a varying thickness can also be done, as the procedure is similar to that of a uniform element; this would be good homework practice for readers after reading this chapter.

Consider now a plate that is represented by a two-dimensional domain in the $x–y$ plane, as shown in Figure 8.1. The plate is divided in a proper manner into a number of *rectangular elements*, as shown in Figure 8.2. Each element will have four nodes and four straight edges. At a node, the DOFs include the deflection w, the rotation about x axis θ_x, and the rotation about y axis θ_y, making the total DOF of each node three. Hence, for a rectangular element with four nodes, the total DOF of the element would be twelve.

Following the Reissner–Mindlin plate theory (see Chapter 2), its shear deformation will force the cross-section of the plate to rotate in the way shown in Figure 8.3. Any straight fiber that is perpendicular to the middle plane of the plate before the deformation rotates, but remains straight after the deformation. The two displacement components that are parallel to the middle surface can then be expressed mathematically as

$$u(x, y, z) = z\theta_y(x, y) \tag{8.1}$$

$$v(x, y, z) = -z\theta_x(x, y)$$

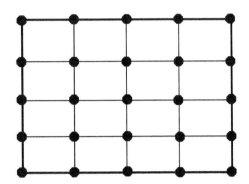

FIGURE 8.2

2D domain of plate meshed by rectangular elements.

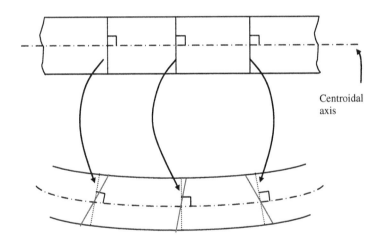

Centroidal
axis

FIGURE 8.3

Shear deformation in a plate. A straight fiber that is perpendicular to the middle plane of the plate before
deformation rotates but remains straight after deformation.

where θ_x and θ_y are, respectively, the rotations of the vertical fiber of the plate with respect to the x and
y axes. The in-plane strains can then be given as

$$
\varepsilon = \begin{Bmatrix} \varepsilon_{xx} \\ \varepsilon_{yy} \\ \varepsilon_{xy} \end{Bmatrix} = -z\chi
\tag{8.2}
$$

where χ is the curvature, given as

$$\chi = \mathbf{L}\theta = \underbrace{\begin{bmatrix} 0 & -\partial/\partial x \\ \partial/\partial y & 0 \\ \partial/\partial x & -\partial/\partial y \end{bmatrix}}_{\mathbf{L}} \left\{ \begin{array}{c} \theta_x \\ \theta_y \end{array} \right\} = \left\{ \begin{array}{c} -\partial\theta_y/\partial x \\ \partial\theta_x/\partial y \\ \partial\theta_x/\partial x - \partial\theta/\partial y \end{array} \right\} \tag{8.3}$$

in which $\mathbf{L}$ is the differential operator defined in Chapter 2, and is re-written as

$$\mathbf{L} = \begin{bmatrix} 0 & -\partial/\partial x \\ \partial/\partial y & 0 \\ \partial/\partial x & -\partial/\partial y \end{bmatrix} \tag{8.4}$$

Since the vertical fiber of the plate (which is initially normal to the mid plane of the plate) rotates and no longer remains normal to the mid plane of the plate, there are off-plane shear strain components and these are given as

$$\gamma = \left\{ \begin{array}{c} \gamma_{xz} \\ \gamma_{yz} \end{array} \right\} = \left\{ \begin{array}{c} \theta_y + \frac{\partial w}{\partial x} \\ -\theta_x + \frac{\partial w}{\partial y} \end{array} \right\} \tag{8.5}$$

Note that Hamilton's principle uses energy functions for derivation of the equation of motion. The potential (strain) energy expression for a thick plate element is

$$U_e = \frac{1}{2} \int_{A_e} \int_{-h/2}^{h/2} \boldsymbol{\varepsilon}^T \boldsymbol{\sigma} \, dA \, dz + \frac{1}{2} \int_{A_e} \int_{-h/2}^{h/2} \boldsymbol{\tau}^T \boldsymbol{\gamma} \, dA \, dz \tag{8.6}$$

The first term on the right-hand side of Eq. (8.6) relates to the in-plane stresses and strains, whereas the second term is for the transverse stresses and strains. $\boldsymbol{\tau}$ is the average shear stresses relating to the shear strain in the form

$$\boldsymbol{\tau} = \left\{ \begin{array}{c} \tau_{xy} \\ \tau_{yz} \end{array} \right\} = \kappa \begin{bmatrix} G & 0 \\ 0 & G \end{bmatrix} \boldsymbol{\gamma} = \kappa \mathbf{c}_s \boldsymbol{\gamma} \tag{8.7}$$

where G is the shear modulus, and κ is a shear correction factor to account for the non-uniformity of the shear stress across the thickness of the plate, while the transverse strains are uniform. This is because the Reissner–Mindlin plate theory assumes that the cross-section remains planar and hence the off-plane strains are uniform, but the actual off-plane shear stresses cannot be uniform: They have to be zero on the surfaces of the plate and maximum in the middle. The value of κ is usually taken to be $\pi^2/12$ or 5/6.

Substituting Eqs. (8.2) and (8.7) into Eq. (8.6), the potential (strain) energy becomes

$$U_e = \frac{1}{2} \int_{A_e} \frac{h^3}{12} \chi^T \mathbf{c}\chi \, dA + \frac{1}{2} \int_{A_e} \kappa h \boldsymbol{\gamma}^T \mathbf{c}_s \boldsymbol{\gamma} \, dA \tag{8.8}$$

The kinetic energy of the thick plate is given by

$$T_e = \frac{1}{2}\int_{V_e} \rho(\dot{u}^2 + \dot{v}^2 + \dot{w}^2)\mathrm{d}V \tag{8.9}$$

which is basically a summation of the contributions of three velocity components in the x, y, and z directions of all the particles in the entire domain of the plate. Substituting Eq. (8.1) into the above equation leads to

$$T_e = \frac{1}{2}\int_{A_e} \rho\left(h\dot{w}^2 + \frac{h^3}{12}\dot{\theta}_x^2 + \frac{h^3}{12}\dot{\theta}_y^2\right)\mathrm{d}A = \frac{1}{2}\int_{A_e} (\dot{\mathbf{d}}^T \mathbf{I}\dot{\mathbf{d}})\mathrm{d}A \tag{8.10}$$

where

$$\dot{\mathbf{d}} = \begin{Bmatrix} \dot{w} \\ \dot{\theta}_x \\ \dot{\theta}_y \end{Bmatrix} \tag{8.11}$$

and

$$\mathbf{I} = \begin{bmatrix} \rho h & 0 & 0 \\ 0 & \rho h^3/12 & 0 \\ 0 & 0 & \rho h^3/12 \end{bmatrix} \tag{8.12}$$

As we can see from Eq. (8.10), the terms related to in-plane displacements caused by the rotations are less important for thin plates, since it is proportional to the cubic of the plate thickness.

8.2.1 Shape functions

It can be seen from the above analysis of the constitutive equations that the rotations, θ_x and θ_y, are independent of the deflection w. Therefore, when it comes to interpolating the generalized displacements, the deflection and rotations can actually be interpolated separately using independent shape functions. Therefore, the procedure of field variable interpolation is the same as that for 2D solid problems, except that there are three instead of two DOFs, for a node.

For four-node rectangular thick plate elements, the deflection and rotations can be summed as

$$w = \sum_{i=1}^{4} N_i w_i, \quad \theta_x = \sum_{i=1}^{4} N_i \theta_{x_i}, \quad \theta_y = \sum_{i=1}^{4} N_i \theta_{y_i} \tag{8.13}$$

where the shape function N_i is the same as the four-node 2D solid element in Chapter 7, i.e.,

$$N_i = \frac{1}{4}(1 + \xi_i\xi)(1 + \eta_i\eta) \tag{8.14}$$

The element constructed will be a conforming element, meaning that w, θ_x, and θ_y are continuous on the edges between elements. Rewriting Eq. (8.13) into a single matrix equation, we have

$$\left\{ \begin{array}{c} w \\ \theta_x \\ \theta_y \end{array} \right\}^h = \mathbf{N}\,\mathbf{d}_e \tag{8.15}$$

where $\mathbf{d}_e$ is the (generalized) displacement vector for all the nodes in the element, arranged in the order

$$\mathbf{d}_e = \left\{ \begin{array}{c} w_1 \\ \theta_{x1} \\ \theta_{y1} \\ w_2 \\ \theta_{x2} \\ \theta_{y2} \\ w_3 \\ \theta_{x3} \\ \theta_{y3} \\ w_4 \\ \theta_{x4} \\ \theta_{y4} \end{array} \right\} \begin{array}{l} \left. \vphantom{\begin{array}{c}a\\a\\a\end{array}} \right\} \text{displacements at node 1} \\ \left. \vphantom{\begin{array}{c}a\\a\\a\end{array}} \right\} \text{displacements at node 2} \\ \left. \vphantom{\begin{array}{c}a\\a\\a\end{array}} \right\} \text{displacements at node 3} \\ \left. \vphantom{\begin{array}{c}a\\a\\a\end{array}} \right\} \text{displacements at node 4} \end{array} \tag{8.16}$$

and the shape function matrix is arranged in the order

$$\mathbf{N} = \begin{bmatrix} N_1 & 0 & 0 & N_2 & 0 & 0 & N_3 & 0 & 0 & N_4 & 0 & 0 \\ 0 & N_1 & 0 & 0 & N_2 & 0 & 0 & N_3 & 0 & 0 & N_4 & 0 \\ 0 & 0 & N_1 & 0 & 0 & N_2 & 0 & 0 & N_3 & 0 & 0 & N_4 \end{bmatrix} \tag{8.17}$$

$$\underbrace{\hphantom{N_1\ 0\ 0}}_{\text{node 1}} \underbrace{\hphantom{N_2\ 0\ 0}}_{\text{node 2}} \underbrace{\hphantom{N_3\ 0\ 0}}_{\text{node 3}} \underbrace{\hphantom{N_4\ 0\ 0}}_{\text{node 4}}$$

8.2.2 Element matrices

Once the shape function and nodal variables have been defined, element matrices can then be formulated following the standard procedure given in Chapter 7 for 2D solid elements. The only difference is that there are three DOFs at one node for plate elements.

To obtain the element mass matrix $\mathbf{m}_e$ and the element stiffness matrix $\mathbf{k}_e$, we have to use the energy functions given by Eqs. (8.8) and (8.9) and Hamilton's principle. Substituting Eq. (8.15) into the kinetic energy function, Eq. (8.9) gives

$$T_e = \frac{1}{2}\dot{\mathbf{d}}_e^T \mathbf{m}_e \, \dot{\mathbf{d}}_e \tag{8.18}$$

where the mass matrix $\mathbf{m}_e$ is given as

$$\mathbf{m}_e = \int_{A_e} \mathbf{N}^T \mathbf{I} \mathbf{N} \, dA \tag{8.19}$$

The above integration can be carried out analytically, but it will be omitted here. Interested readers can refer to Petyt (1990) for details. In practice, we often perform the integration numerically using the Gauss integration scheme, discussed in Chapter 7.

To obtain the stiffness matrix $\mathbf{k}_e$, we substitute Eq. (8.15) into Eq. (8.6), from which we obtain

$$\mathbf{k}_e = \int_{A_e} \frac{h^3}{12}[\mathbf{B}^I]^T \mathbf{c} \mathbf{B}^I dA + \int_{A_e} \kappa h[\mathbf{B}^O]^T \mathbf{c}_s \mathbf{B}^O dA \tag{8.20}$$

The first term in the above equation represents the strain energy associated with the in-plane stress and strain. The strain matrix $\mathbf{B}^I$ has the form of

$$\mathbf{B}^I = \begin{bmatrix} \mathbf{B}_1^I & \mathbf{B}_2^I & \mathbf{B}_3^I & \mathbf{B}_4^I \end{bmatrix} \tag{8.21}$$

where

$$\mathbf{B}_j^I = \begin{bmatrix} 0 & 0 & -\partial N_j \partial x \\ 0 & \partial N_j \partial x & 0 \\ 0 & \partial N_j \partial x & -\partial N_j \partial y \end{bmatrix} \tag{8.22}$$

Using the expression for the shape functions in Eq. (8.14), we obtain

$$\begin{aligned} \frac{\partial N_j}{\partial x} &= \frac{\partial N_j}{\partial \xi}\frac{\partial \xi}{\partial x} = \frac{1}{4a}\xi_i(1 + \eta_i\eta) \\ \frac{\partial N_j}{\partial y} &= \frac{\partial N_j}{\partial \eta}\frac{\partial \eta}{\partial y} = \frac{1}{4b}(1 + \xi_i\xi)\eta_i \end{aligned} \tag{8.23}$$

In deriving Eq. (8.23), the relationship given in Eq. (7.47) has been used.

The second term in Eq. (8.20) relates to the strain energy associated with the off-plane shear stress and strain. The strain matrix $\mathbf{B}^O$ has the form

$$\mathbf{B}^O = [\mathbf{B}_1^O \quad \mathbf{B}_2^O \quad \mathbf{B}_3^O \quad \mathbf{B}_4^O] \tag{8.24}$$

where

$$\mathbf{B}_j^O = \begin{bmatrix} \partial N_j/\partial x & 0 & N_j \\ \partial N_j/\partial y & -N_j & 0 \end{bmatrix} \tag{8.25}$$

The integration in the stiffness matrix $\mathbf{k}_e$, Eq. (8.20) can be evaluated analytically as well. Practically however, the Gauss integration scheme is used to evaluate the integrations numerically. Note that when the thickness of the plate is reduced, the element becomes over-stiff, a phenomenon that relates to so-called "shear locking." The simplest and most practical means to alleviate this problem is to use 2×2 Gauss points for the integration of the first term, and use only one Gauss point for the second term in Eq. (8.20).

As for the force vector, we substitute the interpolation of the generalized displacements, given in Eq. (8.15), into the usual equation, as in Eq. (3.81), assuming that there is a distributed transverse force, f_z, acting on the surface of the plate:

$$\mathbf{f}_e = \int_{A_e} \mathbf{N}^T \begin{Bmatrix} f_z \\ 0 \\ 0 \end{Bmatrix} dA \tag{8.26}$$

If the load is uniformly distributed in the element, f_z is constant, and the above equation becomes

$$\mathbf{f}_e^T = abf_z\{1 \quad 0 \quad 0 \quad 1 \quad 0 \quad 0 \quad 1 \quad 0 \quad 0 \quad 1 \quad 0 \quad 0\} \tag{8.27}$$

Equation (8.27) implies that the distributed force is divided evenly into four concentrated forces of one quarter of the total force.

8.2.3 Higher order elements

For an eight-node rectangular thick plate element, the deflection and rotations can be summed as

$$w = \sum_{i=1}^{8} N_i w_i, \quad \theta_x - \sum_{i=1}^{8} N_i \theta_{xI}, \quad \theta_y - \sum_{i=1}^{8} N_i \theta_{yi} \tag{8.28}$$

where the shape function N_i is the same as the eight-node 2D solid element given by Eq. (7.113). The element constructed will be a conforming element, as w, θ_x, and θ_y are continuous on the edges between elements. The formulation procedure is the same as for the rectangular plate elements.

8.3 Shell elements

A shell structure carries loads in all directions, and therefore undergoes bending and twisting, as well as in-plane deformation. Some common examples would be the dome-like design of the roof of a building with a large volume of space; or a building with special architectural requirements such as a church or

FIGURE 8.4

The fuselage of an aircraft can be considered to be a typical shell structure.

mosque; or structures with a special functional requirement such as cylindrical and hemispherical water tanks; or lightweight structures like the fuselage of an aircraft, as shown in Figure 8.4. Shell elements are suitable for modeling such structures.

8.3.1 The idea of superposition

The simplest yet most widely used shell element can be formulated easily by superposing the 2D solid element formulated in Chapter 7 onto the plate element formulated in the previous section. The 2D solid elements handle the *membrane* or in-plane effects, while the plate elements are used to handle *bending* or off-plane effects. The procedure for developing such an element is very similar to the simple method of superposition used to formulate the frame elements using the truss and beam elements, as discussed in Chapter 6. Of course, the shell element can also be formulated using the usual method of defining shape functions, substituting into the constitutive equations based on shell theory, and thus obtaining the element matrices. However, as you might have guessed, it is going to be very tedious. Bear in mind, however, that the basic concept of deriving the finite element equation still holds, even though in this book, we introduce the simple method based on assembly. The derivation for four-nodal, rectangular shell elements will be outlined using the above-mentioned approach.

Since the plate structure can be treated as a special case of the shell structure, the shell element developed in this section is applicable for modeling plate structures. In fact, it is common practice to use a shell element offered in a commercial FE package to analyze plate structures.

8.3.2 Elements in the local coordinate system

Shell structures are usually curved. In our FEM formulation a shell-like structure is divided into shell elements that are flat. The curvature of the shell structure is then approximated by changing the orientation of the shell elements in space. Therefore, if the curvature of the shell is very large, a fine mesh of elements has to be used. Even though this assumption may sound like a crude approximation, it is very practical and widely used in engineering practice. There are alternatives of more accurately formulated shell elements, but they are used only in academic research and seldom implemented in any commercially available software packages. Therefore, we focus on the formulation of only flat shell elements in this book.

Similar to the frame structure, there are six DOFs at a node for a shell element: Three translational displacements in the x, y, and z directions, and three rotational deformations with respect to the x, y, and z axes. Figure 8.5 shows the middle plane of a rectangular shell element and the DOFs at the nodes. The generalized displacement vector for the element can be written as

$$\mathbf{d_e} = \begin{Bmatrix} \mathbf{d}_1 \\ \mathbf{d}_2 \\ \mathbf{d}_3 \\ \mathbf{d}_4 \end{Bmatrix} \begin{matrix} \text{node 1} \\ \text{node 2} \\ \text{node 3} \\ \text{node 4} \end{matrix} \tag{8.29}$$

where $\mathbf{d}_i$ ($i = 1, 2, 3, 4$) are the displacement vector at node i:

$$\mathbf{d}_i = \begin{Bmatrix} u_i \\ v_i \\ w_i \\ \theta_{xi} \\ \theta_{yi} \\ \theta_{zi} \end{Bmatrix} \begin{matrix} \text{displacement in } x \text{ direction} \\ \text{displacement in } y \text{ direction} \\ \text{displacement in } z \text{ direction} \\ \text{rotation about } x\text{-axis} \\ \text{rotation about } y\text{-}axis \\ \text{rotation about } z\text{-axis} \end{matrix} \tag{8.30}$$

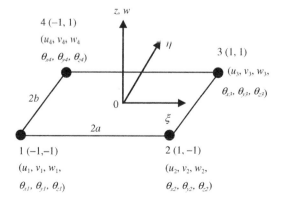

FIGURE 8.5

The middle plane of a rectangular shell element.

The stiffness matrix for a 2D solid, rectangular element is used to account for the membrane effects of the element, which corresponds to DOFs of u and v. The membrane stiffness matrix can thus be expressed in the following form using sub-matrices according to the nodes:

$$
\mathbf{k}_e^m =
\begin{array}{c}

\end{array}
\begin{bmatrix}
\mathbf{k}_{11}^m & \mathbf{k}_{12}^m & \mathbf{k}_{13}^m & \mathbf{k}_{14}^m \\
\mathbf{k}_{21}^m & \mathbf{k}_{22}^m & \mathbf{k}_{23}^m & \mathbf{k}_{24}^m \\
\mathbf{k}_{31}^m & \mathbf{k}_{32}^m & \mathbf{k}_{33}^m & \mathbf{k}_{34}^m \\
\mathbf{k}_{41}^m & \mathbf{k}_{42}^m & \mathbf{k}_{43}^m & \mathbf{k}_{44}^m
\end{bmatrix}
\begin{array}{l}
\text{node 1} \\
\text{node 2} \\
\text{node 3} \\
\text{node 4}
\end{array}
\tag{8.31}
$$

where the superscript m stands for the membrane matrix. Each sub-matrix will have a dimension of 2×2, since it corresponds to the two DOFs u and v at each node. Note again that the matrix above is actually the same as the stiffness matrix of the 2D rectangular, solid element, except it is written in terms of sub-matrices according to the nodes.

The stiffness matrix for a rectangular plate element is used for the bending effects, corresponding to DOFs of w and θ_x, θ_y. The bending stiffness matrix can thus be expressed in the following form using sub-matrices according to the nodes:

$$
\mathbf{k}_e^b =
\begin{bmatrix}
\mathbf{k}_{11}^b & \mathbf{k}_{12}^b & \mathbf{k}_{13}^b & \mathbf{k}_{14}^b \\
\mathbf{k}_{21}^b & \mathbf{k}_{22}^b & \mathbf{k}_{23}^b & \mathbf{k}_{24}^b \\
\mathbf{k}_{31}^b & \mathbf{k}_{32}^b & \mathbf{k}_{33}^b & \mathbf{k}_{34}^b \\
\mathbf{k}_{41}^b & \mathbf{k}_{42}^b & \mathbf{k}_{43}^b & \mathbf{k}_{44}^b
\end{bmatrix}
\begin{array}{l}
\text{node 1} \\
\text{node 2} \\
\text{node 3} \\
\text{node 4}
\end{array}
\tag{8.32}
$$

where the superscript b stands for the bending stiffness. Each bending sub-matrix has a dimension of 3×3.

The stiffness matrix for the shell element in the local coordinate system is then formulated by combining Eqs. (8.31) and (8.32):

$$
\mathbf{k}_e =
\begin{bmatrix}
\mathbf{k}_{11}^m & 0 & 0 & \mathbf{k}_{12}^m & 0 & 0 & \mathbf{k}_{13}^m & 0 & 0 & \mathbf{k}_{14}^m & 0 & 0 \\
0 & \mathbf{k}_{11}^b & 0 & 0 & \mathbf{k}_{12}^b & 0 & 0 & \mathbf{k}_{13}^b & 0 & 0 & \mathbf{k}_{14}^b & 0 \\
0 & 0 & 0 & 0 & 0 & 0 & 0 & 0 & 0 & 0 & 0 & 0 \\
\mathbf{k}_{21}^m & 0 & 0 & \mathbf{k}_{22}^m & 0 & 0 & \mathbf{k}_{23}^m & 0 & 0 & \mathbf{k}_{24}^m & 0 & 0 \\
0 & \mathbf{k}_{21}^b & 0 & 0 & \mathbf{k}_{23}^b & 0 & 0 & \mathbf{k}_{23}^b & 0 & 0 & \mathbf{k}_{24}^b & 0 \\
0 & 0 & 0 & 0 & 0 & 0 & 0 & 0 & 0 & 0 & 0 & 0 \\
\mathbf{k}_{31}^m & 0 & 0 & \mathbf{k}_{32}^m & 0 & 0 & \mathbf{k}_{33}^m & 0 & 0 & \mathbf{k}_{34}^m & 0 & 0 \\
0 & \mathbf{k}_{31}^b & 0 & 0 & \mathbf{k}_{33}^b & 0 & 0 & \mathbf{k}_{33}^b & 0 & 0 & \mathbf{k}_{34}^b & 0 \\
0 & 0 & 0 & 0 & 0 & 0 & 0 & 0 & 0 & 0 & 0 & 0 \\
\mathbf{k}_{41}^m & 0 & 0 & \mathbf{k}_{44}^m & 0 & 0 & \mathbf{k}_{43}^m & 0 & 0 & \mathbf{k}_{44}^m & 0 & 0 \\
0 & \mathbf{k}_{41}^b & 0 & 0 & \mathbf{k}_{43}^b & 0 & 0 & \mathbf{k}_{43}^b & 0 & 0 & \mathbf{k}_{44}^b & 0 \\
0 & 0 & 0 & 0 & 0 & 0 & 0 & 0 & 0 & 0 & 0 & 0
\end{bmatrix}
\tag{8.33}
$$

The stiffness matrix for a rectangular shell matrix has a dimension of 24×24. Note that in Eq. (8.33), the components related to the DOF θ_z are zeros. This is because there is no θ_z in the local coordinate system. If these zero terms are removed, the stiffness matrix would have a reduced dimension of 20×20. However, using the extended 24×24 stiffness matrix will make it more convenient for transforming the matrix from the local coordinate system into the global coordinate system.

Similarly, the mass matrix for a rectangular element can be obtained in the same way as for the stiffness matrix. The mass matrix for the 2D solid element is used for the membrane effects, corresponding to DOFs of u and v. The membrane mass matrix can be expressed in the following form using sub-matrices according to the nodes:

$$
\mathbf{m}_e^m =
\begin{array}{c}
\begin{array}{cccc} \text{node1} & \text{node2} & \text{node3} & \text{node4} \end{array} \\
\begin{bmatrix}
\mathbf{m}_{11}^m & \mathbf{m}_{12}^m & \mathbf{m}_{13}^m & \mathbf{m}_{14}^m \\
\mathbf{m}_{21}^m & \mathbf{m}_{22}^m & \mathbf{m}_{23}^m & \mathbf{m}_{24}^m \\
\mathbf{m}_{31}^m & \mathbf{m}_{32}^m & \mathbf{m}_{33}^m & \mathbf{m}_{34}^m \\
\mathbf{m}_{41}^m & \mathbf{m}_{42}^m & \mathbf{m}_{43}^m & \mathbf{m}_{44}^m
\end{bmatrix}
\begin{array}{l} \text{node1} \\ \text{node2} \\ \text{node3} \\ \text{node4} \end{array}
\end{array}
\tag{8.34}
$$

where the superscript m stands for the membrane matrix. Each membrane sub-matrix has a dimension of 2×2.

The mass matrix for a rectangular plate element is used for the bending effects, corresponding to DOFs of w and θ_x, θ_y. The bending mass matrix can also be expressed in the following form using sub-matrices according to the nodes:

$$
\mathbf{m}_e^b =
\begin{array}{c}
\begin{array}{cccc} \text{node1} & \text{node2} & \text{node3} & \text{node4} \end{array} \\
\begin{bmatrix}
\mathbf{m}_{11}^b & \mathbf{m}_{12}^b & \mathbf{m}_{13}^b & \mathbf{m}_{14}^b \\
\mathbf{m}_{21}^b & \mathbf{m}_{22}^b & \mathbf{m}_{23}^b & \mathbf{m}_{24}^b \\
\mathbf{m}_{31}^b & \mathbf{m}_{32}^b & \mathbf{m}_{33}^b & \mathbf{m}_{34}^b \\
\mathbf{m}_{41}^b & \mathbf{m}_{42}^b & \mathbf{m}_{43}^b & \mathbf{m}_{44}^b
\end{bmatrix}
\begin{array}{l} \text{node1} \\ \text{node2} \\ \text{node3} \\ \text{node4} \end{array}
\end{array}
\tag{8.35}
$$

where the superscript b stands for the bending matrix. Each bending sub-matrix has a dimension of 3×3. The mass matrix for the shell element in the local coordinate system is then formulated by combining Eqs. (8.34) and (8.35):

$$
\mathbf{k}_e =
\begin{bmatrix}
\mathbf{m}_{11}^m & 0 & 0 & \mathbf{m}_{12}^m & 0 & 0 & \mathbf{m}_{13}^m & 0 & 0 & \mathbf{m}_{14}^m & 0 & 0 \\
0 & \mathbf{m}_{11}^b & 0 & 0 & \mathbf{m}_{12}^b & 0 & 0 & \mathbf{m}_{13}^b & 0 & 0 & \mathbf{m}_{14}^b & 0 \\
0 & 0 & 0 & 0 & 0 & 0 & 0 & 0 & 0 & 0 & 0 & 0 \\
\mathbf{m}_{21}^m & 0 & 0 & \mathbf{m}_{22}^m & 0 & 0 & \mathbf{m}_{23}^m & 0 & 0 & \mathbf{m}_{24}^m & 0 & 0 \\
0 & \mathbf{m}_{21}^b & 0 & 0 & \mathbf{m}_{23}^b & 0 & 0 & \mathbf{m}_{23}^b & 0 & 0 & \mathbf{m}_{24}^b & 0 \\
0 & 0 & 0 & 0 & 0 & 0 & 0 & 0 & 0 & 0 & 0 & 0 \\
\mathbf{m}_{31}^m & 0 & 0 & \mathbf{m}_{32}^m & 0 & 0 & \mathbf{m}_{33}^m & 0 & 0 & \mathbf{m}_{34}^m & 0 & 0 \\
0 & \mathbf{m}_{31}^b & 0 & 0 & \mathbf{m}_{33}^b & 0 & 0 & \mathbf{m}_{33}^b & 0 & 0 & \mathbf{m}_{34}^b & 0 \\
0 & 0 & 0 & 0 & 0 & 0 & 0 & 0 & 0 & 0 & 0 & 0 \\
\mathbf{m}_{41}^m & 0 & 0 & \mathbf{m}_{44}^m & 0 & 0 & \mathbf{m}_{43}^m & 0 & 0 & \mathbf{m}_{44}^m & 0 & 0 \\
0 & \mathbf{m}_{41}^b & 0 & 0 & \mathbf{m}_{43}^b & 0 & 0 & \mathbf{m}_{43}^b & 0 & 0 & \mathbf{m}_{44}^b & 0 \\
0 & 0 & 0 & 0 & 0 & 0 & 0 & 0 & 0 & 0 & 0 & 0
\end{bmatrix}
\tag{8.36}
$$

Similarly, it is noted that the terms corresponding to the DOF θ_z are zero for the same reasons as explained for the stiffness matrix.

8.3.3 Elements in the global coordinate system

The matrices for shell elements in the global coordinate system can be obtained by performing the transformations

$$\mathbf{K}_e = \mathbf{T}^T \mathbf{k}_e \mathbf{T} \tag{8.37}$$

$$\mathbf{M}_e = \mathbf{T}^T \mathbf{m}_e \mathbf{T} \tag{8.38}$$

$$\mathbf{F}_e = \mathbf{T}^T \mathbf{f}_e \tag{8.39}$$

where $\mathbf{T}$ is the transformation matrix, given by

$$\mathbf{T} = \begin{bmatrix} \mathbf{T}_3 & 0 & 0 & 0 & 0 & 0 & 0 & 0 \\ 0 & \mathbf{T}_3 & 0 & 0 & 0 & 0 & 0 & 0 \\ 0 & 0 & \mathbf{T}_3 & 0 & 0 & 0 & 0 & 0 \\ 0 & 0 & 0 & \mathbf{T}_3 & 0 & 0 & 0 & 0 \\ 0 & 0 & 0 & 0 & \mathbf{T}_3 & 0 & 0 & 0 \\ 0 & 0 & 0 & 0 & 0 & \mathbf{T}_3 & 0 & 0 \\ 0 & 0 & 0 & 0 & 0 & 0 & \mathbf{T}_3 & 0 \\ 0 & 0 & 0 & 0 & 0 & 0 & 0 & \mathbf{T}_3 \end{bmatrix}_{24 \times 24} \tag{8.40}$$

in which

$$\mathbf{T}_3 = \begin{bmatrix} l_x & m_x & n_x \\ l_y & m_y & n_y \\ l_z & m_z & n_z \end{bmatrix}_{3 \times 3} \tag{8.41}$$

where l_k, m_k and n_k ($k = x, y, z$) are direction cosines, which can be obtained in exactly the same way as described in Section 6.3.2. The difference is that there is no need to define the additional point 3, as there are already four nodes for the shell element. The local coordinates x, y, z can be conveniently defined under the global coordinate system using the four nodes of the shell element.

The global matrices obtained will not have zero columns and rows if the elements joined at a node are not in the same plane. If all the elements joined at a node are in the same plane, then the global matrices will be singular. This kind of case is encountered when using shell elements to model a flat plate. In such situations, special techniques, such as a "stabilizing matrix," have to be used to solve the global system equations.

8.4 Remarks

The direct superposition of the matrices for 2D solid elements and plate elements are performed by assuming that the membrane effects are not coupled with the bending effects at the individual element level. This implies that the membrane forces will not result in any bending deformation, and bending forces will not cause any in-plane displacement in the element. For a shell structure in 3D space, the membrane and bending effects are actually coupled globally, meaning that the membrane force at an element may result in bending deformations in the other elements, and the bending forces in an element may create in-plane displacements in other elements. The coupling effects are more significant for shell structures with a strong curvature. Therefore, for those structures, a finer element mesh should be used. Using the shell elements developed in this chapter implies that the curved shell structure has to be meshed by piecewise flat elements. This simplification in geometry needs to be taken into account when evaluating the results obtained.

8.5 Case study: Natural frequencies of the micro-motor

In this case study, we examine the natural frequencies and mode shapes of the micro-motor described in Section 7.10. Natural frequencies are properties of a system, and it is important to study the natural frequencies and corresponding mode shapes of a system, because if a forcing frequency is applied to the system near to or at the natural frequency, resonance will occur. That is, there will be very large amplitude vibration that might be disastrous in some situations. In this case study, the flexural vibration modes of the rotor of the micro-motor will be analyzed.

8.5.1 Modeling

The geometry of the micro-motor's rotor will be the same as that of Figure 7.22, and the elastic properties will remain unchanged using the properties in Table 7.2. To show the mode shapes more clearly, we model the rotor as a whole, rather than as a symmetrical quarter model. However, using a quarter model is still possible, but one has to take note of symmetrical and anti-symmetrical modes (to be discussed in more detail in Chapter 11). Figure 8.6 shows the finite element model of the micro-motor containing 480 nodes and 384 elements. To study the flexural vibration modes, plate elements discussed in this chapter are used. However, as mentioned earlier in this chapter, most commercial finite element packages, including ABAQUS, do not allow the use of pure plate elements. Therefore, shell elements will be utilized here for meshing up the model of the micro-motor. 2D, four-nodal shell elements (S4) are used. Recall that each shell element has three translational DOFs and three rotational DOFs, and it is actually a superposition of a plate element with a 2D solid element. Hence, to obtain just the flexural modes, we would need to constrain the DOFs corresponding to the x translational displacement and the y translational displacement, as well as the rotation about the z axis. This would leave each node of a shell element with the three DOFs of a plate element. As before, the nodes along the edge of the center hole will be constrained to be fixed. Since we are interested in the natural frequencies, there will be no external forces on the rotor.

8.5.2 **ABAQUS input file**

The ABAQUS input file for the problem described is shown below. Note that some parts are not shown due to limitations of space available in this text.

```
*HEADING, SPARSE
EIGENVALUE ANALYSIS OF MICRO MOTOR
**
*NODE
1, 8., 0.
2, 7.99238, 0.348994
3, 7.96955, 0.697324
4, 7.93155, 1.04427
5, 7.87846, 1.38919
 ⋮
997, −8.68241, −49.2404
998, −6.52629, −49.5722
999, −4.35774, −49.8097
1000, −2.1809, −49.9524
**
**
*ELEMENT, TYPE = S4, ELSET = MOTOR
1, 1, 6, 7, 2
2, 2, 7, 8, 3
3, 3, 8, 9, 4
4, 4, 9, 10, 5
 ⋮
830, 994, 998, 999, 995
831, 995, 999, 1000, 996
832, 996, 1000, 760, 755
**
**
**
```

Nodal cards
These define the coordinates of the nodes in the model. The first entry being the node ID, while the second and third are the *x* and *y* coordinates of the position of the node, respectively.

Element (connectivity) cards
These define the element type and what nodes make up the element. S4 represents that it is a four-nodal, shell element. The "ELSET = MOTOR" statement is simply for naming this set of elements so that it can be referenced when defining the material properties. In the subsequent data entry, the first entry is the element ID, and the following four entries are the nodes making up the element. The order of the nodes for all elements must be consistent and counter-clockwise.

```
*SHELL SECTION, ELSET = MOTOR, MATERIAL=POLYSI
13.
**
** PolySi
**
*MATERIAL, NAME = POLYSI
**
*DENSITY
```

Property cards
These define properties to the elements of set "MOTOR." It will have the material properties defined under "POLYSI." The thickness of the elements is also defined in the data line.

```
2.3E-15,
**
*ELASTIC, TYPE = ISO
169000., 0.262
**
**
*BOUNDARY, OP = NEW
1, 1,, 0.
1, 2,, 0.
1, 3,, 0.
2, 1,, 0.
2, 2,, 0.
2, 3,, 0.
3, 1,, 0.
⋮
903, 4,, 0.
903, 5,, 0.
903, 6,, 0.
**
** fixedxy
**
*BOUNDARY, OP = NEW
6, 1,, 0.
6, 2,, 0.
7, 1,, 0. 7, 2,, 0.
⋮
997, 6,, 0.
998, 6,, 0.
999, 6,, 0.
1000, 6,, 0.
**
**
** Step 1, freq
** LoadCase, Default
**
*STEP, NLGEOM
```
This load case is the default load case that always appears
```
*FREQUENCY
8, , , , 30
**
**
**
*NODE PRINT, FREQ = 1
```

Material cards
These define material properties under the name "POLYSI." Density and elastic properties are defined. TYPE = ISO represents isotropic properties.

BC cards
These define boundary conditions. In this case, all the nodes along the center circle are constrained to zero displacements. To simulate plate elements, DOFs 1, 2, and 6 are constrained for all the nodes in the model.

Output control cards
These define the output required. In this case, the nodal displacement components (U) are requested.

U,
*NODE FILE, FREQ=1
U,
**
**
**
*END STEP

> *Control cards*
> These indicate the analysis step. In this case it is a "FREQUENCY" analysis, which extracts the eigenvalues for the problem.

8.5.3 Solution process

Looking at the mesh in Figure 8.6, one can see that quadrilateral shell elements are used. Therefore, the equations for a linear, quadrilateral shell element must be formulated by ABAQUS. As before, the formulation of the element matrices would require information from the nodal cards and the element connectivity cards. The element type used here is S4, representing four-nodal shell elements. There are other types of shell elements available in the ABAQUS element library.

After the nodal and element cards, next to be considered would be the property and material cards. The properties for the shell element used here must be defined, which in this case includes the material used and the thickness of the shell elements. The material cards are similar to those of the case study in Chapter 7, except that here the density of the material must be included, since we are not carrying out a static analysis, as in Chapter 7.

The boundary (BC) cards then define the boundary conditions on the model. In this problem, we would like to obtain only the flexural vibration modes of the motor, hence the components of

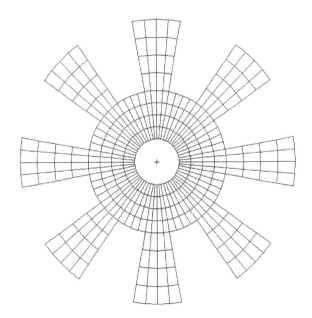

FIGURE 8.6

Finite element mesh using 2D, four-nodal, shell elements.

displacements in the plane of the motor are not actually required. As mentioned, this is very much the characteristic of the plate elements. Therefore, DOFs 1, 2, and 6 corresponding to the x and y displacements, and rotation about the z axis, are constrained. The other boundary condition would be the constraining of the displacements of the nodes at the center of the motor.

Without the need to define any external loadings, the control cards then define the type of analysis ABAQUS would carry out. ABAQUS uses the sub-space iteration scheme by default to evaluate the eigenvalues of the equation of motion. This method is a very effective method of determining a number of lowest eigenvalues and corresponding eigenvectors for a very large system of several thousand DOFs. The procedure is outlined in the case study in Chapter 5. Finally, the output control cards define the necessary output required by the analyst.

8.5.4 **Results and discussion**

Using the input file above, an eigenvalue extraction is carried out in ABAQUS. The output is extracted from the ABAQUS results file showing the first eight natural frequencies and these are tabulated in Table 8.1. The table also shows results obtained from using triangular elements as well as a finer mesh of quadrilateral elements. It is interesting to note that for certain modes, the eigenvalues and hence the frequencies are repetitive with the previous one. This is due to the symmetry of the circular rotor structure. For example, modes 1 and 2 have the same frequency, and looking at their corresponding mode shapes in Figures 8.7 and 8.8, respectively, one would notice that they are actually of the same shape but bending at a plane 90° from each other. As such, many consider this as one single mode. Therefore, though eight eigenmodes are extracted, it is effectively equivalent to only five eigenmodes. However, to be consistent with the result file from ABAQUS, all the modes extracted will be shown here. Figures 8.9–8.14 show the other mode shapes from this analysis. Remember that, since the in-plane displacements are already constrained, these modes are only the flexural modes of the rotor.

Comparing the natural frequencies obtained using 768 triangular elements with those obtained using the quadrilateral elements, one can see that the frequencies are generally higher using the triangular elements. Note that for the same number of nodes, using the quadrilateral elements requires half the number of elements. The results obtained using 384 quadrilateral elements do not differ much

Table 8.1   Natural frequencies obtained from analyses.

	Natural Frequencies (MHz)		
Mode	768 Triangular Elements with 480 Nodes	384 Quadrilateral Elements with 480 Nodes	1280 Quadrilateral Elements with 1472 Nodes
1	7.67	5.08	4.86
2	7.67	5.08	4.86
3	7.87	7.44	7.41
4	10.58	8.52	8.30
5	10.58	8.52	8.30
6	13.84	11.69	11.44
7	13.84	11.69	11.44
8	14.86	12.45	12.17

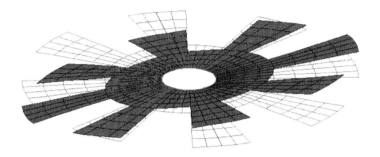

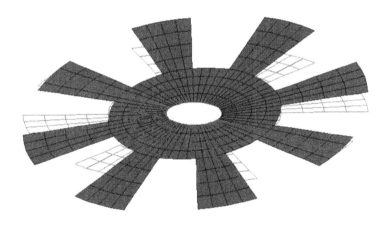

FIGURE 8.7

Mode 1.

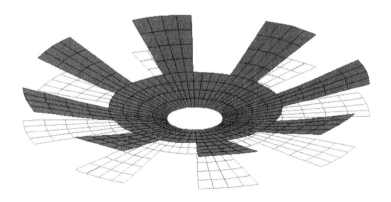

FIGURE 8.8

Mode 2.

FIGURE 8.9

Mode 3.

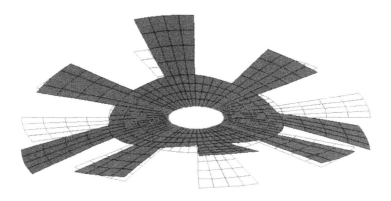

FIGURE 8.10

Mode 4.

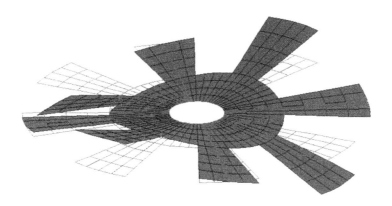

FIGURE 8.11

Mode 5.

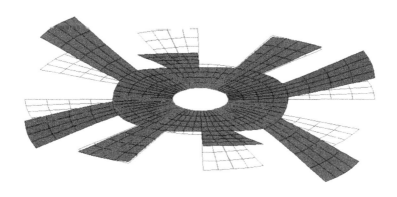

FIGURE 8.12

Mode 6.

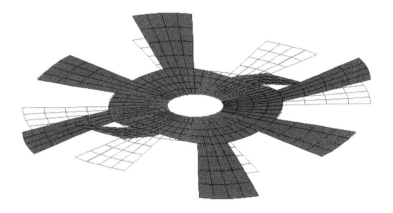

FIGURE 8.13

Mode 7.

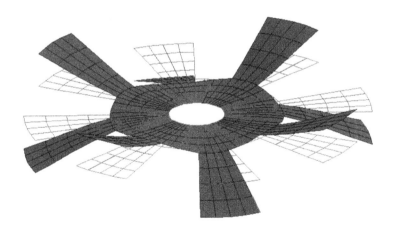

FIGURE 8.14

Mode 8.

from those that use 1280 elements. This again shows that the triangular elements are less accurate than the quadrilateral elements. Note that the mode shapes obtained in the three analyses are the same. It is interesting to also note that the natural frequencies are higher (stiffer) when less elements are used, which shows the over-stiffening effect of the FEM. Using more elements relaxes the constrains from the shape functions more and hence the structure becomes softer.

8.6 Case study: Transient analysis of a micro-motor

While analyzing the micro-motor, another case study is included here to illustrate an example of a transient analysis using ABAQUS. The same micro-motor shown in Chapter 7 will be analyzed here.

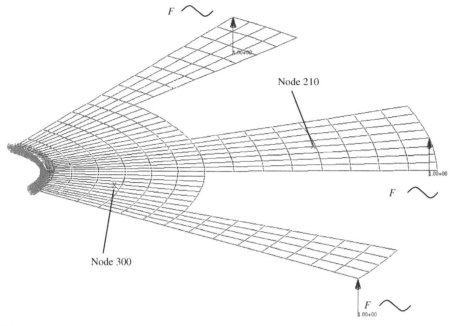

FIGURE 8.15

Quarter model of micro model with sinusoidal forces applied.

The rotor of the micro-motor rotates due to the electrostatic force between the rotor and the stator poles of the motor. Let's assume a hypothetical case where there is a misalignment between the rotor and the stator poles in the motor. As such, there might be other force components acting on the rotor. The actual analysis of such a problem can be very complex, so in this case study we simply analyze a very simple case of the problem with loading conditions as shown in Figure 8.15. It can be seen that symmetrical conditions are used, resulting in a quarter model. The transient response of the transverse displacement components of the various parts of the rotor is to be calculated here.

8.6.1 **Modeling**

Since we are analyzing the same structure as that in Chapter 7, the meshing aspects of the geometry will not be discussed again. It should be noted that an optimum number of elements (nodes) should be used for every finite element analysis. The same treatment of using the shell elements and constraining the necessary DOFs (1, 2, and 6) is carried out to simulate plate elements. The difference here is that there will be loadings in the form of a sinusoidal function with respect to time,

$$F = A \sin \bar{\omega} t \qquad (8.42)$$

applied as concentrated loadings at the positions shown in Figure 8.15.

8.6.2 ABAQUS input file

The ABAQUS input file for the problem described is shown below. Note that some parts are not shown due to limitations of space available in this text.

```
*HEADING
TRANSIENT ANALYSIS OF MICRO MOTOR
**
*NODE
1, 5.46197E-7, 50.
2, 2.1809, 49.9524
3, 4.35774, 49.8097
⋮
380, 46.8304, 2.04468
381, 49.5722, 6.52638
382, 49.9524, 2.18099
383, 49.8097, 4.35783
**
**
*ELEMENT, TYPE=S4, ELSET=MOTOR
1, 343, 342, 347, 348
2, 342, 341, 346, 347
3, 341, 340, 345, 346
⋮
317, 74, 65, 67, 75
318, 65, 55, 57, 67
319, 55, 43, 45, 57
320, 43, 29, 31, 45
**
**
*NSET, NSET=EDGE1
283, 298, 311, 322, 331, 338, 343, 348,
353, 358, 363, 368, 373, 376, 377
**
**
*NSET, NSET=EDGE2, GENERATE
1, 1, 1
6, 6, 1
11, 11, 1
16, 16, 1
21, 31, 1
**
**
```

Nodal cards
These define the coordinates of the nodes in the model. The first entry is the node ID, while the second and third are the x and y coordinates of the position of the node, respectively.

Element (connectivity) cards
These define the element type and what nodes make up the element. S4 represents that it is a four-nodal, shell element. The "ELSET=MOTOR" statement is simply for naming this set of elements so that it can be referenced when defining the material properties. In the subsequent data entry, the first entry is the element ID, and the following four entries are the nodes making up the element. The order of the nodes for all elements must be consistent and counter-clockwise

Node sets
Sets of nodes are defined to be used for referencing when defining boundary conditions.

```
*NSET, NSET=CENTER
21, 36, 49, 60, 69, 76, 83, 90,
97, 104, 111, 118, 127, 149, 169, 188,
205, 220, 233, 244, 253, 260, 267, 274,
275, 276, 277, 278, 279, 280, 281, 282,
283,
**
**
*SHELL SECTION, ELSET=MOTOR, MATERIAL=POLYSI
13.,
**
**
*MATERIAL,
NAME=POLYSI
*DENSITY
2.3E-15,
**
*ELASTIC, TYPE=ISO
169000., 0.262
**
**
**
**
**
*BOUNDARY, OP=NEW
DOF, 1,, 0.
DOF, 2,, 0.
DOF, 6,, 0.
EDGE1, YSYMM
EDGE2, XSYMM
CENTER, ENCASTRE
**
*AMPLITUDE, NAME=SINE, DEFINITION–PERIODIC
1,12.566,0,0
0,10
**
**
**
**
**
**
*STEP, INC=1000
**
```

Material cards
These define material properties under the name "POLYSI." Density and elastic properties are defined. TYPE=ISO represents isotropic properties.

Property cards
These define properties to the elements of set "MOTOR." It will have the material properties defined under "POLYSI." The thickness of the elements is also defined in the data line.

BC cards
These define boundary conditions. In this case, all the nodes along the center circle are constrained to zero displacements. To simulate plate elements, DOFs 1, 2, and 6 are constrained for all the nodes in the model. Symmetrical conditions are also applied.

Amplitude curve
This defines an amplitude curve that can be a function of time or frequency. Loads or boundary conditions can then be made to follow the defined amplitude curve. In this case, a periodic function of the Fourier series is defined. The name of this amplitude curve is given as "SINE."

```
*DYNAMIC, DIRECT
NOHAF
0.1, 1.0
**
*NSET, NSET=DOF,
GENERATE
1, 383, 1
*NSET, NSET=FORCE
1, 143, 377
**
**
** FORCE
**
*CLOAD, OP=NEW, AMPLITUDE=SINE
FORCE, 3, 1.
**
**
*NODE PRINT, FREQ=1
U,
V,
A,
*NODE FILE, FREQ=1
U,
V,
A,
**
*END STEP
```

Control cards
These indicate the analysis step. In this case it is a "DYNAMIC" analysis, which performs a direct integration step to determine the transient response. The parameters following the keyword, DYNAMIC, specify various parameters for the algorithm. The first entry in the data line specifies the duration of each time step and the second specifies the total time step.

Load cards
"CLOAD" defines concentrated loading on the node set "FORCE" defined earlier. The load follows the amplitude curve, "SINE," defined earlier.

Output control cards
These define the output required. In this case, the nodal displacement components (U), velocity components (V), and acceleration components (A) are requested.

8.6.3 Solution process

The significance of the information provided in the above input file is very similar to the previous case study. Therefore, this section will highlight the differences that are mainly used for the transient analysis.

The definition of the amplitude curve is important here as it enables the load (or boundary condition) to be defined as a function of time. In this case the load will follow the sinusoidal function defined in the amplitude curve block. The sinusoidal function is defined as a periodic function whereby the formula used is actually the Fourier series. The data lines in the amplitude curve block basically define the angular frequency and the other constants in the Fourier series.

The control card specifies that the analysis is a direct integration, transient analysis. In ABAQUS, Newmarks's method (Section 3.7.2) together with the Hilber–Hughes–Taylor operator (1978) applied on the equilibrium equations, is used as the implicit solver for direct integration analysis. The time increment is specified to be 0.1 s, and the total time of the step is 1.0 s. As mentioned in Chapter 3, implicit methods involve the solving of the matrix equation at each individual increment in time, therefore the analysis can be rather computationally expensive. The algorithm used by ABAQUS is quite complex, involving the capabilities of having automatic deduction of the required time increments. Further details are, however, beyond the scope of this book.

8.6.4 Results and discussion

Upon the analysis of the problem defined by the input file above, the displacement, velocity, and acceleration components throughout each individual time increment can be obtained until the final time step specified. Therefore, we have what is known as the displacement–time history, the velocity–time history and the acceleration–time history, as shown in Figures 8.16, 8.17, and 8.18, respectively. The plots show the displacement, velocity, and acceleration histories of nodes 210 and 300.

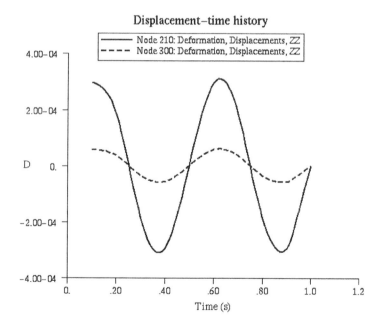

FIGURE 8.16

Displacement–time history at nodes 210 and 300.

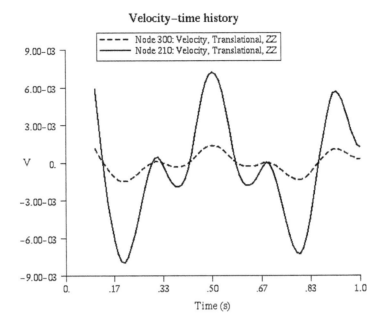

FIGURE 8.17

Velocity–time history at nodes 210 and 300.

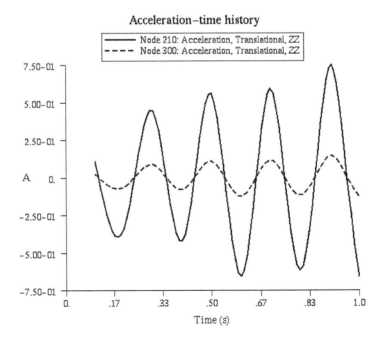

FIGURE 8.18

Acceleration–time history at nodes 210 and 300.

8.7 **Review questions**

1. If the plate were not homogenous but laminated, how would the finite element equation be different?
2. State the procedure to develop a triangular plate element.
3. Is it possible to develop plate elements using the classical plate theory (for thin plates)? What could be possible issues when doing so?
4. What could be a problem, if the plate elements developed based on the Mindlin plate theory (or other higher order theory) used for thin plates?
5. How should one develop a four-node quadrilateral element? How should one develop an eight-node element with curved edges?
6. How many Gauss points are required to obtain the exact results for Eqs. (8.19) and (8.20)?
7. Consider a uniform, isotropic square plate with a dimension of 1 m by 1 m. The plate is made of a material with Young's modulus $E = 70$ GPa and Poisson ratio $\nu = 0.3$. Using any FEM software package, compute the deflection, distribution of moments, shear forces, and a normal stress on the top surface of the plate. Using square elements with nodal density of 2×2, 4×4, 8×8, and 16×16 to perform the analysis, compare the FEM solution with the analytical solution (based on thin plate theory) for the deflection at the center of the plate. You are requested to perform the following studies:
   a. The plate is of thickness 0.001 m and clamped on all edges (CCCC);
   b. The plate is of thickness 0.01 m and clamped on all edges (CCCC);
   c. The plate is of thickness 0.1 m and simply-supported on all edges (SSSS);
   d. The plate is of thickness 0.001 m and simply-supported on all edges (SSSS);
   e. The plate is of thickness 0.01 m and simply-supported on all edges (SSSS);
   f. The plate is of thickness 0.1 m and simply-supported on all edges (SSSS);
   g. The plate is of thickness 0.001 m, clamped on two opposite edges and simply-supported on the other two edges (CSCS);
   h. The plate is of thickness 0.01 m, clamped on two opposite edges and simply-supported on the other two edges (CSCS);
   i. The plate is of thickness 0.1 m, clamped on two opposite edges and simply-supported on the other two edges (CSCS).

FEM for 3D Solid Elements

9

9.1 Introduction

A three-dimensional (3D) solid element can be considered to be the most general of all solid finite elements because the field variables are described fully in terms of all three physical coordinates: x, y, and z. An example of a 3D solid structure under loading is shown in Figure 9.1, where the force vectors are described arbitrarily in any spatial direction. A 3D solid can also have any arbitrary shape,

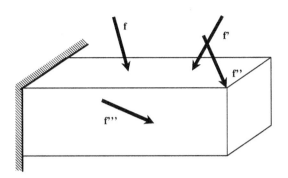

FIGURE 9.1

Example of a 3D solid under loadings.

material properties, and boundary conditions in space. As such, there are a total of six possible stress components, three normal and three shear, that need to be taken into consideration. Typically, a 3D solid element can be tetrahedron or hexahedron in shape with either flat or curved surfaces. Each node of the element will have three translational degrees of freedom (DOFs). The element can thus deform in all three directions in space.

Since the 3D element is considered the most general solid element, the truss, beam, plate, 2D solid and shell elements can all be considered to be special cases of the 3D element. So, why is there a need to develop all the other elements covered in previous chapters? Why not just use the 3D elements to model everything? Theoretically, yes, the 3D element can actually be used to model all kinds of structural components including trusses, beams, plates, shells, and so on. However, it can be very tedious in geometry creation and meshing. Furthermore, it is also the most demanding on computer resources. Hence, the general rule of thumb is, when a structure can be assumed within acceptable tolerances to be simplified into a 1D (trusses, beams, and frames) or 2D (2D solids and plates) structure, we shall always do so. The creation of 1D or 2D FEM models is very much easier and more efficient. We use 3D solid elements only when we have no other choices. The formulation of 3D solid elements is straightforward, because it is basically an extension of 2D solid elements. All the techniques used in 2D solids can be utilized, except that all the variables are now functions of x, y, and z. The basic concepts, procedures, and formulations for 3D solid elements may be found in many existing books (see, for example, Washizu, 1981; Rao, 1999; Zienkiewicz and Taylor, 2000).

9.2 Tetrahedron element

9.2.1 Strain matrix

Consider the same 3D solid structure as Figure 9.1, whose domain is divided in a proper manner into a number of *tetrahedron* elements with four nodes and four surfaces, as shown in Figure 9.2. A tetrahedron element has four nodes and each node has three DOFs (u, v, and w), which makes the total DOFs

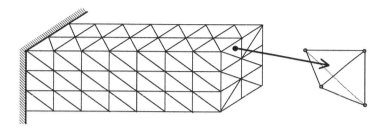

FIGURE 9.2

Solid block divided into 4-node tetrahedron elements.

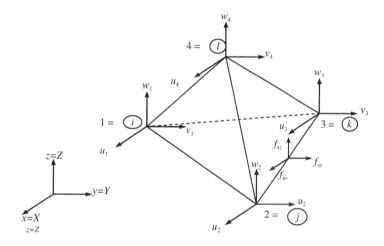

FIGURE 9.3

A tetrahedron element.

in a tetrahedron element twelve, as shown in Figure 9.3. The nodes are numbered 1, 2, 3, and 4 by the right- hand rule. The local Cartesian coordinate system for a tetrahedron element can usually be the same as the global coordinate system, as there are no advantages in having a separate local Cartesian coordinate system. In an element, the displacement vector $\mathbf{U}$ is a function of the coordinate x, y, and z, and is interpolated by shape functions in the following form, which should be, by now, part and parcel of the finite element method.

$$\mathbf{U}^h(x, y, z) = \mathbf{N}(x, y, z)\mathbf{d}_e \qquad (9.1)$$

where the nodal displacement vector, $\mathbf{d}_e$, is given as

$$\mathbf{d}_e = \left\{ \begin{array}{l} \left. \begin{array}{l} u_1 \\ v_1 \\ w_1 \end{array} \right\} \text{displacements at node 1} \\ \left. \begin{array}{l} u_2 \\ v_2 \\ w_2 \end{array} \right\} \text{displacements at node 2} \\ \left. \begin{array}{l} u_3 \\ v_3 \\ w_3 \end{array} \right\} \text{displacements at node 3} \\ \left. \begin{array}{l} u_4 \\ v_4 \\ w_4 \end{array} \right\}_{12 \times 1} \text{displacements at node 4} \end{array} \right. \tag{9.2}$$

and the matrix of shape functions has the form

$$\mathbf{N} = \begin{bmatrix} \overbrace{N_1 \quad 0 \quad 0}^{\text{node 1}} & \overbrace{N_2 \quad 0 \quad 0}^{\text{node 2}} & \overbrace{N_3 \quad 0 \quad 0}^{\text{node 3}} & \overbrace{N_4 \quad 0 \quad 0}^{\text{node 4}} \\ 0 \quad N_1 \quad 0 & 0 \quad N_2 \quad 0 & 0 \quad N_3 \quad 0 & 0 \quad N_4 \quad 0 \\ 0 \quad 0 \quad N_1 & 0 \quad 0 \quad N_2 & 0 \quad 0 \quad N_3 & 0 \quad 0 \quad N_4 \end{bmatrix}_{3 \times 12} \tag{9.3}$$

To develop the shape functions, we make use of what are known as the *volume coordinates*, which are a natural extension from the area coordinates for 2D solids. The use of the volume coordinates makes it more convenient for shape function construction and element matrix integration. The volume coordinates for node 1 are defined as

$$L_1 = \frac{V_{P234}}{V_{1234}} \tag{9.4}$$

where V_{P234} and V_{1234} denote, respectively, the volumes of the tetrahedrons P234 and 1234, as shown in Figure 9.4. The volume coordinate for node 2–4 can also be defined in the same manner.

$$L_2 = \frac{V_{P134}}{V_{1234}}, \quad L_3 = \frac{V_{P124}}{V_{1234}}, \quad L_4 = \frac{V_{P123}}{V_{1234}} \tag{9.5}$$

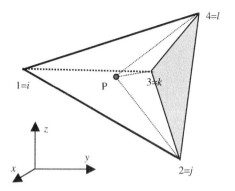

FIGURE 9.4

Volume coordinates for tetrahedron elements.

The volume coordinate can also be viewed as the ratio between the distance of the point P and point 1 to the plane 234.

$$L_1 = \frac{d_{P-234}}{d_{1-234}}, \quad L_2 = \frac{d_{P-134}}{d_{2-134}}, \quad L_3 = \frac{d_{P-124}}{d_{3-214}}, \quad L_4 = \frac{d_{P-123}}{d_{4-231}} \tag{9.6}$$

It can easily be confirmed that

$$L_1 + L_2 + L_3 + L_4 = 1 \tag{9.7}$$

since

$$V_{P234} + V_{P134} + V_{P124} + V_{P123} = V_{1234} \tag{9.8}$$

It can also easily be confirmed that

$$L_i = \begin{cases} 1 & \text{at the home node } i \\ 0 & \text{at the remote nodes } jkl \end{cases} \tag{9.9}$$

Using Eq. (9.9) above, the relationship between the volume coordinates and the Cartesian coordinates can be easily derived.

$$\begin{aligned} x &= L_1 x_1 + L_2 x_2 + L_3 x_3 + L_4 x_4 \\ y &= L_1 y_1 + L_2 y_2 + L_3 y_3 + L_4 y_4 \\ z &= L_1 z_1 + L_2 z_2 + L_3 z_3 + L_4 z_4 \end{aligned} \tag{9.10}$$

Equations (9.7) and (9.10) can then be expressed as a single matrix equation as follows:

$$
\begin{Bmatrix} 1 \\ x \\ y \\ z \end{Bmatrix} = \begin{bmatrix} 1 & 1 & 1 & 1 \\ x_1 & x_2 & x_3 & x_4 \\ y_1 & y_2 & y_3 & y_4 \\ z_1 & z_2 & z_3 & z_4 \end{bmatrix} \begin{Bmatrix} L_1 \\ L_2 \\ L_3 \\ L_4 \end{Bmatrix}
\tag{9.11}
$$

The inversion of Eq. (9.11) above will give

$$
\begin{Bmatrix} L_1 \\ L_2 \\ L_3 \\ L_4 \end{Bmatrix} = \frac{1}{6V} \begin{bmatrix} a_1 & b_1 & c_1 & d_1 \\ a_2 & b_2 & c_2 & d_2 \\ a_3 & b_3 & c_3 & d_3 \\ a_4 & b_4 & c_4 & d_4 \end{bmatrix} \begin{Bmatrix} 1 \\ x \\ y \\ z \end{Bmatrix}
\tag{9.12}
$$

where

$$
a_i = (-1)^{j+k+l+1} \det \begin{bmatrix} x_j & y_j & z_j \\ x_k & y_k & z_k \\ x_l & y_l & z_l \end{bmatrix}, \quad b_i = (-1)^{j+k+l} \det \begin{bmatrix} 1 & y_j & z_j \\ 1 & y_k & z_k \\ 1 & y_l & z_l \end{bmatrix}
$$

$$
c_i = (-1)^{j+k+l} \det \begin{bmatrix} x_j & 1 & z_j \\ x_k & 1 & z_k \\ x_l & 1 & z_l \end{bmatrix}, \quad d_i = (-1)^{j+k+l} \det \begin{bmatrix} x_j & y_j & 1 \\ x_k & y_k & 1 \\ x_l & y_l & 1 \end{bmatrix}
\tag{9.13}
$$

in which the subscript i varies from 1 to 4, and j, k, and l are determined by a cyclic permutation in the order of i, j, k, l. For example, if $i=1$, then $j=2$, $k=3$, $l=4$. When $i=2$, then $j=3$, $k=4$, $l=1$. The volume of the tetrahedron element V can be obtained by

$$
V = \frac{1}{6} \times \det \begin{bmatrix} 1 & x_i & y_i & z_i \\ 1 & x_j & y_j & z_j \\ 1 & x_k & y_k & z_k \\ 1 & x_l & y_l & z_l \end{bmatrix}
\tag{9.14}
$$

The properties of L_i as depicted in Eqs. (9.6) to (9.9) show that L_i can be used as the shape function of a 4-nodal tetrahedron element.

$$N_i = L_i = \frac{1}{6V}(a_i + b_i x + c_i y + d_i z)$$

(9.15)

It can be seen from above that the shape function is a linear function of x, y, and z, hence, the 4-nodal tetrahedron element is a linear element. Note that from Eq. (9.14), the moment matrix of the linear basis functions will never be singular, unless the volume of the element is zero (or the four nodes of the element are in a plane). Based on Lemma 2 and 3 in Chapter 3, we can ensure that the shape functions given by Eq. (9.15) satisfy sufficiently the requirement of FEM shape functions.

It was mentioned that there are a total of six stress components in a 3D element. The stress components can be written in a vector form as

$$\left\{ \sigma_{xx} \; \sigma_{yy} \; \sigma_{zz} \; \sigma_{yz} \; \sigma_{xz} \; \sigma_{xy} \right\}^T$$

To get the corresponding vector of strain components,

$$\left\{ \varepsilon_{xx} \; \varepsilon_{yy} \; \varepsilon_{zz} \; \gamma_{yz} \; \gamma_{xz} \; \gamma_{xy} \right\}^T$$

we can substitute Eq. (9.1) into Eq. (2.5) from Chapter 2.

$$\varepsilon = \mathbf{LU} = \mathbf{LNd}_e = \mathbf{Bd}_e$$

(9.16)

where the strain matrix $\mathbf{B}$ is given by

$$\mathbf{B}_{6\times 12} = \mathbf{LN} = \begin{bmatrix} \partial/\partial x & 0 & 0 \\ 0 & \partial/\partial y & 0 \\ 0 & 0 & \partial/\partial z \\ 0 & \partial/\partial z & \partial/\partial y \\ \partial/\partial z & 0 & \partial/\partial x \\ \partial/\partial y & \partial/\partial x & 0 \end{bmatrix}_{6\times 3} \mathbf{N}_{3\times 12}$$

(9.17)

Substituting Eq. (9.3) into the above, the strain matrix **B** can be obtained as

$$\mathbf{B} = \frac{1}{6V} \begin{bmatrix} b_1 & 0 & 0 & b_2 & 0 & 0 & b_3 & 0 & 0 & b_4 & 0 & 0 \\ 0 & c_1 & 0 & 0 & c_2 & 0 & 0 & c_3 & 0 & 0 & c_4 & 0 \\ 0 & 0 & d_1 & 0 & 0 & d_2 & 0 & 0 & d_3 & 0 & 0 & d_4 \\ c_1 & b_1 & 0 & c_2 & b_2 & 0 & c_3 & b_3 & 0 & c_4 & b_4 & 0 \\ 0 & d_1 & c_1 & 0 & d_2 & c_2 & 0 & d_3 & c_3 & 0 & d_4 & c_4 \\ d_1 & 0 & b_1 & d_2 & 0 & b_2 & d_3 & 0 & b_3 & d_4 & 0 & b_4 \end{bmatrix}_{6 \times 12} \tag{9.18}$$

It can be seen that the strain matrix for a linear tetrahedron element is a constant matrix. This implies that the strain within a linear tetrahedron element is constant, and thus the stress is also constant. Therefore, the linear tetrahedron elements are also often referred to as a *constant strain element* or *constant stress element*, similar to the case of 2D linear triangular elements.

9.2.2 Element matrices

Once the strain matrix is obtained, the stiffness matrix $\mathbf{k}_e$ for 3D solid elements can be obtained by substituting Eq. (9.18) into Eq. (3.71) from Chapter 3. Since the strain is constant, the element stiffness matrix is obtained as

$$\mathbf{k}_e = \int_{V_e} \mathbf{B}^T \mathbf{c} \mathbf{B} dV = V_e \left[\mathbf{B}^T \mathbf{c} \mathbf{B} \right]_{12 \times 12} \tag{9.19}$$

Note that the material constant matrix **c** is given generally by Eq. (2.9) in Chapter 2.
   The mass matrix can similarly be obtained using Eq. (3.75) from Chapter 3.

$$\mathbf{m}_e = \int_{V_e} \rho \mathbf{N}^T \mathbf{N} dV = \int_{V_e} \rho \begin{bmatrix} \mathbf{N}_{11} & \mathbf{N}_{12} & \mathbf{N}_{13} & \mathbf{N}_{14} \\ \mathbf{N}_{21} & \mathbf{N}_{22} & \mathbf{N}_{23} & \mathbf{N}_{24} \\ \mathbf{N}_{31} & \mathbf{N}_{32} & \mathbf{N}_{33} & \mathbf{N}_{34} \\ \mathbf{N}_{41} & \mathbf{N}_{42} & \mathbf{N}_{43} & \mathbf{N}_{44} \end{bmatrix}_{12 \times 12} dV \tag{9.20}$$

where

$$\mathbf{N}_{ij} = \begin{bmatrix} N_i N_j & 0 & 0 \\ 0 & N_i N_j & 0 \\ 0 & 0 & N_i N_j \end{bmatrix}_{3 \times 1} \tag{9.21}$$

Using the following formula (Eisenberg and Malvern, 1973) for volume coordinates,

$$\int_{V_e} L_1^m L_2^n L_3^p L_4^q \, dV = \frac{m!\,n!\,p!\,q!}{(m+n+p+q+3)!} 6V_e \tag{9.22}$$

we can conveniently evaluate the integral in Eq. (9.20) to give

$$\mathbf{m}_e = \frac{\rho V_e}{20}
\begin{bmatrix}
2 & 0 & 0 & 1 & 0 & 0 & 1 & 0 & 0 & 1 & 0 & 0 \\
  & 2 & 0 & 0 & 1 & 0 & 0 & 1 & 0 & 0 & 1 & 0 \\
  &   & 2 & 0 & 0 & 1 & 0 & 0 & 1 & 0 & 0 & 1 \\
  &   &   & 2 & 0 & 0 & 1 & 0 & 0 & 1 & 0 & 0 \\
  &   &   &   & 2 & 0 & 0 & 1 & 0 & 0 & 1 & 0 \\
  &   &   &   &   & 2 & 0 & 0 & 1 & 0 & 0 & 1 \\
  &   &   &   &   &   & 2 & 0 & 0 & 1 & 0 & 0 \\
  &   &   &   &   &   &   & 2 & 0 & 0 & 1 & 0 \\
  &   &   &   &   &   &   &   & 2 & 0 & 0 & 1 \\
  & sy. & &   &   &   &   &   &   & 2 & 0 & 0 \\
  &   &   &   &   &   &   &   &   &   & 2 & 0 \\
  &   &   &   &   &   &   &   &   &   &   & 2 \\
\end{bmatrix}_{12\times 12} \tag{9.23}$$

An alternative way to calculate the mass matrix for 3D solid elements is to use a special natural coordinate system, which is defined as shown in Figures 9.5, 9.6 and 9.7. In Figure 9.5, the plane of $\xi=$ constant is defined in such a way that the edge P-Q stays parallel to the edge 2–3 of the element, and the point 4 coincides with point 4 of the element. When P moves to point 1, $\xi=0$, and when P moves to point 2, $\xi=1$. In Figure 9.6, the plane of $\eta=$ constant is defined in such a way that the edge 1–4 on the triangle coincides with the edge 1–4 of the element, and point P stays on the edge 2 3 of the element. When P moves to point 2, $\eta=0$, and when P moves to point 3, $\eta=1$. The plane of $\zeta=$ constant is defined in Figure 9.7, in such a way that the plane P-Q-R stays parallel to the plane 1–2–3 of the element, and when P moves to point 4, $\zeta=0$, and when P moves to point 2, $\zeta=1$. In addition, the plane 1–2–3 on the element sits on the x–y plane. Therefore, the relationship between xyz and $\xi\eta\zeta$ can be obtained in the following steps:

In Figure 9.8, the coordinates at point P are first interpolated using the x, y, and z coordinates at points 2 and 3:

$$
\begin{aligned}
x_P &= \eta(x_3 - x_2) + x_2 \\
y_P &= \eta(y_3 - y_2) + y_2 \\
z_P &= 0
\end{aligned}
\tag{9.24}
$$

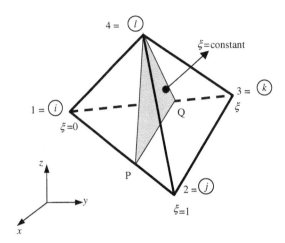

FIGURE 9.5

Natural coordinate, where ξ=constant.

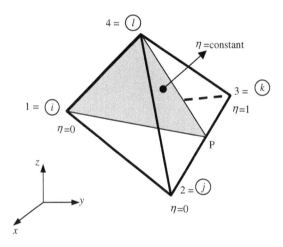

FIGURE 9.6

Natural coordinate, where η=constant.

The coordinates at point B are then interpolated using the x, y, and z coordinates at points 1 and P:

$$x_B = \xi(x_P - x_1) + x_1 = \xi\eta(x_3 - x_2) + \xi(x_2 - x_1) + x_1$$
$$y_B = \xi(y_P - y_1) + y_1 = \xi\eta(y_3 - y_2) + \xi(y_2 - y_1) + y_1 \qquad (9.25)$$
$$z_B = 0$$

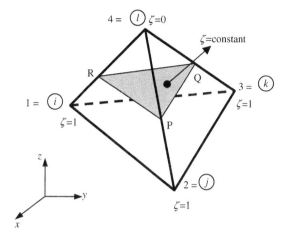

FIGURE 9.7

Natural coordinate, where ζ=constant

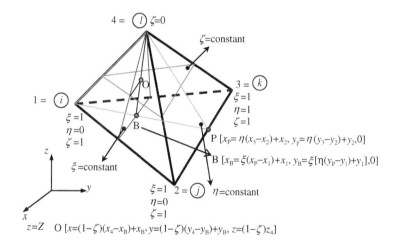

FIGURE 9.8

Cartesian coordinates xyz of point O in term of $\xi\eta\zeta$.

The coordinates at point O are finally interpolated using the x, y, and z coordinates at points 4 and B:

$$
\begin{aligned}
x &= x_4 - \zeta(x_4 - x_B) = x_4 - \zeta(x_4 - x_1) + \xi\zeta(x_2 - x_1) - \xi\zeta\eta(x_2 - x_3) \\
y &= y_4 - \zeta(y_4 - y_B) = y_4 - \zeta(y_4 - y_1) + \xi\zeta(y_2 - y_1) - \xi\zeta\eta(y_2 - y_3) \\
z &= (1 - \zeta)z_4
\end{aligned}
\tag{9.26}
$$

With this special natural coordinate system, the shape functions in the matrix of Eq. (9.3) can be written by inspection as

$$
\begin{aligned}
N_1 &= (1 - \xi)\zeta \\
N_2 &= \xi\zeta(1 - \eta) \\
N_3 &= \xi\eta\zeta \\
N_4 &= (1 - \zeta)
\end{aligned}
\tag{9.27}
$$

The Jacobian matrix between xyz and $\xi\eta\zeta$ is required and is given as

$$
\mathbf{J} =
\begin{bmatrix}
\frac{\partial x}{\partial \xi} & \frac{\partial y}{\partial \xi} & \frac{\partial z}{\partial \xi} \\[4pt]
\frac{\partial x}{\partial \eta} & \frac{\partial y}{\partial \eta} & \frac{\partial z}{\partial \eta} \\[4pt]
\frac{\partial x}{\partial \zeta} & \frac{\partial y}{\partial \zeta} & \frac{\partial z}{\partial \zeta}
\end{bmatrix}
\tag{9.28}
$$

Using Eqs. (9.26) and (9.27), the determinate of the Jacobian can be found to be

$$
\det (\mathbf{J}) = -6V\xi\zeta^2
\tag{9.29}
$$

The mass matrix can now be obtained as

$$
\mathbf{m}_e = \int_{V_e} \rho \mathbf{N}^T \mathbf{N} dV = \int_0^1 \int_0^1 \int_0^1 \rho \mathbf{N}^T \mathbf{N} \det[\mathbf{J}] d\xi \, d\eta \, d\zeta
\tag{9.30}
$$

which gives

$$
\mathbf{m}_e = -6V_e\rho \int_0^1 \int_0^1 \int_0^1 \xi\zeta^2
\begin{bmatrix}
\mathbf{N}_{11} & \mathbf{N}_{12} & \mathbf{N}_{13} & \mathbf{N}_{14} \\
\mathbf{N}_{21} & \mathbf{N}_{22} & \mathbf{N}_{23} & \mathbf{N}_{24} \\
\mathbf{N}_{31} & \mathbf{N}_{32} & \mathbf{N}_{33} & \mathbf{N}_{34} \\
\mathbf{N}_{41} & \mathbf{N}_{42} & \mathbf{N}_{43} & \mathbf{N}_{44}
\end{bmatrix}_{12 \times 12}
d\xi \, d\eta \, d\zeta
\tag{9.31}
$$

where $\mathbf{N}_{ij}$ is given by Eq. (9.21), but in which the shape functions should be defined by Eq. (9.27). Evaluating the integrals in Eq. (9.31) would give the same mass matrix as in Eq. (9.23).

The nodal force vector for 3D solid elements can be obtained using Eqs. (3.78), (3.79) and (3.81). Suppose the element is loaded by a distributed force $\mathbf{f}_s$ on the edge 2–3 of the element as shown in Figure 9.3, the nodal force vector becomes

$$
\mathbf{f}_e = \int_l [\mathbf{N}]^T \Big|_{2-3}
\begin{Bmatrix}
f_{sx} \\
f_{sy} \\
f_{sz}
\end{Bmatrix}
dl
\tag{9.32}
$$

If the load is uniformly distributed, f_{sx}, f_{sy}, and f_{sz} are constants and the above equation becomes

$$\mathbf{f}_e = \frac{1}{2}l_{2-3}\left\{\begin{array}{c} \{0\}_{3\times 1} \\ \left\{\begin{array}{c} f_{sx} \\ f_{sy} \\ f_{sz} \end{array}\right\} \\ \left\{\begin{array}{c} f_{sx} \\ f_{sy} \\ f_{sz} \end{array}\right\} \\ \{0\}_{3\times 1} \end{array}\right\}_{12\times 1} \tag{9.33}$$

where l_{2-3} is the length of the edge 2–3. Eq. (9.33) implies that the distributed forces are equally divided and applied at the two nodes. This conclusion applies also to evenly distribute surface forces applied on any face of the element, and to evenly distribute body force applied on the entire body of the element. Finally, the stiffness matrix, $\mathbf{k}_e$, the mass matrix, $\mathbf{m}_e$, and the nodal force vector, $\mathbf{f}_e$, can be used directly to assemble the global FE equation, Eq. (3.96), without going through a coordinate transformation.

9.3 Hexahedron element
9.3.1 Strain matrix

Consider now a 3D domain, which is divided in a proper manner into a number of *hexahedron elements* with eight nodes and six surfaces, as shown in Figure 9.9. Each hexahedron element has nodes numbered 1, 2, 3, 4 and 5, 6, 7, 8 in a counter-clockwise manner, as shown in Figure 9.10.

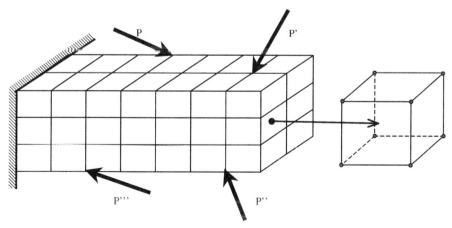

FIGURE 9.9

Solid block divided into 8-nodal hexahedron elements.

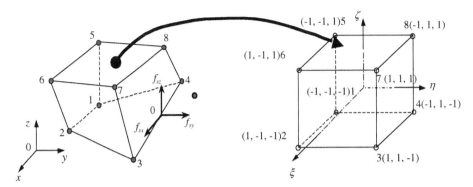

FIGURE 9.10

An 8-nodal hexahedron element and the coordinate systems.

As there are three DOFs at one node, there are now a total of twenty-four DOFs in a hexahedron element. It is again useful to define a *natural coordinate system* (ξ, η, ζ) with its origin at the center of the transformed cube, as this makes it easier to construct the shape functions and to evaluate the matrix integration. The coordinate mapping is performed in a similar manner to that of quadrilateral elements in Chapter 7. Like the quadrilateral element, shape functions are also used to interpolate the coordinates from the nodal coordinates:

$$x = \sum_{i=1}^{8} N_i(\xi, \eta, \zeta)x_i$$

$$y = \sum_{i=1}^{8} N_i(\xi, \eta, \zeta)y_i \qquad (9.34)$$

$$z = \sum_{i=1}^{8} N_i(\xi, \eta, \zeta)z_i$$

The shape functions are given in the local natural coordinate system as

$$N_1 = \frac{1}{8}(1 - \xi)(1 - \eta)(1 - \zeta)$$

$$N_2 = \frac{1}{8}(1 + \xi)(1 - \eta)(1 - \zeta)$$

$$N_3 = \frac{1}{8}(1 + \xi)(1 + \eta)(1 - \zeta)$$

$$N_4 = \frac{1}{8}(1 - \xi)(1 + \eta)(1 - \zeta)$$

$$N_5 = \frac{1}{8}(1 - \xi)(1 - \eta)(1 + \zeta) \qquad (9.35)$$

$$N_6 = \frac{1}{8}(1 + \xi)(1 - \eta)(1 + \zeta)$$

$$N_7 = \frac{1}{8}(1 + \xi)(1 + \eta)(1 + \zeta)$$

$$N_8 = \frac{1}{8}(1 - \xi)(1 + \eta)(1 + \zeta)$$

or in a concise form of

$$N_i = \frac{1}{8}(1 + \xi\xi_i)(1 + \eta\eta_i)(1 + \zeta\zeta_i) \tag{9.36}$$

where (ξ_i, η_i, ζ_i) denotes the natural coordinates of node i.

From Eq. (9.36), it can be seen that the shape functions vary linearly in each of the ξ, η, and ζ directions. Therefore, these shape functions are sometimes called trilinear functions. The shape function N_i is a three-dimensional analogy of that given in Eq. (7.54). It can be shown through direct observation that the trilinear elements possess delta function property. In addition, since all these shape functions can be formed using the common set of 8 basis functions of

$$1, \xi, \eta, \zeta, \xi\eta, \xi\zeta, \eta\zeta, \xi\eta\zeta \tag{9.37}$$

which contain both constant and linear basis functions, these shape functions can expect to possess both partitions of unity property as well as the linear reproduction property (see Lemma 2 and 3 described in Chapter 3).

In a hexahedron element, the displacement vector $\mathbf{U}$ is a function of the coordinates x, y, and z, and like before, it is interpolated using the shape functions:

$$\mathbf{U} = \mathbf{Nd}_e \tag{9.38}$$

where the nodal displacement vector, $\mathbf{d}_e$, is given by

$$\mathbf{d}_e = \begin{Bmatrix} \mathbf{d}_1 \\ \mathbf{d}_2 \\ \mathbf{d}_3 \\ \mathbf{d}_4 \\ \mathbf{d}_5 \\ \mathbf{d}_6 \\ \mathbf{d}_7 \\ \mathbf{d}_8 \end{Bmatrix} \begin{matrix} \text{displacement components at node 1} \\ \text{displacement components at node 2} \\ \text{displacement components at node 3} \\ \text{displacement components at node 4} \\ \text{displacement components at node 5} \\ \text{displacement components at node 6} \\ \text{displacement components at node 7} \\ \text{displacement components at node 8} \end{matrix} \tag{9.39}$$

in which,

$$\mathbf{d}_i = \begin{Bmatrix} u_1 \\ v_1 \\ w_1 \end{Bmatrix} \quad (i = 1, 2, \cdots, 8) \tag{9.40}$$

is the displacement at node i. The matrix of shape functions is given by

$$\mathbf{N} = [\mathbf{N}_1 \ \mathbf{N}_2 \ \mathbf{N}_3 \ \mathbf{N}_4 \ \mathbf{N}_5 \ \mathbf{N}_6 \ \mathbf{N}_7 \ \mathbf{N}_8] \tag{9.41}$$

in which each sub-matrix, $\mathbf{N}_i$, is given as

$$\mathbf{N}_i = \begin{bmatrix} N_i & 0 & 0 \\ 0 & N_i & 0 \\ 0 & 0 & N_i \end{bmatrix} \quad (i = 1, 2, \cdots, 8) \tag{9.42}$$

In this case, the strain matrix defined by Eq. (9.17) can be expressed as

$$\mathbf{B} = [\mathbf{B}_1 \ \mathbf{B}_2 \ \mathbf{B}_3 \ \mathbf{B}_4 \ \mathbf{B}_5 \ \mathbf{B}_6 \ \mathbf{B}_7 \ \mathbf{B}_8] \tag{9.43}$$

whereby

$$\mathbf{B}_i = \mathbf{L} \mathbf{N}_i = \begin{bmatrix} \partial N_i/\partial x & 0 & 0 \\ 0 & \partial N_i/\partial y & 0 \\ 0 & 0 & \partial N_i/\partial z \\ 0 & \partial N_i/\partial z & \partial N_i/\partial y \\ \partial N_i/\partial z & 0 & \partial N_i/\partial x \\ \partial N_i/\partial y & \partial N_i/\partial x & 0 \end{bmatrix} \tag{9.44}$$

As the shape functions are defined in terms of the natural coordinates, ξ, η, and ζ, to obtain the derivatives with respect to x, y, and z in the strain matrix, the chain rule of partial differentiation needs to be used:

$$\begin{aligned} \frac{\partial N_i}{\partial \xi} &= \frac{\partial N_i}{\partial x}\frac{\partial x}{\partial \xi} + \frac{\partial N_i}{\partial y}\frac{\partial y}{\partial \xi} + \frac{\partial N_i}{\partial z}\frac{\partial z}{\partial \xi} \\ \frac{\partial N_i}{\partial \eta} &= \frac{\partial N_i}{\partial x}\frac{\partial x}{\partial \eta} + \frac{\partial N_i}{\partial y}\frac{\partial y}{\partial \eta} + \frac{\partial N_i}{\partial z}\frac{\partial z}{\partial \eta} \\ \frac{\partial N_i}{\partial \zeta} &= \frac{\partial N_i}{\partial x}\frac{\partial x}{\partial \zeta} + \frac{\partial N_i}{\partial y}\frac{\partial y}{\partial \zeta} + \frac{\partial N_i}{\partial z}\frac{\partial z}{\partial \zeta} \end{aligned} \tag{9.45}$$

which can be expressed in the matrix form of

$$\begin{Bmatrix} \frac{\partial N_i}{\partial \xi} \\ \frac{\partial N_i}{\partial \eta} \\ \frac{\partial N_i}{\partial \zeta} \end{Bmatrix} = \mathbf{J} \begin{Bmatrix} \frac{\partial N_i}{\partial x} \\ \frac{\partial N_i}{\partial y} \\ \frac{\partial N_i}{\partial z} \end{Bmatrix} \tag{9.46}$$

where $\mathbf{J}$ is the *Jacobian matrix* defined by

$$\mathbf{J} = \begin{bmatrix} \dfrac{\partial x}{\partial \xi} & \dfrac{\partial y}{\partial \xi} & \dfrac{\partial z}{\partial \xi} \\[2mm] \dfrac{\partial x}{\partial \eta} & \dfrac{\partial y}{\partial \eta} & \dfrac{\partial z}{\partial \eta} \\[2mm] \dfrac{\partial x}{\partial \zeta} & \dfrac{\partial y}{\partial \zeta} & \dfrac{\partial z}{\partial \zeta} \end{bmatrix} \tag{9.47}$$

Recall that the coordinates, x, y, and z are interpolated by the shape functions from the nodal coordinates. Hence, substituting the interpolation of the coordinates, Eq. (9.34) into Eq. (9.47) above gives

$$\mathbf{J} = \begin{bmatrix} \dfrac{\partial N_1}{\partial \xi} & \dfrac{\partial N_2}{\partial \xi} & \dfrac{\partial N_3}{\partial \xi} & \dfrac{\partial N_4}{\partial \xi} & \dfrac{\partial N_5}{\partial \xi} & \dfrac{\partial N_6}{\partial \xi} & \dfrac{\partial N_7}{\partial \xi} & \dfrac{\partial N_8}{\partial \xi} \\[2mm] \dfrac{\partial N_1}{\partial \eta} & \dfrac{\partial N_2}{\partial \eta} & \dfrac{\partial N_3}{\partial \eta} & \dfrac{\partial N_4}{\partial \eta} & \dfrac{\partial N_5}{\partial \eta} & \dfrac{\partial N_6}{\partial \eta} & \dfrac{\partial N_7}{\partial \eta} & \dfrac{\partial N_8}{\partial \eta} \\[2mm] \dfrac{\partial N_1}{\partial \zeta} & \dfrac{\partial N_2}{\partial \zeta} & \dfrac{\partial N_3}{\partial \zeta} & \dfrac{\partial N_4}{\partial \zeta} & \dfrac{\partial N_5}{\partial \zeta} & \dfrac{\partial N_6}{\partial \zeta} & \dfrac{\partial N_7}{\partial \zeta} & \dfrac{\partial N_8}{\partial \zeta} \end{bmatrix} \begin{bmatrix} x_1 & y_1 & z_1 \\ x_2 & y_2 & z_2 \\ x_3 & y_3 & z_3 \\ x_4 & y_4 & z_4 \\ x_5 & y_5 & z_5 \\ x_6 & y_6 & z_6 \\ x_7 & y_7 & z_7 \\ x_8 & y_8 & z_8 \end{bmatrix} \tag{9.48}$$

or

$$\mathbf{J} = \begin{bmatrix} \sum_{i=1}^{8} x_i \dfrac{\partial N_i}{\partial \xi} & \sum_{i=1}^{8} y_i \dfrac{\partial N_i}{\partial \xi} & \sum_{i=1}^{8} z_i \dfrac{\partial N_i}{\partial \xi} \\[3mm] \sum_{i=1}^{8} x_i \dfrac{\partial N_i}{\partial \eta} & \sum_{i=1}^{8} y_i \dfrac{\partial N_i}{\partial \eta} & \sum_{i=1}^{8} z_i \dfrac{\partial N_i}{\partial \eta} \\[3mm] \sum_{i=1}^{8} x_i \dfrac{\partial N_i}{\partial \zeta} & \sum_{i=1}^{8} y_i \dfrac{\partial N_i}{\partial \zeta} & \sum_{i=1}^{8} z_i \dfrac{\partial N_i}{\partial \zeta} \end{bmatrix} \tag{9.49}$$

Equation (9.46) can be re-written as

$$\begin{Bmatrix} \dfrac{\partial N_i}{\partial x} \\[2mm] \dfrac{\partial N_i}{\partial y} \\[2mm] \dfrac{\partial N_i}{\partial z} \end{Bmatrix} = \mathbf{J}^{-1} \begin{Bmatrix} \dfrac{\partial N_i}{\partial \xi} \\[2mm] \dfrac{\partial N_i}{\partial \eta} \\[2mm] \dfrac{\partial N_i}{\partial \zeta} \end{Bmatrix} \tag{9.50}$$

which is then used to compute the strain matrix, $\mathbf{B}$, in Eqs. (9.43) and (9.44), by replacing all the derivatives of the shape functions with respect to x, y, and z to those with respect to ξ, η, and ζ.

9.3.2 Element matrices

Once the strain matrix, **B**, is computed, the stiffness matrix, $\mathbf{k}_e$, for 3D solid elements can be obtained by substituting **B** into Eq. (3.71) from Chapter 3.

$$\mathbf{k}_e = \int_{V_e} \mathbf{B}^T \mathbf{c} \mathbf{B} dV = \int_{-1}^{+1} \int_{-1}^{+1} \int_{-1}^{+1} \mathbf{B}^T \mathbf{c} \mathbf{B} \det[\mathbf{J}] d\xi d\eta d\zeta \tag{9.51}$$

Note that the matrix of material constant, **c**, is given by Eq. (2.9) from Chapter 2. As the strain matrix, **B**, is a function of ξ, η, and ζ, evaluating the integrations in Eq. (9.51) can be tedious. Therefore, the integrals are performed using a numerical integration scheme. The Gauss integration scheme discussed in Section 7.3.4 of Chapter 7 is often used to carry out the integral. For three-dimensional integrations, the Gauss integration is sampled in three directions as follows:

$$I = \int_{-1}^{+1} \int_{-1}^{+1} \int_{-1}^{+1} f(\xi, \eta) d\xi d\eta = \sum_{i=1}^{n} \sum_{j=1}^{m} \sum_{k=1}^{l} w_i w_j w_k f(\xi_i, \eta_j, \zeta_k) \tag{9.52}$$

To obtain the mass (inertia) matrix for the hexahedron element, substitute the shape function matrix, Eq. (9.41), into Eq. (3.75) from Chapter 3.

$$\mathbf{m}_e = \int_{V_e} \rho \mathbf{N}^T \mathbf{N} dV = \int_{-1}^{1} \int_{-1}^{1} \int_{-1}^{1} \rho \mathbf{N}^T \mathbf{N} \det[\mathbf{J}] d\xi d\eta d\zeta \tag{9.53}$$

The above integral is also usually carried out using the Gauss integration. If the hexahedron is rectangular with dimensions of $2a \times 2b \times 2c$, the determinate of the Jacobian matrix is simply given by

$$\det[\mathbf{J}] = abc \tag{9.54}$$

and the mass matrix can be explicitly obtained as

$$\mathbf{m}_e = \begin{bmatrix} m_{11} & m_{12} & m_{13} & m_{14} & m_{15} & m_{16} & m_{17} & m_{18} \\ & m_{22} & m_{23} & m_{24} & m_{25} & m_{26} & m_{27} & m_{28} \\ & & m_{33} & m_{24} & m_{25} & m_{26} & m_{27} & m_{28} \\ & & & m_{44} & m_{45} & m_{46} & m_{47} & m_{48} \\ & & & & m_{55} & m_{56} & m_{57} & m_{58} \\ & & & & & m_{66} & m_{67} & m_{68} \\ & sy. & & & & & m_{77} & m_{78} \\ & & & & & & & m_{88} \end{bmatrix} \tag{9.55}$$

where

$$\mathbf{m}_{ij} = \int_{-1}^{1}\int_{-1}^{1}\int_{-1}^{1} \rho abc \mathbf{N}_i \mathbf{N}_j d\xi d\eta d\zeta$$

$$= \rho abc \int_{-1}^{1}\int_{-1}^{1}\int_{-1}^{1} \begin{bmatrix} N_i & 0 & 0 \\ 0 & N_i & 0 \\ 0 & 0 & N_i \end{bmatrix}\begin{bmatrix} N_j & 0 & 0 \\ 0 & N_j & 0 \\ 0 & 0 & N_j \end{bmatrix} d\xi d\eta d\zeta$$

$$= \rho abc \int_{-1}^{1}\int_{-1}^{1}\int_{-1}^{1} \begin{bmatrix} N_i N_j & 0 & 0 \\ 0 & N_i N_j & 0 \\ 0 & 0 & N_i N_j \end{bmatrix} d\xi d\eta d\zeta$$

$$\text{(9.56)}$$

or

$$\mathbf{m}_{ij} = \begin{bmatrix} m_{ij} & 0 & 0 \\ 0 & m_{ij} & 0 \\ 0 & 0 & m_{ij} \end{bmatrix} \tag{9.57}$$

in which,

$$m_{ij} = \rho abc \int_{-1}^{+1}\int_{-1}^{+1}\int_{-1}^{+1} N_i N_j d\xi d\eta d\zeta$$

$$= \frac{\rho abc}{64}\int_{-1}^{+1}(1+\xi_i\xi)(1+\xi_j\xi)d\xi \int_{-1}^{+1}(1+\eta_i\eta)(1+\eta_j\eta)d\eta \int_{-1}^{+1}(1+\zeta_i\zeta)(1+\zeta_j\zeta)d\zeta$$

$$= \frac{\rho abc}{8}(1+\frac{1}{3}\xi_i\xi_j)(1+\frac{1}{3}\eta_i\eta_j)(1+\frac{1}{3}\zeta_i\zeta_j)$$

$$\text{(9.58)}$$

As an example, m_{33} is calculated as follows:

$$m_{33} = \frac{\rho abc}{8}(1+\frac{1}{3}\times 1\times 1)(1+\frac{1}{3}\times 1\times 1)(1+\frac{1}{3}\times 1\times 1) = 8\times\frac{\rho abc}{27} \tag{9.59}$$

The other components of the mass matrix for a rectangular hexahedron element are thus

$$m_{11} = m_{22} = m_{33} = m_{44} = m_{55} = m_{66} = m_{77} = m_{88} = \frac{8\rho abc}{27}$$

$$m_{12} = m_{23} = m_{34} = m_{56} = m_{67} = m_{78} = m_{14} = m_{58} = m_{15} = m_{26} = m_{37} = m_{48} = \frac{4\rho abc}{27}$$

$$m_{13} = m_{24} = m_{16} = m_{25} = m_{36} = m_{47} = m_{57} = m_{68} = m_{27} = m_{38} = m_{45} = m_{18} = \frac{2\rho abc}{27}$$

$$m_{17} = m_{28} = m_{35} = m_{46} = \frac{1\rho abc}{27}$$

$$(9.60)$$

Note that the equalities in the above equation can be easily figured out by observing the relative geometric positions of the nodes in the cube element. For example the relative geometric positions of nodes 1–2 are equivalent to the relative geometric positions of nodes 2–3, and the relative geometric positions of nodes 1–7 are equivalent to the relative geometric positions of nodes 2–8. If we write the portion of the mass matrix corresponding to only one translational direction, say the x direction, we have

$$\mathbf{m}_e = \frac{\rho abc}{27} \begin{bmatrix} 8 & 4 & 2 & 4 & 4 & 2 & 1 & 2 \\ & 8 & 4 & 2 & 2 & 4 & 2 & 1 \\ & & 8 & 4 & 1 & 2 & 4 & 2 \\ & & & 8 & 2 & 1 & 2 & 4 \\ & & & & 8 & 4 & 2 & 4 \\ & & & & & 8 & 4 & 2 \\ & sy. & & & & & 8 & 4 \\ & & & & & & & 8 \end{bmatrix} \qquad (9.61)$$

The mass matrices corresponding to only the y and z directions are exactly the same as $\mathbf{m}_e$.

The nodal force vector for a rectangular hexahedron element can be obtained using Eqs. (3.78), (3.79), and (3.81). Suppose the element is loaded by a distributed force $\mathbf{f}_s$ on the edge 3–4 of the element, as shown in Figure 9.10, the nodal force vector becomes

$$\mathbf{f}_e = \int_l [\mathbf{N}]^T \Big|_{3-4} \begin{Bmatrix} f_{sx} \\ f_{sy} \\ f_{sz} \end{Bmatrix} dl \qquad (9.62)$$

If the load is uniformly distributed and f_{sx}, f_{sy}, and f_{sz} are constants, then the above equation becomes:

$$\mathbf{f}_e = \frac{1}{2}l_{3-4}\begin{Bmatrix} \{\mathbf{0}\}_{3\times 1} \\ \{\mathbf{0}\}_{3\times 1} \\ \begin{Bmatrix} f_{sx} \\ f_{sy} \\ f_{sz} \end{Bmatrix} \\ \begin{Bmatrix} f_{sx} \\ f_{sy} \\ f_{sz} \end{Bmatrix} \\ \{\mathbf{0}\}_{3\times 1} \\ \{\mathbf{0}\}_{3\times 1} \\ \{\mathbf{0}\}_{3\times 1} \\ \{\mathbf{0}\}_{3\times 1} \end{Bmatrix} \qquad (9.63)$$

where l_{3-4} is the length of the edge 3–4. Eq. (9.63) implies that the distributed forces are equally divided and applied at the two nodes. This conclusion also applies to evenly distributed surface forces applied on any face of the element, and to evenly distributed body force applied on the entire body of the element.

9.3.3 Using tetrahedrons to form hexahedrons

An alternative method of formulating hexahedron elements is to make use of tetrahedron elements. This is built upon the fact that a hexahedron can be said to be made up of numerous tetrahedrons. Figure 9.11 shows how a hexahedron can be made up of five tetrahedrons. Of course, the way a hexahedron can be made up of five tetrahedrons is not unique and, in fact, it can also be made up of six tetrahedrons as shown in Figure 9.12. Similarly, there is more than one way of dividing a hexahedron into six tetrahedrons. In this way, the element matrices for a hexahedron can be formed by assembling all the matrices for the tetrahedron elements, each of which is developed in Section 9.2. The assembly is done in a similar way as the assembly between elements.

9.4 Higher order elements

9.4.1 Tetrahedron elements

Two higher order tetrahedron elements with 10 nodes and 20 nodes are shown in Figure 9.13a and b, respectively. The 10-node tetrahedron element is a quadratic element. Compared with the linear tetrahedron element (4-nodal) developed earlier, six additional nodes are added at the middle of the edges

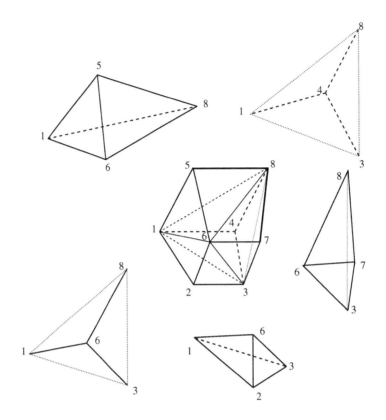

FIGURE 9.11

A hexahedron broken up into 5 tetrahedrons.

of the element. In developing the 10-nodal tetrahedron element, a complete polynomial up to second order can be used. The shape functions for this quadratic tetrahedron element in the volume coordinates are given as follows:

$$N_i = (2L_i - 1)L_i \text{ for corner nodes } i = 1, 2, 3, 4$$

$$
\left.
\begin{aligned}
N_5 &= 4L_2L_3 \\
N_6 &= 4L_1L_3 \\
N_7 &= 4L_1L_2 \\
N_8 &= 4L_1L_4 \\
N_9 &= 4L_2L_4 \\
N_{10} &= 4L_3L_4
\end{aligned}
\right\} \text{ for mid-edge nodes}
\qquad (9.64)
$$

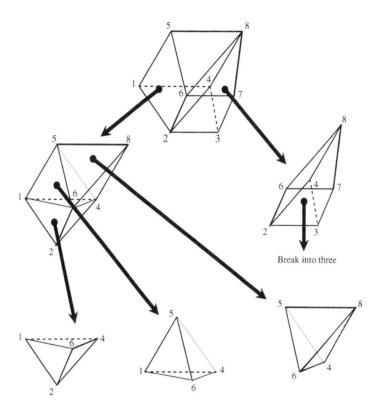

FIGURE 9.12

A hexahedron broken up into 6 tetrahedrons.

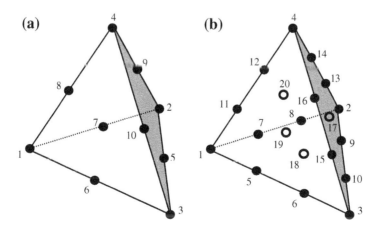

FIGURE 9.13

Higher order 3D tetrahedron elements. (a) 10-node tetrahedron element; (b) 20-node tetrahedron element.

where L_i is the volume coordinate, which is the same as the shape function for the linear tetrahedron elements given by Eq. (9.15).

The 20-node tetrahedron element is a cubic element. Compared with the linear tetrahedron element (4-nodal) developed earlier, two additional nodes are added evenly on each edge of the element, and 4 central-face nodes are added at the geometry center of each triangular surface of the element. In developing the 20-nodal tetrahedron element, a complete polynomial up to third order can be used. The shape functions for this cubic tetrahedron element in the volume coordinates are given as follows:

$$N_i = \frac{1}{2}(3L_i - 1)(3L_i - 2)L_i \text{ for corner nodes } i = 1, 2, 3, 4$$

$$\left.\begin{aligned}
N_5 &= \tfrac{9}{2}(3L_1 - 1)L_1L_3 \quad N_{11} = \tfrac{9}{2}(3L_1 - 1)L_1L_4 \\
N_6 &= \tfrac{9}{2}(3L_3 - 1)L_1L_3 \quad N_{12} = \tfrac{9}{2}(3L_4 - 1)L_1L_4 \\
N_7 &= \tfrac{9}{2}(3L_1 - 1)L_1L_2 \quad N_{13} = \tfrac{9}{2}(3L_2 - 1)L_2L_4 \\
N_8 &= \tfrac{9}{2}(3L_2 - 1)L_1L_2 \quad N_{14} = \tfrac{9}{2}(3L_4 - 1)L_2L_4 \\
N_9 &= \tfrac{9}{2}(3L_2 - 1)L_2L_3 \quad N_{15} = \tfrac{9}{2}(3L_3 - 1)L_3L_4 \\
N_{10} &= \tfrac{9}{2}(3L_3 - 1)L_2L_3 \quad N_{16} = \tfrac{9}{2}(3L_4 - 1)L_3L_4
\end{aligned}\right\} \text{ for edge nodes}$$

$$\left.\begin{aligned}
N_{17} &= 27L_2L_3L_4 \\
N_{18} &= 27L_1L_2L_3 \\
N_{19} &= 27L_1L_3L_4 \\
N_{20} &= 27L_1L_2L_4
\end{aligned}\right\} \text{ for center surface nodes}$$

(9.65)

where L_i is the volume coordinate, which is the same as the shape function for the linear tetrahedron elements given by Eq. (9.15).

9.4.2 Brick elements

9.4.2.1 Lagrange type elements

The Lagrange type brick elements can be developed in precisely the same manner as the 2D rectangular elements described in Chapter 7. A brick element with $n_d = (n+1)(m+1)(p+1)$ nodes is shown in Figure 9.14. The element is defined in the domain of $(-1 \leq \xi \leq 1, -1 \leq \eta \leq 1, -1 \leq \zeta \leq 1)$ in the natural coordinates ξ, η, and ζ. Due to the regularity of the nodal distribution along ξ, η, and ζ directions, the shape function of the element can be simply obtained by multiplying one-dimensional shape functions with respect to ξ, η, and ζ directions using the Lagrange interpolants defined in Eq. (4.73) (Zienkiewicz et al., 2000).

$$N_i = N_I^{1D}N_J^{1D}N_K^{1D} = l_I^n(\xi)l_J^m(\eta)l_K^p(\varsigma)$$

(9.66)

Due to the delta function proper of the 1D shape functions given in Eq. (4.74), it is easy to confirm that N_i given by Eq. (9.66) also possesses the delta function property.

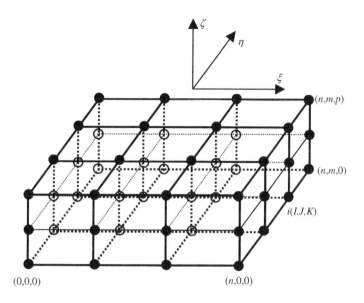

FIGURE 9.14

Rectangular element of arbitrary high orders.

9.4.2.2 *Serendipity type elements*

The method used in constructing the Lagrange type of elements is very systematic. However, the Lagrange type of elements is not very widely used, due the presence of the interior nodes. Serendipity types of brick elements without interior nodes are created by inspective construction methods as described in Chapter 7 for 2D rectangular elements.

Figure 9.15a shows a 20-nodal tri-quadratic element. The element has eight corner nodes and twelve mid-side nodes. The shape functions in terms of natural coordinates for the quadratic rectangular element are given as follows:

$$N_j = \frac{1}{8}(1 + \xi_j\xi)(1 + \eta_j\eta)(1 + \varsigma_j\varsigma)(\xi_j\xi + \eta_j\eta + \varsigma_i\varsigma - 2)$$

$$\text{for corner nodes } j = 1, \cdots, 8$$

$$N_j = \frac{1}{4}(1 - \xi^2)(1 + \eta_j\eta)(1 + \varsigma_j\varsigma) \quad \text{for mid-side nodes } j = 10, 12, 14, 16$$

$$N_j = \frac{1}{4}(1 - \eta^2)(1 + \xi_j\xi)(1 + \varsigma_j\varsigma) \quad \text{for mid-side nodes } j = 9, 11, 13, 15$$

$$N_j = \frac{1}{4}(1 - \varsigma^2)(1 + \xi_j\xi)(1 + \eta_j\eta) \quad \text{for mid-side nodes } j = 17, 18, 19, 20 \quad (9.67)$$

where (ξ_j, η_j) are the natural coordinates of node j. It is very easy to observe that the shape functions possess the delta function property. The shape function is constructed by simple inspections,

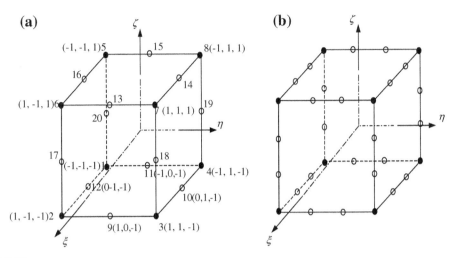

(a)

(b)

FIGURE 9.15

High order 3D serendipity elements. (a) 20-node quadratic element; (b) 32-node cubic element.

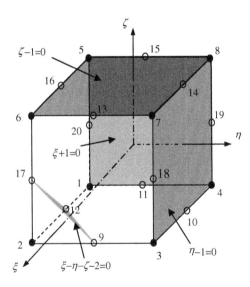

FIGURE 9.16

Construction of a 20-node serendipity element. Four flat planes passing through the remote nodes of node 9 are used.

making use of the shape function properties. For example, for the corner node 2 (where $\xi_2 = 1$, $\eta_2 = -1$, $\zeta_2 = -1$), the shape function N_2 has to pass the following four planes as shown in Figure 9.16 to ensure it vanishes (i.e. $N_2 = 0$) at remote nodes.

$$1 + \xi = 0 \qquad\qquad \Rightarrow \text{vanishes at nodes } 1,4,5,8,11,15,19,20$$
$$\eta - 1 = 0 \qquad\qquad \Rightarrow \text{vanishes at nodes } 3,4,7,8,10,14,18,19$$
$$\varsigma - 1 = 0 \qquad\qquad \Rightarrow \text{vanishes at nodes } 5,6,7,8,13,14,15,16 \qquad (9.68)$$
$$\xi - \eta - \varsigma - 2 = 0 \quad \Rightarrow \text{vanishes at nodes } 9,12,17$$

The shape N_2 can immediately be written as

$$N_2 = C(1 + \xi)(1 - \eta)(1 - \varsigma)(\xi - \eta - \varsigma - 2) \qquad (9.69)$$

where C is a constant to be determined using the condition that it has to be unity at node 2 ($\xi_2 = 1$, $\eta_2 = -1$, $\varsigma_2 = -1$), which gives

$$C = \frac{1}{(1+1)(1-(-1))(1-(-1))(1-(-1)-(-1)-2)} = \frac{1}{8} \qquad (9.70)$$

We finally have

$$N_2 = \frac{1}{8}(1 + \xi_2\xi)(1 + \eta_2\eta)(1 + \varsigma_2\varsigma)(\xi_2\xi + \eta_2\eta + \varsigma_2\varsigma - 2) \qquad (9.71)$$

which is the first equation in Eq. (9.67) for $j = 2$.

Shape functions at all the other corner nodes can be constructed in the same manner. As for the mid-side nodes, say node 9, we enforce the shape function passing through the following three planes as shown in Figure 9.17.

$$1 + \xi = 0 \qquad \Rightarrow \text{vanishes at nodes } 1, 4, 5, 8, 11, 15, 19, 20$$
$$\eta - 1 = 0 \qquad \Rightarrow \text{vanishes at nodes } 3, 4, 7, 8, 10, 14, 18, 19$$
$$\varsigma - 1 = 0 \qquad \Rightarrow \text{vanishes at nodes } 5, 6, 7, 8, 13, 14, 15, 16 \qquad (9.72)$$
$$\eta + 1 = 0 \qquad \Rightarrow \text{vanishes at nodes } 1, 2, 5, 6, 12, 13, 16, 17$$

The shape function, N_9, can then immediately be written as

$$N_9 = C(1 - \eta^2)(1 + \xi)(1 - \varsigma) \qquad (9.73)$$

where C is a constant to be determined using the condition that it has to be unity at node 9 ($\xi_9 = ,1$, $\eta_9 = 0$, $\varsigma_9 = -1$), which gives

$$C = \frac{1}{(1 - \eta^2)(1 + \xi)(1 - \varsigma)} = \frac{1}{(1 - 0^2)(1 + 1)(1 - (-1))} = \frac{1}{4} \qquad (9.74)$$

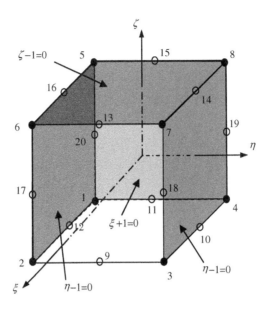

FIGURE 9.17

Construction of a 20-node serendipity element. Four flat planes passing through the remote nodes of node 9 are used.

We then have

$$N_9 = \frac{1}{4}(1 - \eta^2)(1 + \xi_9\xi)(1 + \varsigma_9\varsigma)$$

(9.75)

which is the third equation in Eq. (9.67) for $j = 9$.

Because the delta function property is used for the construction of shape functions given in Eq. (9.67), they naturally possess the delta function property. It can be easily seen that all the shape functions can be formed using the following common set of basis functions.

$$1, \xi, \eta, \varsigma\xi\eta, \eta\varsigma, \xi\varsigma, \xi^2, \eta^2, \varsigma^2,$$

(9.76)

$$\xi\eta\varsigma, \xi\eta^2, \xi\varsigma^2, \eta\xi^2, \eta\varsigma^2, \varsigma\xi^2, \varsigma\eta^2, \xi^2\eta\varsigma, \eta^2\xi\varsigma, \xi\eta\varsigma^2$$

that are linearly-independent and contains all the linear terms. From Lemma 2 and 3, we confirm that the shape functions are partitions of unity, and at least linear field reproduction. Hence, they satisfy the sufficient requirements for FEM shape functions.

Following a similar procedure, the shape functions for the 32-node tri-cubic element shown in Figure 9.15b can be written as

$$N_j = \frac{1}{64}(1 + \xi_j\xi)(1 + \eta_j\eta)(1 + \varsigma_j\varsigma)(9\xi^2 + 9\eta^2 + 9\varsigma^2 - 19)$$

for corner nodes $j = 1, \cdots, 8$

$$N_j = \frac{9}{64}(1 - \xi^2)(1 + 9\xi_j\xi)(1 + \eta_j\eta)(1 + \varsigma_j\varsigma)$$

for side nodes with $\xi_j = \pm\frac{1}{3}, \eta_j = \pm 1$ and $\varsigma_j = \pm 1$

$$N_j = \frac{9}{64}(1 - \eta^2)(1 + 9\eta_j\eta)(1 + \xi_j\xi)(1 + \varsigma_j\varsigma) \qquad (9.77)$$

for side nodes with $\eta_j = \pm\frac{1}{3}, \xi_j = \pm 1$ and $\varsigma_j = \pm 1$

$$N_j = \frac{9}{64}(1 - \varsigma^2)(1 + 9\varsigma_j\varsigma)(1 + \xi_j\xi)(1 + \eta_j\eta)$$

for side nodes with $\varsigma_j = \pm\frac{1}{3}, \xi_j = \pm 1$ and $\eta_j = \pm 1$

The reader is encouraged to identify the planes that are used to form these shape functions listed in Eq. (9.77). When $\zeta = \zeta_I = 1$, the above equations reduce to the two dimensional serendipity quadratic and cubic element defined by Eqs. (7.113) and (7.123) respectively.

9.5 Elements with curved surfaces

Using high order elements, elements with curved surfaces can be used in the modeling. Two frequently used higher order elements of curved edges are shown in Figure 9.18a. In formulating these types of elements, the same mapping technique used for the linear quadrilateral elements (Section 9.3) can be used. In the physical coordinate system, elements with curved edges are first formed in the problem domain as shown in Figure 9.18a. These elements are then mapped to the natural coordinate system using Eq. (9.34). The elements mapped in the natural coordinate system will have straight edges as shown in Figure 9.18b.

Higher order elements of curved surfaces are often used for modeling curved boundaries. Note that elements with excessively curved edges may cause problems in the numerical integration. Therefore, more elements should be used where the curvature of the boundary is large. In addition, it is recommended that in the internal portion of the domain, elements with straight edges should be used whenever possible.

9.6 Case study: Stress and strain analysis of a quantum dot heterostructure

Quantum dots are clusters of atoms nanometers in size, usually made from semiconducting materials like silicon, cadmium selenide or gallium arsenide. What makes quantum dots interesting is that they have unusual electrical and optical properties, hence they have the potential for use in a wide variety of

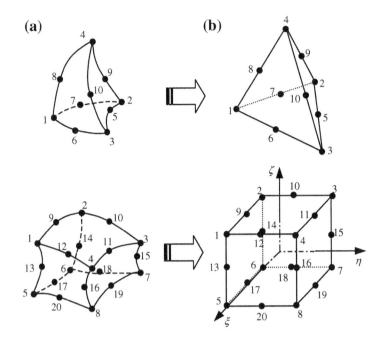

FIGURE 9.18

3D solid elements with curved surfaces. (a) Elements with curved surfaces in the physical coordinate system; (b) Brick elements obtained by mapping.

novel electronic devices, including light emitting diodes, photovoltaic cells, and quantum semiconductor lasers. An interesting way of fabricating such quantum dot structures is to actually grow the dots directly by depositing a thin film layer of material on a substrate under appropriate growth conditions. Usually, the thin film layer is of a different material with the substrate material and, thus, such structures are also known generally as heterostructures. This growth mode is due to the mismatch in the lattice parameters of the different materials and is known as the Stranski-Krastanow (SK) growth mode.

The growth of the quantum dot structures is partly driven by the strain energy in the thin film, which makes studying the stress distribution in the film a crucial part of understanding the growth mechanism. In this case study, an example of modeling a 3D finite element model to analyze the stress distribution in and around such structures will be shown. The stress distribution also affects the electrical and optical properties of the quantum dot structure.

Figure 9.19 shows a schematic representation of a quantum dot grown on top of the substrate and embedded in a cap layer. This is just a single quantum dot and can probably be considered as a single basic unit of the heterostructure. In reality, there could be many of such quantum dots distributed on top of a layer of substrate. It can also be seen from Figure 9.19 that the quantum dot is usually pyramidal or trapezoidal in shape. The pyramidal shape of the quantum dot is often approximated by many analysts by using a 2D axisymmetric model of a cone. However, it should be noted that using a 2D axisymmetric model is not fully representative of the pyramidal shape. For the purposes of this chapter, this case study will use the 3D solid element to model the structure.

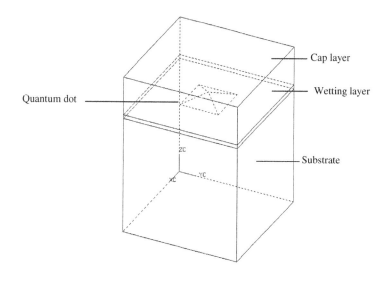

FIGURE 9.19

Schematic representation of a quantum dot heterostructure.

9.6.1 **Modeling**
Meshing

As mentioned, the modeling of any 3D structure is generally more complex and tedious. In this case, 8-nodal, hexahedron elements are being used for the meshing of the 3D geometry. It can be seen from Figure 9.19 that the problem domain is very much symmetrical. Therefore, in order to work on a more manageable problem, a quarter of the model is being modeled using mirror symmetry. Note that it is also possible to use a one-eighth model, which then requires the use of multiple point constraints (MPC) equations (see Chapter 11).

Proper meshing in this case is very important as it has been found that a poor mesh usually yields bad results. The 3D mesh of the heterostructure is shown in Figures 9.20 and 9.21. The model is generally divided into two main parts geometrically for the analysts to distinguish them more conveniently. The parts of the heterostructure comprising the substrate and the cap layer, as shown in Figure 9.19, are grouped together as the matrix; and the parts of the heterostructure comprising the wetting layer and the quantum dot itself are grouped together as the island. Figure 9.22 also shows the plan view of the mesh of the island (or matrix). It can be seen from this figure how smaller elements are concentrated at and around the pyramidal quantum dot. To generate the mesh here, the analyst has employed the aid of automatic mesh generators that can still mesh the relatively complex shape of the pyramid with hexahedron elements. Some mesh generators may not be able to achieve this and one may end up with either tetrahedron elements or a mixture of both hexahedron and tetrahedron elements.

Material properties

In this case study, the heterostructure system of indium arsenide (InAs) quantum dots embedded in gallium arsenide (GaAs) substrate and cap layer is being analyzed. Therefore, the matrix part of the

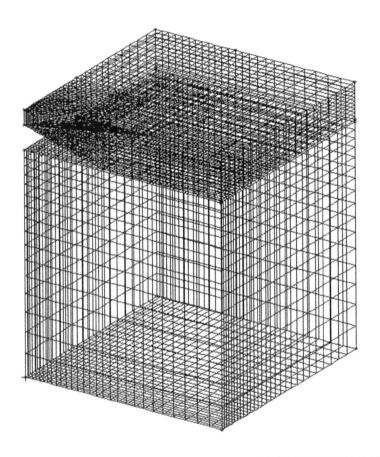

FIGURE 9.20

3D mesh of the matrix.

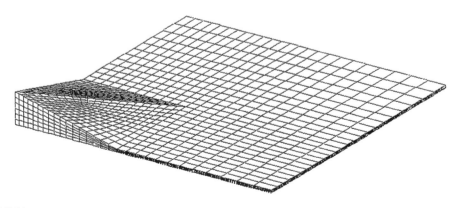

FIGURE 9.21

3D mesh of the island.

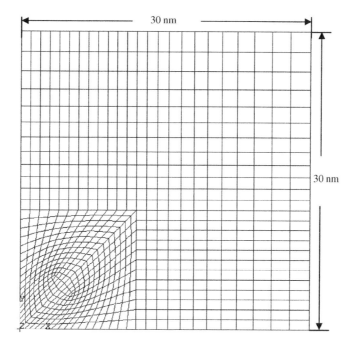

FIGURE 9.22

Plan view of finite element mesh of quantum dot heterostructure.

model will be of the material GaAs and the island part of the model will be of the material InAs. This is an example of the convenience of dividing the model into these two parts. It is assumed here that the materials have isotropic properties and they are listed in Table 9.1.

Constraints and boundary conditions

As the model is a symmetric quarter model, symmetrical boundary conditions must be applied. Here it means that the nodes on the planes corresponding to $x = 0$ nm, $y = 0$ mm, $x = 30$ mm, and $y - 30$ nm have their displacement components normal to their respective planes constrained. Another displacement boundary condition would be the base of the matrix, whereby all the displacement components of the nodes are constrained.

There is also a contact constraint condition imposed between the outer surfaces of the island and the surfaces of the cap layer and the substrate in the matrix. Contact modeling is a relatively advanced

Table 9.1	Material properties of GaAs and InAs.	
Material	**E (GPa)**	**v**
GaAs	86.96	0.31
InAs	51.42	0.35

technique and will not be covered in this book. Basically contact modeling is used to model the sliding or the movement between two surfaces. ABAQUS offers a "tied" contact condition whereby the two surfaces in contact are actually tied to one another. This "tied" contact condition is used here to model the bonding between the island (InAs) and the matrix (GaAs).

There is actually no load acting on this model. Rather, thermal expansion is being made use of to simulate the strain induced due to the lattice mismatch between GaAs and InAs. The strain induced due to the lattice mismatch can be calculated from the lattice parameters to be -0.067. To represent this lattice mismatch, a corresponding thermal expansion coefficient of $\alpha_T = 0.067$ is applied to the elements in the island and the temperature is raised by 1 K. This would effectively result in an expansion of the island and because it is constrained by the matrix, thermal strain corresponding to the lattice mismatch strain is induced. Note that this thermal expansion does not take place in the physical case, but is just used to produce the mismatch strain. This thermal strain actually contributes to the force vector in the finite element equations.

9.6.2 ABAQUS input file

Parts of the ABAQUS input file for the problem defined above are shown below. As the problem is a rather large one, the full input file would consist of a large amount of data defining the nodes, elements, and so on. As such, the full data will not be included here and some parts of the input file that have been explained in previous case studies will not be explained again here.

```
*HEADING, SPARSE
Calculation of stress distribution in quantum dot structure
**
*NODE
**
**Elements are divided into two main parts: ISLAND and MATRIX
**Elements used are 8-nodal, hexahedral elements (C3D8)
**
*ELEMENT, TYPE=C3D8, ELSET=ISLAND
*ELEMENT, TYPE=C3D8, ELSET=MATRIX
**
**
*NSET, NSET=ISLAND
**
*NSET, NSET=BASE
**
*NSET, NSET=FIXED_X
**
*NSET, NSET=FIXED_Y
**
*SOLID SECTION, ELSET=ISLAND, MATERIAL=INAS
1.,
```

Nodal cards
Node I.D., x-coordinate, y-coordinate, z-coordinate

Element (connectivity) cards
Element I.D., node 1, node 2, node 3, ... , node 8

```
**
**
*SOLID SECTION, ELSET=MATRIX, MATERIAL=GAAS
1.,
**
** GaAs
**
*MATERIAL, NAME=GAAS
**
*ELASTIC, TYPE=ISO
 86.96, 0.31
**
** InAs
**
*MATERIAL, NAME=INAS
**
*ELASTIC, TYPE=ISO
 51.42, 0.35
**
*EXPANSION, TYPE=ISO
 0.067,
**
** Displacement boundaries
**
*BOUNDARY, OP=NEW
BASE, ENCASTRE
FIXED_X, XSYMM
FIXED_Y, YSYMM
**
** contact1
**
*SURFACE DEFINITION, NAME=M20

*SURFACE DEFINITION, NAME=S20

**
*CONTACT PAIR, INTERACTION=I20,
 ADJUST=0.0001, TIED S20, M20
*SURFACE INTERACTION, NAME=I20
**
** contact2
**
*SURFACE DEFINITION, NAME=M21
*SURFACE DEFINITION, NAME=S21
**
```

Node set
Nodes in ISLAND are grouped in a node set named "ISLAND."

Node set
Nodes on base surface grouped in a node set named "BASE."

Node set
Nodes to be constrained in *x* direction grouped in node set, "FIXED_X."

Node set
Nodes to be constrained in *y* direction grouped in node set, "FIXED_Y."

Property cards
Define properties to the elements of sets "ISLAND" and "MATRIX." It will have the material properties defined under "INAS" and "GAAS" respectively.

Material cards
Define material properties under the name "GAAS" and "INAS." Elastic properties are defined. TYPE=ISO represents isotropic properties. Note that for "INAS." the thermal expansion coefficient is defined under *EXPANSION.

BC cards
The nodes grouped under BASE, FIXED_X and FIXED_Y are given the corresponding constraints. ENCASTRE represents a fully clamped in boundary; XSYMM represents conditions symmetrical to a plane *x* = constant; YSYMM represents conditions symmetrical to a plane *y* = constant.

```
*CONTACT PAIR, INTERACTION=I21, ADJUST=0.0001,
  TIED S21, M21
*SURFACE INTERACTION, NAME=I21
**
**
*INITIAL CONDITIONS, TYPE=TEMPERATURE
ISLAND, 300.
**
**
*STEP, AMPLITUDE=RAMP
Linear Static Analysis
**
**
*STATIC
**
**
**
*TEMPERATURE, OP=NEW
ISLAND, 301.
**
**
*NODE PRINT, FREQ=1
U,
*NODE FILE, FREQ=1
U,
**
*EL PRINT, POS=INTEG, FREQ=1
S,
E,
*EL FILE, POS=INTEG, FREQ=1
S,
E,
**
*END STEP
```

Defines contact surfaces

Defines conatact surfaces

Contact cards
Contact conditions are being defined here. The surfaces to be in contact with are defined in *SURFACE DEFINITION. The contact conditions are then specified in *CONTACT PAIR and *SURFACE INTERACTION. Details will not be shown here since it is beyond the scope of this book.

IC cards
Initial temperature conditions are being defined and nodes in ISLAND are set at a temperature of 300K.

Control cards
Indicate the STATIC analysis procedure.

Load cards
The load here is temperature and the nodes in the ISLAND are given a temperature of 301K, which implies a raise of 1K from the initial conditions.

Output control cards
Define the output requested. In this case, the displacements (U), the stresses (S), and the strains (E).

The input file would provide the information ABAQUS would need to perform tasks like forming the stiffness matrix and the force vector (no mass matrix since this is a static analysis). The full input file may consist of many pages, which is common for large problems.

9.6.3 Solution process

The information provided in the input file is very similar to previous case studies. The nodal and element connectivity information is read for the formulation of the element matrices. The element type used here is C3D8, which represents a 3D, hexahedral element with 8 nodes. More 3D element types are also available in the ABAQUS element library. The material properties provided in the input file will be used to formulate the element stiffness matrix (Eq. (9.51)) as well. Recall that the integration in the stiffness matrix is usually carried out using the Gauss integration scheme sampled in three directions and in ABAQUS, the default number of integration points per face of the hexahedral element is 4, making the total number of integration points per element 24. All the element matrices will be assembled together using the connectivity information provided. The application of the boundary conditions and the thermal strain induced by the thermal expansion is carried out by the specifications in the boundary cards and the load card. Finally, the finite element equation will be solved using the algorithm for static analysis as discussed in Chapter 6.

9.6.4 Results and discussion

Running the problem in ABAQUS, we are able to get the stress distribution and the strain distribution as requested in the input file. Figures 9.23 and 9.24 shows the stress distribution obtained in the matrix and island, respectively, of a particular plane of $\theta = 45°$, where θ is measured from the x–z plane counter-clockwise. Despite the extra effort in the meshing of a 3D model, the advantage of it is that it enables the analyst to view, in this case, the stress distribution in any arbitrary plane in the entire model. This would be difficult to achieve if a 2D, axisymmetric approximation was carried out instead. From the stress distribution, one would observe that there are compressive stresses in the island and tensile stresses in the matrix area above the island. In the island, there is also stress relaxation in the quantum dot with the maximum stress relaxation at the tip of the pyramidal quantum dot. This actually verifies the thermodynamics aspect of quantum dot formation, since the formation of a quantum dot results in a

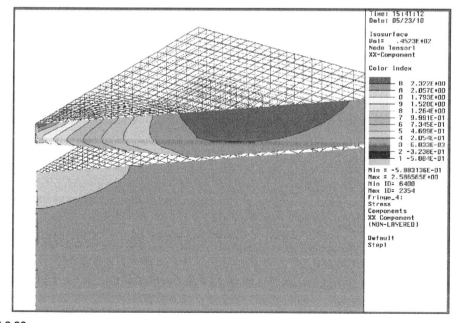

FIGURE 9.23

Stress σ_{xx} distribution of plane $\theta = 45°$ in matrix.

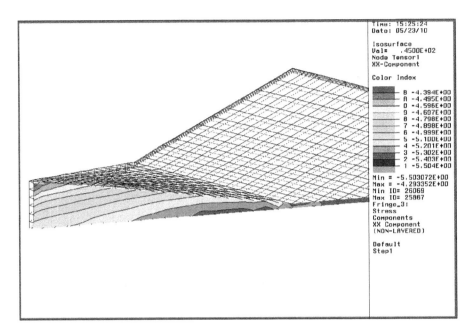

FIGURE 9.24

Stress σ_{xx} distribution of plane $\theta=45°$ in island.

lower energy level (lower elastic strain energy). The tensile stress in the matrix area above the quantum dot is also important as this stress actually causes subsequent quantum dots to be formed directly above the buried quantum dot when a subsequent InAs layer is deposited.

9.7 Review questions

1. Can 3D solid elements be used for solving 2D plane stress and plane strain problems? Give a justification for your answer.
2. What is the difference between using tetrahedron elements and hexahedron elements derived using assemblage of tetrahedron elements? Can they give the same results for the same problem? Give a justification for your answer.
3. Give shape functions N_1, N_2, N_3, and N_4 for the element given in Figure 9.25.
4. Evaluate the strain matrix **B** for the tetrahedral solid element shown in Figure 9.25.
5. For the element shown in Figure 9.25, assume the nodal displacements have been found as (in cm):

$$
\begin{array}{lll}
u_1 = 0.005 & v_1 = 0.0 & w_1 = 0.0 \\
u_2 = 0.001 & v_2 = 0.0 & w_2 = 0.001 \\
u_3 = 0.005 & v_3 = 0.0 & w_3 = 0.0 \\
u_4 = -0.001 & v_4 = 0.0 & w_4 = 0.005
\end{array}
$$

Determine the strains and then the stresses in the element. Let $E=210$ GPa and $v=0.3$.

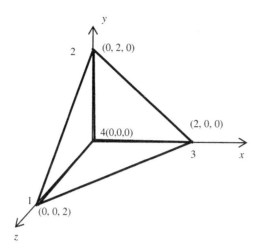

FIGURE 9.25

A tetrahedral element for 3D solids.

6. Can one develop pentahedron elements? How?
7. How many Gauss points should be used for evaluating mass and stiffness matrices for 4-node tetrahedron elements? Give a justification for your answer.
8. How many Gauss points should be used for evaluating mass and stiffness matrices for 8-node hexahedron elements? Give a justification for your answer.
9. If a higher order shape function is used, do Eqs. (9.30) and (9.63) still hold? Give a justification for your answer.
10. Consider a uniform, isotropic square "plate" with an in-plane dimension of 1 m by 1 m. The plate is clamped on all edges (CCCC), made of a material with Young's modulus $E = 70$ GPa and Poisson ratio $\upsilon = 0.3$. Using 3D block element with a density of 16×16 in the plate-plane to perform the analysis; compare the 3D FEM solution with the FEM solution using plate elements with a density of 16×16, and the analytical solution (based on thin plate theory) in terms of the deflection at the center of the plate. You are requested to perform the following studies.
    a. The plate is of thickness 0.01 m;
    b. The plate is of thickness 0.1 m;
    c. The plate is of thickness 0.2 m;
    d. The plate is of thickness 0.5 m.

Special Purpose Elements

CHAPTER OUTLINE HEAD

10.1 Introduction

This chapter introduces some special purpose elements and methods that are designed for specific circumstances. They are used to either simplify meshing and calculation or to obtain better accuracy, which the usual elements cannot obtain. These include crack tip elements, infinite elements, finite strip elements, and strip elements. In addition, a brief introduction is provided on the recent meshfree and the smoothed finite element methods (S-FEMs). The characteristics of these elements and recent advanced models are summarized in Table 10.1.

10.2 Crack tip elements

In fracture mechanics, the region around the tip of the crack is usually of much interest to analysts since the tip is a singularity point where the stress level becomes mathematically infinite. When modeled with the conventional, polynomial-based finite elements discussed in previous chapters, the finite element approximations are usually not accurate around the crack tip. Refinement is a simple way to improve the solution *near* the crack tip, but this may not prove feasible at times and is highly inefficient when computational resources are limited. Even if the elements are refined around the crack tip, it is

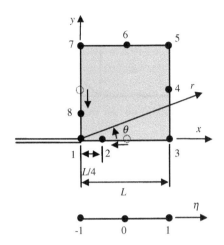

FIGURE 10.3

Eight-node, isoparametric, quadratic crack tip element.

by a quarter of the edge-length towards the crack tip. The following explains how the stress singularity is created by this simple modification.

Consider the element edge joining nodes 1, 2, and 3 of the isoparametric quadratic element shown in Figure 10.3, with node 1 at the crack tip. Following the formulation of the conventional isoparametric 8-nodal element, the coordinate x and the displacement u are both interpolated by shape functions as follows:

$$x = -0.5\eta(1 - \eta)x_1 + (1 + \eta)(1 - \eta)x_2 + 0.5\eta(1 + \eta)x_3 \tag{10.3}$$

$$u = -0.5\eta(1 - \eta)u_1 + (1 + \eta)(1 - \eta)u_2 + 0.5\eta(1 + \eta)u_3 \tag{10.4}$$

Let both x and u be measured from node 1, and let the mid-edge node 2 be moved to the quarter-point node 2. For an edge of length L, we have

$$x_1 = 0, x_2 = L/4, x_3 = L, u_1 = 0 \tag{10.5}$$

Substitution of Eq. (10.5) into Eqs. (10.3) and (10.4) leads to

$$x = 0.25(1 + \eta)(1 - \eta)L + 0.5\eta(1 + \eta)L \tag{10.6}$$

$$u = (1 + \eta)(1 - \eta)u_2 + 0.5\eta(1 + \eta)u_3 \tag{10.7}$$

Simplifying the above equations will give us

$$x = 0.25(1 + \eta)^2 L \tag{10.8}$$

$$u = (1 + \eta)[(1 - \eta)u_2 + 0.5\eta u_3] \tag{10.9}$$

Now, we know that along the x axis, $x = r$. Therefore,

$$r = 0.25(1 + \eta)^2 L \text{ or } (1 + \eta) = 2\sqrt{\frac{r}{L}} \tag{10.10}$$

Substitution of Eq. (10.10) into Eq. (10.9) leads to

$$u = 2(\sqrt{r}/\sqrt{L})[(1 - \eta)u_2 + 0.5\eta u_3] \tag{10.11}$$

Notice that by shifting the middle node to the quarter position, the displacement now follows a behavior that is proportional to $\sqrt{r}$. Furthermore, the strain relates directly to derivatives of the displacement:

$$\frac{\partial u}{\partial x} = \frac{\partial u}{\partial \eta}\frac{\partial \eta}{\partial x} \tag{10.12}$$

where from Eqs. (10.8) and (10.10),

$$\frac{\partial x}{\partial \eta} = 0.5(1 + \eta)L = \sqrt{r}\sqrt{L} \tag{10.13}$$

Thus, by using Eqs. (10.9), (10.12), and (10.13),

$$\frac{\partial u}{\partial x} = \frac{1}{\sqrt{r}}\frac{1}{\sqrt{L}}[-2\eta u_2 + (\frac{1}{2} + \eta)u_3] \tag{10.14}$$

It is noted that the strain will have a behavior given by Eq. (10.14): proportional to $1/\sqrt{r}$. Since the stress is proportional to the strain, this is also true for the stresses. Therefore it can be seen that by shifting the middle node 2 to the quarter position, we are able to obtain an approximation that follows the behavior of the stresses and displacements near the crack tip as predicted by linear fracture mechanics. Similar procedures can be applied to the other side consisting of nodes 1, 7, and 8.

By the same argument, other types of crack tip elements with different shapes can also be obtained and some examples are shown in Figure 10.4.

10.3 Methods for infinite domains

There are many problems in real life that actually involve an infinite or semi-infinite domain, for example, the radiation of heat from a point source into space, the propagation of waves on the surface of the ground and under the ocean, and so on. For the above problems, the strength of the heat radiation and the amplitude of the waves vanish at infinity. So far, the finite element methods we have discussed in this book all include finite enclosed boundaries. If these conventional elements are used to model an infinite domain, the boundary will affect the results obtained. For the propagation of waves, any finite

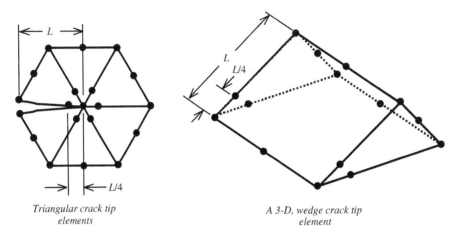

Triangular crack tip
elements

A 3-D, wedge crack tip
element

FIGURE 10.4

Other examples of crack tip elements.

boundary will reflect the waves and this will result in the superposition of the transmitted and reflected waves. Similarly for other problems, the approximations using the conventional finite elements will thus be inaccurate. Intuitively, one might think that one solution to modeling the infinite or semi-infinite domain is to place the finite boundary far away from the area of interest. The question of "how far" will be far enough will then arise and, besides, this method would usually require excessively large numbers of elements to model regions that the analyst has little interest in. To overcome such difficulties caused by infinite or semi-infinite domains, many methods have been proposed, one of the most effective and efficient of which is the use of *infinite elements*.

10.3.1 Infinite elements formulated by mapping

An infinite element is created by using shape functions to approximate a sequence of decaying terms

$$\frac{C_1}{r} + \frac{C_2}{r^2} + \frac{C_3}{r^3} + \cdots \tag{10.15}$$

where C_i are arbitrary constants and r is the radial distance from the origin or *pole*, which is usually the "central" point of the problem with the infinite domain. Consider the 1D mapping of the line OPQ, which coincides with the x axis, as shown in Figure 10.5. Like the finite element formulation for isoparametric elements, the coordinates are interpolated from the nodal coordinates and thus let us propose that

$$x = -\frac{\xi}{1-\xi}x_O + \left(1 + \frac{\xi}{1-\xi}\right)x_Q \tag{10.16}$$

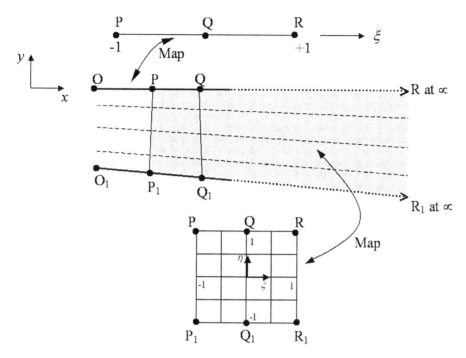

FIGURE 10.5

Infinite line and 2D element mapping.

From Eq. (10.16), it is observed that $\xi=0$ corresponds to $x=x_Q$, $\xi=1$ corresponds to $x=\alpha$, and $\xi=-1$ corresponds to $x = \frac{x_Q-x_O}{2} = x_P$. As mentioned, r is the distance measured from the origin or pole and we assume that the pole is at O. Therefore,

$$r = x - x_O \tag{10.17}$$

Solving Eq. (10.16) for ξ would give

$$\xi = 1 - \frac{x_Q - x_O}{x - x_O} = 1 - \frac{x_Q - x_O}{r} \tag{10.18}$$

If the unknown variable, say the displacement u, is approximated by a polynomial function such as

$$u = \alpha_0 + \alpha_1 \xi + \alpha_2 \xi^2 + \alpha_3 \xi^3 + \cdots \tag{10.19}$$

Then, substituting Eq. (10.18) into (10.19) would give us a series of decaying terms in the form of Eq. (10.15) with the linear shape function in ξ corresponding to $1/r$ terms, quadratic to $1/r^2$, and so on.

A generalization to 2D or 3D can be achieved by simple products of the 1D, infinite, mapping shown above with a standard type of shape function in η (and ζ for 3D) direction in the manner shown in Figure 10.5. First, we generalize the interpolation of Eq. (10.16) for any straight line in the x, y, and z space:

$$x = -\frac{\xi}{1-\xi}x_{O_1} + \left(1 + \frac{\xi}{1-\xi}\right)x_{Q_1}$$

$$y = -\frac{\xi}{1-\xi}y_{O_1} + \left(1 + \frac{\xi}{1-\xi}\right)y_{Q_1} \qquad (10.20)$$

$$z = -\frac{\xi}{1-\xi}z_{O_1} + \left(1 + \frac{\xi}{1-\xi}\right)z_{Q_1}$$

Then we complete the interpolation and map the whole domain of $\xi\eta(\zeta)$ by adding a standard interpolation in the $\eta(\zeta)$ directions. Thus, for element $PP_1QQ_1RR_1$ in Figure 10.5, we can write

$$x = N_1(\eta)\left[-\frac{\xi}{1-\xi}x_O + \left(1 + \frac{\xi}{1-\xi}\right)x_Q\right] + N_0(\eta)\left[-\frac{\xi}{1-\xi}x_{O_1} + \left(1 + \frac{\xi}{1-\xi}\right)x_{Q_1}\right] \quad (10.21)$$

with

$$N_1(\eta) = \frac{1+\eta}{2}, \quad N_0(\eta) = \frac{1-\eta}{2} \qquad (10.22)$$

and map the points as shown. In a similar manner, quadratic interpolations could be used as well. These infinite elements can be joined to a standard finite element mesh as shown in Figure 10.6.

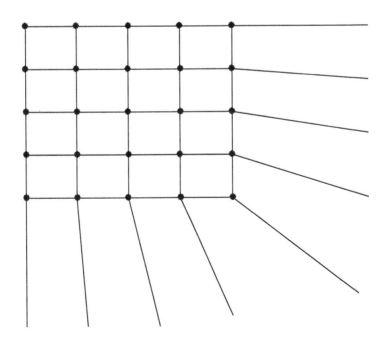

FIGURE 10.6

Infinite elements attached to the boundary of a standard finite element mesh.

10.3.2 **Gradual damping elements**

Using elements with gradually increased artificial damping elements attached on the regular finite element mesh is a very efficient way to model vibration and wave propagation problems with infinite boundaries. This method was proposed by Liu (1994) and Liu and Quek (2003) for modeling such problems using discrete numerical methods. One of the applications of this technique is in the study of lamb wave propagation in infinite plates or beams. Lamb waves are dispersive waves that involve multiple characteristic reflections with the top and bottom surface of the plate as it progresses along the plate as shown schematically in Figure 10.7. This method uses the conventional finite elements and the infinite domain is approximated by adding additional elements with a gradual increase in damping to damp down the amplitude of the propagating waves, as shown in Figure 10.8. A detailed analysis of the method was given by Liu (1994) and Liu and Quek (2003).

10.3.3 **Coupling of FEM and the boundary element method**

Another effective method of dealing with infinite domains is to use the FEM coupled with the boundary element method (BEM). The FEM is used in the interior portions of the problem domain where the problem is very complex (non-linear, inhomogeneous, etc), and BEM is used for exterior portion that

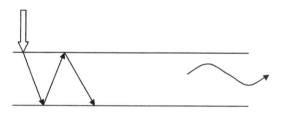

FIGURE 10.7

Dispersive characteristic of lamb wave propagation.

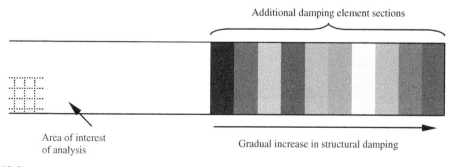

FIGURE 10.8

Damping element sets attached outside area of interest.

can extend to infinity. Much research has been done in this area. An example can be found in a paper by Liu et al. (1992) for wave propagation problems.

10.3.4 Coupling of FEM and the strip element method

Coupling of FEM with the strip element method (SEM; see Section 10.5) can also effectively handle infinite domains. In such a combination, the FEM is used in the interior portions of the problem domain where the problem is very complex (anisotropy, non-linear, inhomogeneous, complex geometry, etc), and SEM is used for exterior portions that can extend to infinity. This combination is applicable for domains of anisotropic materials (see Liu, 2002).

10.4 Finite strip elements

Using finite strip elements, instead of the conventional finite elements, can be a very effective method for solving structural problems involving regular geometry and simple boundary conditions, as shown in Figure 10.9. This method was developed by Y. K. Cheung in 1968. In his method, the structure is divided into 2D strips or 3D prisms or layers of sub-domains. This method usually requires the geometry of the structure to be constant along one or two coordinate axes so that the width of the strip or the cross-section of the prism or layer will not change from one end to the other.

10.5 Strip element method

The strip element method (SEM) was proposed by Kausel and Roësset (1977) and Tassoulas and Kausel (1983) for solids of isotropic materials, and Liu and co-workers (Liu and Achenbach, 1994; Liu and Achenbach, 1995; Liu and Xi, 2001) for solids of anisotropic materials. The SEM is a semi-analytic method for stress analysis of solids and structures. It has been mainly applied for solving wave

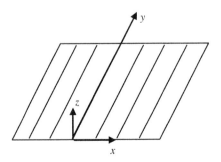

FIGURE 10.9

Finite strip elements used in a plate.

propagating in composite laminates. The SEM is a semi-exact method that discretizes the problem domain in one or two directions. Polynomial shape functions are then used in these directions together with the weak forms of system equation to produce a set of dimension-reduced spatial differential equations. These differential equations are then solved analytically. The dimension of the final discretized system equations would therefore be reduced by one order. The details can be found in a monograph by Liu and Xi (2001). Due to the semi-analytic nature of the SEM, it is applicable for problems of arbitrary boundary conditions including the infinite boundary conditions.

The coupling of SEM with FEM has also been proposed by Liu (2002). In such a combination, the FEM is used for small domains of complex geometry and SEM is used for bulky domains of regular geometry.

10.6 Meshfree methods

The FEM uses elements, and all the numerical operations are based on elements, strictly following the Galerkin weakform or the standard variational/energy principles. The FEM has dominated the area of numerical methods during the last half a century. However, shortcomings for the FEM are also becoming evident. We have seen that the FEM model is a stiff model, meaning that it behaves stiffer than the actual real solids/structures. We have also found that FEM models using T-meshes (triangles for 2D and tetrahedron for 3D) are particularity stiff, and the stress solutions are often inaccurate. However, we engineers prefer to use such a simple T-mesh, because it can be generated automatically for complicated geometries, which is extremely important for automation in computations. It is believed that automation is the future in computer modeling and simulation. Therefore, efforts have been made in improving the FEM, which led to the development of a class of methods called meshfree methods in the past few decades (Liu, 2009).

In meshfree methods, numerical operations (interpolation, integration, etc.) are conducted beyond the elements and based on nodes and/or particles. Some meshfree methods, such as the smoothed particle hydrodynamics (SPH) (Liu and Liu, 2003), use only particles. The SPH method is powerful enough to simulate highly dynamic problems, including high velocity impact, penetration, and explosions. Some meshfree methods, such as the Smoothed Point Interpolation Methods (S-PIM) and the Element Free Galerkin (EFG) methods can use T-meshes (triangles for 2D and tetrahedra for 3D), and therefore are good candidates for adaptive analyses, highly non-linear problems, and crack propagation problems. The formulation of S-PIM is based on smoothing the weakened weak (W2) form, and that of EFG is still based on the Galerkin weakform. Details on this new development can be found in *Meshfree Methods: Moving Beyond the Finite Element Method*, by Liu (2009), where major meshfree methods are introduced in great detail.

10.7 S-FEM

The meshfree method can be computationally more expensive compared to the well-established FEM. The idea of combining the meshfree techniques with the FEM therefore seems like an attractive proposition. This has led to the recent development of the S-FEM that combines the existing

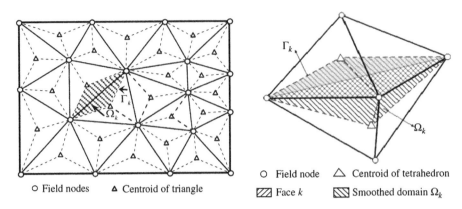

○ Field node   △ Centroid of tetrahedron

▨ Face k ▨ Smoothed domain Ω_k

○ Field nodes   ▲ Centroid of triangle

FIGURE 10.10

Smoothing domains used in S-FEM for 2D and 3D problems. Left: edge-based smoothing domains in 2D problems; Right: face-based smoothing domains for 3D problems.

standard FEM and the existing strain smoothing techniques used in meshfree methods (Liu and Trung, 2010). In S-FEM, the field function interpolation is based on elements (as in the FEM), but it uses smoothed strains to construct numerical models, and hence no integration is needed for the weakform. The formulation of S-FEM can be based either on smoothed weakform or the W2 form. The key to the S-FEM lies in how one performs the strain operations. The currently proven strain smoothing operations are based on node-based, edge-based, cell-based, and face-based smoothing domains, or even their combinations. Therefore, there is a family of models called ES-FEM; NS-FEM; CS-FEM; FS-FEM; and αFEM. Figure 10.10 shows an ES-FEM for 2D problems and an FS-FEM for 3D problems. It works very well for T-meshes that can be generated automatically for complicated geometries.

The S-FEM has been found to have the following significant features: (1) S-FEM models are always "softer" than the standard FEM, promising to provide more effective numerical solutions; (2) S-FEM models give more freedom and convenience in constructing shape functions for special purposes or enrichments (e.g., various degrees of singular fields near the crack tip, highly oscillating fields, etc.); (3) S-FEM models allow the use of distorted elements and general n-sided polygonal elements; (4) NS-FEM offers a much simpler tool to estimate the quality of the solution (global error, bounds of solutions, etc) for many types of problems; (5) the αFEM, a combination of various S-FEM models, can offer solutions of very high accuracy. Interested readers are referred to the S-FEM book by Liu and Trung (2010).

Modeling Techniques

11.1 **Introduction**

In this chapter, various modeling techniques for creating FEM models will be introduced, many of which have been discussed in NAFEMS (1986). These techniques are necessary when carrying out finite element analysis to help to ensure the reliability and accuracy of the results obtained. With the development in the power of computer hardware and software, a FEM analysis can now be performed with ease. In fact, FEM software packages are very often used as a routine tool in the design process by analysts who may not have a proper background of finite element analysis. To many analysts, the FEM software is like a "black box," into which inputs are prescribed, and out of which results are computed and visualized. However, it should be stressed that improper use of commercial software can lead to erroneous results, which are often obscured by colorful stress or deformation plots, movies or other post-processed results. The earlier chapters of this book aim to imbue readers with a clear idea of what goes on in a commercial FE software package. The primary objective of this chapter is to present practical modeling techniques that ensure proper implementations of the FEM theory, and to avoid unnecessary mistakes in creating a FEM model when using a commercial package.

Another reason to learn some of these modeling techniques is to improve both the efficiency of the computation as well as the accuracy of the results. An experienced analyst should be able to obtain accurate results with as little effort in modeling and using as little computer resources as possible. The efficiency of the FE analysis is measured by the effort to accuracy ratio as shown in Figure 11.1. For example, the use of a symmetrical model to simulate a problem with symmetrical geometry can greatly reduce the modeling and computation time with possibly even more accurate numerical results. Therefore, a good FEM model requires more than just discretizing the problem

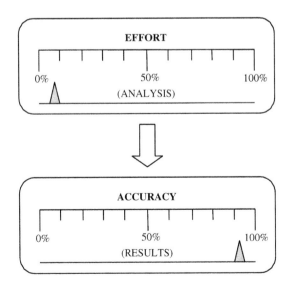

FIGURE 11.1

Minimum effort to yield maximum accuracy.

domain with elements. To come up with a good finite element model, the following factors need to be considered:

- Computational and manpower resources that limit the scale of the FEM model
- Requirements for results that define the purpose and hence the approaches for the analysis
- Mechanical characteristics of the geometry that determine the types of elements to use
- Boundary conditions
- Loading and initial conditions

11.2 CPU time estimation

Despite continuous advances in the computer industry, computer resource can still be one of the decisive factors limiting how complex a finite element model can be built. The CPU time required for a static analysis can be roughly estimated using the following simple relation (it is termed as *complexity* of a linear algebraic system):

$$t_{CPU} \propto n_{dof}^{\alpha} \tag{11.1}$$

where n_{dof} is the number of the total degrees of freedom (DOFs) in the finite element equation system, and α is a constant that is in general in the range of 1.7 to 3.0 depending on different solvers used in the FEM package and the structure of the stiffness matrix.

One of the very important factors that affect α is the bandwidth of the stiffness matrix as illustrated in Figure 11.2. The smaller the bandwidth the smaller the value of α, and hence a faster computation. From the direct assembly procedure described in Example 4.2, it is clear that bandwidth depends on the difference of the global node number assigned to the elements. The element that has the biggest difference in node number controls is the bandwidth of the global stiffness matrix. The bandwidth can be changed even for the same FEM model by changing the global numbering of the nodes.

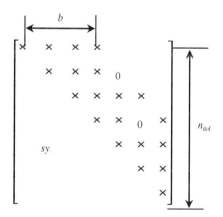

FIGURE 11.2

Schematic of the sparse and banded structure of the stiffness matrix.

Therefore, tools have been developed for minimizing the bandwidth through re-numbering of nodes. Most of the FEM packages have one or more of such tools built-in. The user can simply invoke the tool in the FEM software to minimize the bandwidth after meshing the problem domain. This simple operation can sometimes reduce the CPU time drastically. A very simple method for minimizing the difference of node numbers and hence the bandwidth can be found in the book by Liu (2009, Chapter 15).

Equation (11.1) clearly indicates that a finer mesh, with a large number of DOFs, results in an exponentially increasing computational time. This implies the importance of the reduction of the DOFs. Many techniques discussed in this chapter are related to the reduction of DOFs. Our aims are therefore:

1. To create a FEM model with minimum DOFs by using elements of as low a dimension as possible, and
2. To use as coarse a mesh as possible, and use fine meshes only for important areas.

These have to be done without sacrificing the desired accuracy of the results.

11.3 Geometry modeling

Actual physical structures are usually very complex. The analyst should decide on the ways, where possible, to reduce a complex geometry to a manageable one. The first issue the analyst needs to consider is the type of elements that should be used: 3D elements? 2D (2D solids, plates, and shells) elements? Or 1D (truss and beam) elements? This requires a good understanding of the mechanics of the problem and the behavior of the structure. As we have mentioned in Chapter 9, 3D elements can be used for modeling all types of structures. However, it can be extremely expensive if 3D elements are used everywhere in the entire problem domain, because it will definitely lead to a huge number of DOFs. Therefore, for complex problems, the mesh is often a combination of different types of elements created by taking full geometrical advantage of the problem domain. The analyst should analyze the problem in hand, examine the geometry of the problem domain, and try to make use of 2D and 1D elements for areas or parts of the structure that satisfy the assumptions that lead to the formulation of 2D or 1D elements. Usually, 2D elements should be used for areas/parts that have a plate- or shell-like geometry, and 1D elements should be used for areas/parts that have a bar- or arch-like geometry. 3D elements are used only for bulky parts of the structure to which 2D or 1D elements cannot apply. This process is very important because the use of 2D and 1D elements can significantly reduce the number of DOFs.

As shown in Figure 11.3, for the modeling of parts where 3D elements are to be used, 3D objects that have the same geometrical shapes of the structure have to be created. For parts where 2D elements are to be used, only the neutral surfaces that are often the geometrical mid-surfaces need to be created. For parts where 1D elements are to be used, only the neutral axes that are often the geometrical mid-axes need to be created. Therefore, one can imagine that by using 2D and 1D elements, the task of creating the geometry is significantly simplified.

When a mix of different types of elements are used in an FEM model, the interfaces where the different element types meet requires additional attention. At these interfaces, techniques of modeling joints can be used, which will be discussed in detail in Section 11.9. These techniques are required because the type of DOFs at a node that is shared by different element types is different for different

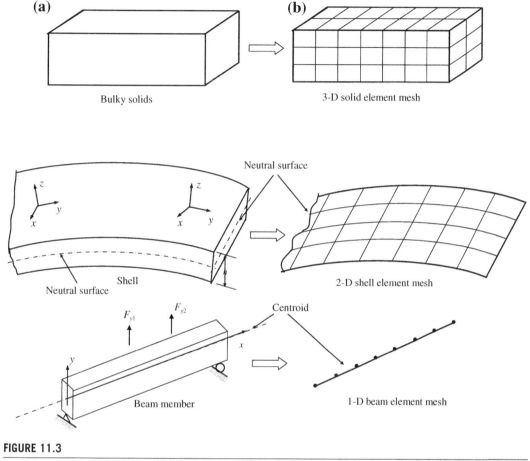

FIGURE 11.3

Geometrical modeling. (a) Physical geometry of the structural parts; (b) Geometry created in FEM models.

element types. Recall that the DOFs for any particular element type arise from the theories of mechanics discussed in Chapter 2 and depends on the mechanics of the problem (e.g., truss, beam, 2D solids, plate, etc.). Table 11.1 lists the number of DOFs for some different types of elements.

Understanding the requirements needed for the analysis is important and has a direct bearing on the generation of the simulation domain. For example, analysts will usually give a detailed modeling of the geometry for areas where critical results are expected. Note that many structures are now designed using so-called computer aided design (CAD) packages. Therefore, the geometry of the structure would have already been created electronically. Most commercial pre-processors of FEM software packages can read CAD files in certain formats. Making use of these CAD files can reduce the effort in creating the geometry of the structure, but it requires a certain amount of effort to modify the CAD geometry for them to be suitable for FEM meshing. There are also ongoing research activities to automatically

Table 11.1 Type of elements and number of DOFs at a node.

No.	Description	DOFs at a Node
1	2D frame analysis (using 2D frame element)	3 (2 translations and 1 rotation)
2	3D frame analysis (using 3D frame element)	6 (3 translations and 3 rotations)
3	2D analysis for plane strain or plane stress analysis	2 (translational displacements)
4	3D analysis for solids with general geometries and loading conditions	3 (translational displacements)
5	2D analysis for axisymmetric solids with axisymmetric or asymmetric loading	2 (translational displacements)
6	Plate bending analysis for out-of-plane loading (bending effects only)	3 (1 translation and 2 rotations)
7	General plate and assembled plate analysis with general loading conditions (combined membrane and bending effects)	5 or 6 (3 translations and 2 or 3 rotations)
8	General shell analysis for shell structures (coupled membrane and bending effects)	5 (or 6) (3 translations and 2 or 3 rotations)
9	1D analysis for axisymmetric shells with axisymmetric loading (membrane and bending effects)	3 (2 translations and 1 rotation)

convert proper 3D geometries to 2D and 1D geometries for FEM mesh, but at the time of writing, such tools have not matured sufficiently for commercial applications.

11.4 Meshing

11.4.1 Mesh density

To minimize the DOFs, it is common to create a mesh of varying density. We usually only need a fine mesh in areas of importance, such as areas of analytical interest, and expected zones of stress concentration, such as at re-entrant corners; holes; slots; notches; or cracks. An example of a finite element mesh exhibiting mesh density transition is shown in Figure 11.4. In this example of the sprocket-chain system of a motorcycle, the focus of the analysis is the contact forces between the sprocket and the chain. Hence, the region at the center of the sprocket is relatively less critical and the mesh used at that region is therefore coarser.

In using FEM packages, the control of mesh density is often performed by creating what are commonly termed "mesh seeds." Mesh seeds are placed on a given or created geometry before actual meshing. These mesh seeds indicate the positions of the nodes of the actual finite element mesh where the node density needs to be controlled. To have more elements in areas of importance, denser mesh seeds are simply placed in these areas before meshing. When the mesh is being generated, nodes will be created at the locations of these seeds.

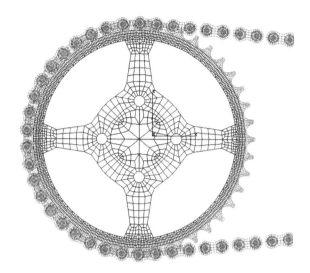

FIGURE 11.4

Finite element mesh for a sprocket-chain system. (Courtesy of the Institute of High Performance Computing.)

11.4.2 **Element distortion**

It is not always possible to fit regularly shaped elements into irregular geometries. Irregular or *distorted* elements are acceptable in FEM, but there are limitations and one needs to control the degree of element distortion in the process of mesh generation. The element distortions are measured against the *basic shape* of the element which are:

- Square $\Rightarrow$ Quadrilateral elements
- Isosceles triangle $\Rightarrow$ Triangle elements
- Cube $\Rightarrow$ Hexahedron elements
- Isosceles tetrahedron $\Rightarrow$ Tetrahedron elements

Five possible forms of element distortions and their rule-of-thumb limits are listed as follows:

1. *Aspect ratio* distortion (elongation of element) as shown in Figure 11.5.
2. *Angular* distortion of the element (Figure 11.6), where any included angle between edges approaches either 0° or 180° (skew and taper).
3. *Curvature* distortion of element (Figure 11.7), where the straight edges from an element are distorted into curves when matching the nodes to the geometric points.

Volumetric distortion occurs in concave elements. As discussed in Chapter 6, in calculating the element stiffness matrix, a mapping is performed to transfer the irregular shape of the element in the *physical coordinate system* into a regular one in the non-dimensional *natural coordinate system*. For concave elements, there are areas outside the elements (see the shadowed area in Figure 11.8) that will

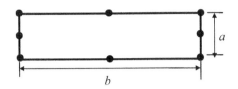

$$\frac{b}{a} \leq \begin{cases} 3 & \text{for stress analysis} \\ 10 & \text{for displacement analysis} \end{cases}$$

FIGURE 11.5

Aspect distortion.

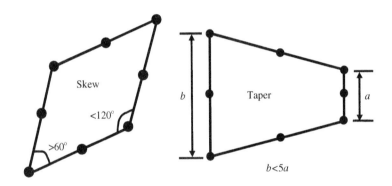

FIGURE 11.6

Angular distortion.

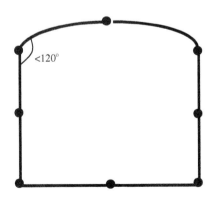

FIGURE 11.7

Curvature distortion.

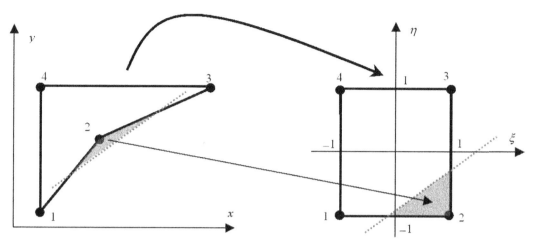

FIGURE 11.8

Mapping between the physical coordinate $(x-y)$ and the natural coordinate $(\xi-\eta)$ for a heavily volumetrically distorted elements leads to mapping of an area outside of the physical element into an interior area in the natural coordinates.

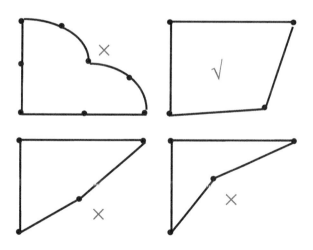

FIGURE 11.9

Unacceptable shapes of quadrilateral elements.

be transformed into an internal area in the natural coordinate system. The element volume integration for the shadowed area based on the natural coordinate system will thus result in a negative value. A few unacceptable shapes of quadrilateral elements are shown in Figure 11.9.

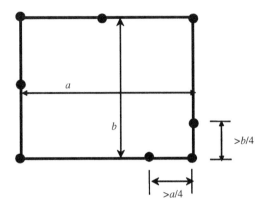

FIGURE 11.10

The limit for mid-node displacing away from the middle edge of the element.

Mid-node position distortion occurs with higher order elements where there are mid-nodes. The mid-node should be placed as close as possible to the middle of the element edge. The limit for mid-node displacing away from the middle edge of the element is a quarter of the element edge, as shown in Figure 11.10. The reason behind this is that shifting of mid-nodes can result in undesirable stress distribution and even singular stress field in the elements as discussed in Section 10.2.

Many pre-processors of the FEM package provide a tool for analyzing the element distortion rate for a created mesh. It is good practice prior to submitting a finite element model for computation to invoke this built-in tool to check for element distortion. A report of distortion rates will be generated for the analyst's examination or heavily distorted elements can be highlighted for any necessary actions.

11.5 Mesh compatibility

In Chapter 3, when the Hamilton's principle is used for deriving the FEM equation, it is required that the displacement has to be admissible, which demands the continuity of the displacement field in the entire problem domain. A mesh is said to be compatible if the displacements are continuous along all the edges between all the elements in the mesh. The use of different types of elements in the same mesh or improper connection of elements can result in an incompatible mesh. The following two subsections will discuss the reasons for mesh incompatibility and suggest methods for fixing or avoiding mesh incompatibility issues. Note that the compatibility conditions can be relaxed if the S-PIM (Liu and Zhang, 2013) is used. This is because the S-PIM is based on the so-called weakened weak form, and the stability and convergence are ensured by the G-space theory.

11.5.1 Different order of elements

Mesh incompatibility issues can arise when we have a transition between different mesh densities or when we have meshes comprising of different element types. When a quadratic element is joined with one or more linear elements as shown in Figure 11.11, incompatibility arises due to the difference in the orders of shape functions used. The 8-nodal quadratic element in Figure 11.11 has quadratic shape functions, which implies that the deformation along the edge follows a quadratic function. On the other hand, the linear shape function used in the 4-nodal linear element in Figure 11.11 will result in a linear deformation along each element edge. For the case shown in Figure 11.11a, the displacements at nodes 1 and 3 for the quadratic element and the linear elements are the same, but deformation of the edges between nodes 1 and 3 will be different. Assuming that nodes 1 and 2 have zero displacements, while node 3 gets displaced, the deformations of these edges can then be shown by the dotted lines in Figure 11.11. A crack-like behavior is clearly observed which can lead to severe erroneous results. For the case shown in Figure 11.11b, the displacements at nodes 1, 2 and 3 for the quadratic element and the two linear elements are the same, but the deformation of the edges between nodes 1 and 2, and nodes 2 and 3 will again be different. If nodes 1 and 3 have zero displacements, while node 2 gets displaced, the deformations of these edges are shown by the dotted lines in Figure 11.11. Again, a crack-like behavior is clearly observed.

Here are some solutions to avoid or solve these mesh incompatibility issues:

1. Use the same type of elements throughout the entire problem domain. This is the simplest solution and is a usual practice, as complete compatibility is automatically satisfied if the same elements are used as those shown in Figure 11.12.

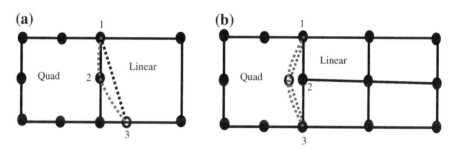

FIGURE 11.11

Incompatible mesh caused by the different shape functions along the common edge of the quadratic and linear elements. (a) A quadratic element connected to one linear element. (b) A quadratic element connected to two linear elements.

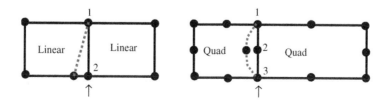

FIGURE 11.12

Use of elements of the same type with complete edge to edge connection automatically ensures the mesh compatibility.

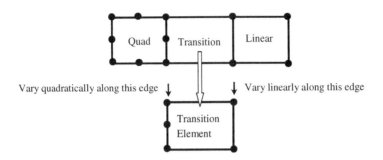

FIGURE 11.13

A transition element with 5 nodes used to connect linear and quadratic elements to ensure mesh compatibility.

2. When elements of different orders of shape functions have to be used for some reason, such as in *p*-adaptive analysis, use *transition elements* whose shape functions have different orders on different edges. An example of a transition element is shown in Figure 11.13. The 5-node element shown can behave in a quadratic fashion on the left edge and linearly on the other edges. In this way, the compatibility of the mesh can be guaranteed.

3. Another method used to enforce mesh compatibility is to use *multipoint constraints* (MPC) equations. MPCs can be used to enforce compatibility for cases shown in Figure 11.11a. This method is more complicated, and requires the knowledge of creating MPC equations, which will be covered in Section 11.10.

11.5.2 Straddling elements

Straddling elements can also result in mesh incompatibility as illustrated in Figure 11.14. Although the order of the shape functions of these connected elements are the same, the straddling can result in an incompatible deformation of edges between nodes 1 and 2, and 2 and 3, as indicated by the dotted lines

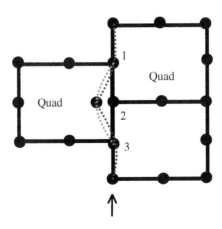

FIGURE 11.14

Incompatible mesh caused by the straddling along the common edge of the same order of elements.

in Figure 11.14. This is because in the assembly of elements, the FEM requires only the continuity of the *displacements* (not the derivatives) at nodes between elements.

The solution to this problem is simply to make sure there are no straddling elements in the mesh. Most mesh generators are designed not to produce such an element in the mesh. However, care needs to be taken if the mesh is created manually.

11.6 **Use of symmetry**

Many structures and objects exhibit some form of symmetry. Figure 11.15 shows some examples of the different types of common structural symmetry. Objects such as a can of drink exhibits axial symmetry and even huge structures such as the Eiffel Tower of Paris exhibits mirror symmetry. An experienced analyst will take full advantage of such symmetries in structures to simplify their modeling process as well as to reduce the DOFs and hence computational time required for the analysis. Imagine a full finite element model of the Eiffel Tower consisting of, say, 100,000 elements. Because of the mirror

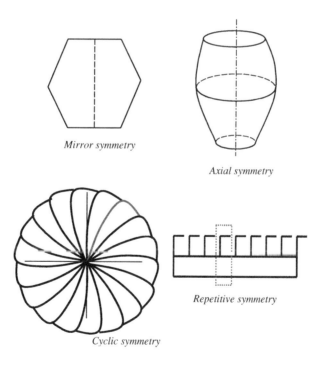

Mirror symmetry

Axial symmetry

Cyclic symmetry

Repetitive symmetry

FIGURE 11.15

Different types of structural symmetry.

symmetry, one can actually perform the analysis by just modeling a quarter of the whole structure and the number of elements will be reduced to a quarter at 25,000 elements. The total DOFs of the system will also be reduced to a quarter. Using Eq. (11.1) and assuming that $\alpha = 3$, it can be found that the CPU time will be reduced to $(1/4)^3 = (1/64)$th of that required for solving the full model. The significance is astonishing! Furthermore, since only a quarter of the model is required, the time for the analyst to create the model is also reduced. In addition, the accuracy of the analysis can be improved as the equation system becomes much smaller and the numerical errors in the computation will reduce. However, proper techniques have to be used to make the full use of the structural symmetry. This section will deal with some of these techniques.

11.6.1 Mirror symmetry or plane symmetry

Mirror symmetry is the symmetry about a particular plane, and it is the most prevailing case of symmetry. A half of the structure is the mirror image of another. The position of the mirror is called the plane of symmetry. A structure is said to have mirror structural symmetry if there is symmetry of geometry, support conditions, and material properties. Many actual structures exhibit this type of symmetry. Some of these structures are actually symmetrical about a particular plane, while others are symmetric with respect to multiple planes. Take for example a cubic block as shown in Figure 11.16. One can actually use the property of single-plane-symmetry and model a half-model or one can use that of two-plane-symmetry to further reduce the finite element model to a quarter of the original structure. In fact, more planes of symmetry can also be used to just model one-eighth of the model and in that case, it would be similar to the case of cyclic symmetry, which will be discussed later.

Consider a symmetric 2D solid shown in Figure 11.17. The 2D solid is symmetric with respect to an axis of symmetry of $x=c$. The left half of the domain is modeled with impositions of the following symmetric boundary conditions at nodes on the symmetric axis.

$$u_1 = 0$$
$$u_2 = 0 \tag{11.2}$$
$$u_3 = 0$$

where u_i $(i=1,2,3)$ denotes the displacements in the x direction at node i. Equation (11.2) gives a set of single point constraint (SPC) equations, because in each equation, there is only one unknown

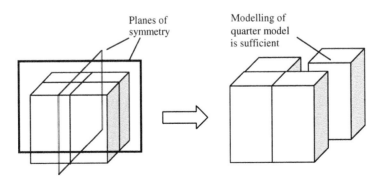

FIGURE 11.16

Modeling a cubic block with two planes of symmetry.

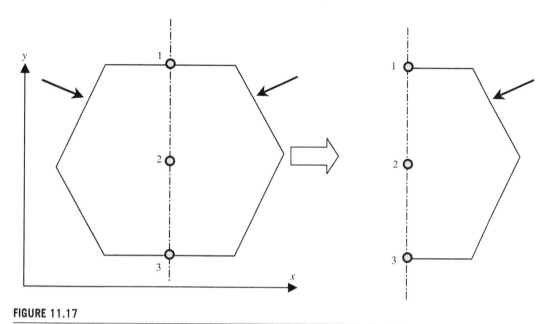

FIGURE 11.17

2D solid with an axis of symmetry at $x = c$. The right half of the domain is modeled with impositions of symmetric boundary conditions at nodes on the symmetric axis.

(or one DOF) involved. This kind of SPC can be simply imposed by removing the corresponding row and column in the global system equations as demonstrated in Examples 4.1 and 4.2.

Loading conditions on a symmetrical structure must also be taken into consideration. A loading is considered symmetric if the loading can also be "reflected" off a particular plane as shown in Figure 11.18. In this case, the problem is symmetric because the whole structure, its support conditions, as well as its loading, is symmetrical about the plane $x = 0$. An analysis of a half of the whole beam structure using the symmetrical boundary condition at $x = 0$ would yield a complete solution as that of the full model with at least less than a quarter of the effort.

A problem can also be anti-symmetric, if the structure is symmetric but the loading is anti-symmetric, as shown in Figure 11.19. Again, modeling half of the structure can also yield a complete solution using an *anti-symmetric* boundary condition, which would be different on the plane of symmetry. In the simple example shown in Figure 11.19, the anti-symmetric boundary condition is that the deformation at the plane of symmetry should be zero. Note that the rotation at the plane of symmetry need not be zero in contrast with the case of symmetric loading.

The following general rules can be applied when deciding the boundary conditions at the plane of symmetry. If the problem is symmetric as shown in Figure 11.18, then:

1. There are no translational displacement components normal to the plane of symmetry.
2. There are no rotational displacement components with respect to the axis that are parallel to the plane of symmetry.

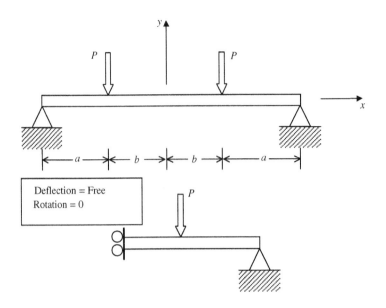

FIGURE 11.18

Simply supported symmetric beam structure.

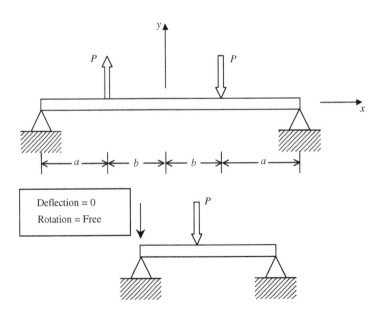

FIGURE 11.19

Simply supported anti-symmetric beam structure.

Table 11.2 Boundary conditions for symmetric loading.

Plane of Symmetry	u	v	w	θ_x	θ_y	θ_z
xy	Free	Free	Fix	Fix	Fix	Free
yz	Fix	Free	Free	Free	Fix	Fix
zx	Free	Fix	Free	Fix	Free	Fix

Table 11.3 Boundary conditions for anti-symmetric loading.

Plane of Symmetry	u	v	w	θ_x	θ_y	θ_z
xy	Fix	Fix	Free	Free	Free	Fix
yz	Free	Fix	Fix	Fix	Free	Free
zx	Fix	Free	Fix	Free	Fix	Free

If the problem is anti-symmetric as shown in Figure 11.19, then:

1. There are no translational displacement components parallel to the plane of symmetry.
2. There are no rotational displacement components with respect to the axis that is normal to the plane of symmetry.

Tables 11.2 and 11.3 give the complete list of conditions for symmetry and anti-symmetry for general three-dimensional cases.

Any load can be decomposed into a symmetric load and an anti-symmetric load. Therefore, as long as the structure is symmetric (in geometry, material, and boundary conditions), one can always take advantage of the symmetry. Consider now a case shown in Figure 11.20a where the simply supported beam structure is symmetric structurally but the loading is asymmetric (neither symmetric nor anti-symmetric). The structure can always be treated as a combination of (a) the same structure with symmetric loading and (b) the same structure with anti-symmetric loading. In this case, one needs to solve two problems with each problem having half the number of DOFs if the whole structure is modeled. One of the problems is symmetrical while the other is anti-symmetric.

Figure 11.21 shows a more complex example of how a framework with asymmetric loading conditions can be analyzed using a half of the framework with symmetric and anti-symmetric conditions. Adding up one problem of the same structure subjected to a symmetric loading with another of the same structure subject to anti-symmetric loading is equivalent to that of analyzing the full frame structure with the asymmetric loading. In this example, there is actually a frame member that is at the plane of symmetry. The properties of this frame member on the symmetric plane need to be halved as well for the two half models. This means that all the properties that contribute to the stiffness matrix for this member need to be halved.

Dynamic problems can also be solved in a similar manner using the symmetric or anti-symmetric properties, for example, if symmetric boundary conditions are imposed on a simple beam structure and a natural frequency analysis is carried out. The natural frequencies obtained will correspond only to the symmetrical modes. To obtain the anti-symmetrical modes, anti-symmetrical boundary conditions will need to be applied to the model. Figure 11.22 shows the symmetric and anti-symmetric conditions for vibration analysis in a simply supported beam.

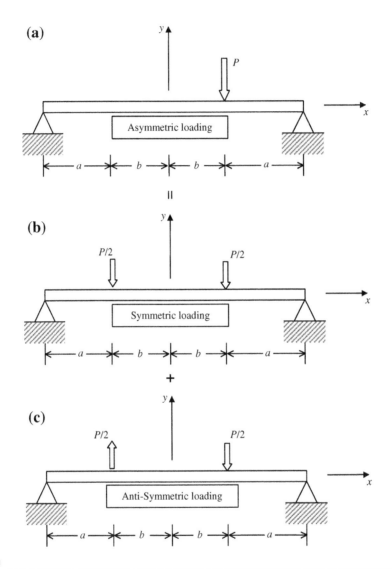

FIGURE 11.20

Simply supported symmetric beam structure subject to an asymmetric load. (a) A structure with an asymmetric load. (b) The same structure with symmetric load. (c) The same structure with anti-symmetric load.

11.6.2 Axial symmetry

A solid or structure has axial symmetry when it can be generated by rotating a planar geometry (1D or 2D) about an axis. To take advantage of axial symmetry, such a solid can be modeled by simply using 1D or 2D axisymmetric elements, as discussed in Chapter 7. In doing so, there is a dimension reduction

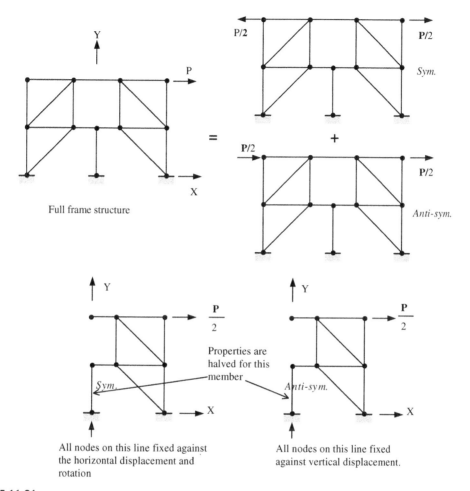

FIGURE 11.21

Using symmetry to analyze symmetrical framework with asymmetric loading.

that will greatly reduce the modeling and computational effort. For example, a cylindrical shell structure can be modeled using 1D axisymmetric beam elements as shown in Figure 11.23. Figure 11.24 shows an example of a 3D solid under axially symmetric loads, which can be modeled using 2D axisymmetric elements.

The formulation of 2D axisymmetric elements is detailed in Section 7.5 and it has been shown that it is very similar to the formulation of 2D plane strain elements. Similarly, the formulation of 1D axisymmetric elements can be easily developed using similar concepts to those used in earlier chapters, but expressed in the polar coordinate system. The shapes for 2D axisymmetric elements are the same as that of regular 2D elements described in Chapter 7. Generally speaking, the use of axisymmetric elements requires much less computational resources compared to a full 3D discretization. Axisymmetric

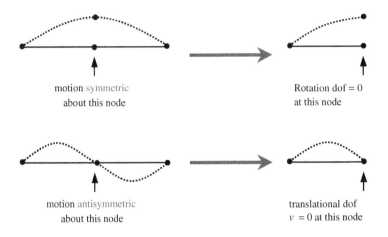

FIGURE 11.22

Using symmetric and anti-symmetric conditions for free vibration analysis.

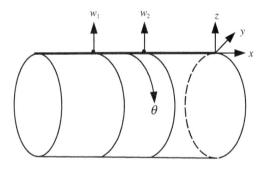

FIGURE 11.23

A cylindrical shell structure modeled using 1D axisymmetric elements.

elements are readily available in most finite element software packages, and the usage of these elements is similar to the regular 1D or 2D elements.

Similar to the plane symmetry problems, the loadings applied on an axisymmetric structure do not have to be symmetric or anti-symmetric about its axis. Any axially asymmetric load can be expressed in a Fourier superimposition of both axial symmetric and axial anti-symmetric components in the θ direction. Therefore, the problem can always be decomposed into two sets of problems of axial symmetric and axial anti-symmetric, as long as the structure is axisymmetric (in geometry, material and boundary support).

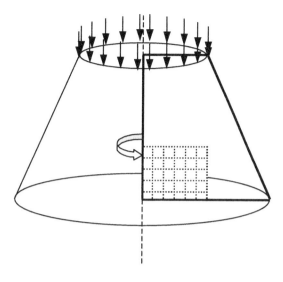

FIGURE 11.24

A 3D structure modeled using 2D axisymmetric elements.

11.6.3 **Cyclic symmetry**

Cyclic symmetry exists in problems whereby both geometry and loading appear as repeated sectors in a cyclic manner. In such a case, a complete solution can be obtained by analyzing only one sector as a *representative cell* with a set of cyclic boundary conditions on the boundaries of the cell, as shown in Figure 11.25. The cyclic symmetric boundary condition for the problem shown in Figure 11.25 requires that all the variables along side A must match exactly those on side B. Constraint equations at all the corresponding nodes along side A and side B can therefore be written as

$$u_{An} = u_{Bn} \tag{11.3}$$

$$u_{At} = u_{Bt} \tag{11.4}$$

Note that in Eqs. (11.3) and (11.4) both u_{An} and u_{Bn} (or u_{At} and u_{Bt}) are unknowns. Thus, Eqs. (11.3) and (11.4) are constraint equations that involve more than one DOF in one equation. These types of constraint equations are termed as Multiple Point Constraints (MPC) equations, which have to be imposed by modifying the global system equations. The imposition of MPCs is, however, more tedious than that of SPC detailed in Chapter 4. To impose an SPC all one needs to do is to remove (or modify) the corresponding row and column in the system equations (see Examples 4.1 and 4.2). The imposition of a MPC requires the use of either penalty method or the method of Lagrange multipliers that will be detailed in Section 11.11.

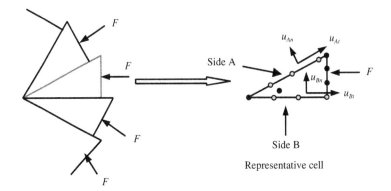

FIGURE 11.25

Representative cell isolated from a cyclic symmetric structure and the cyclic symmetrical conditions on the cell.

11.6.4 Repetitive symmetry

Repetitive symmetry exists in structures consisting of continuously repeating sections under certain loading conditions (usually in the direction of repeating section), as shown in Figure 11.26. In such a case, only one section needs to be modeled and analyzed. Similar to cyclic symmetry, constraint equations are used for the corresponding nodes at the sectioned surface such that

$$u_{Ax} = u_{Bx} \tag{11.5}$$

which is again an MPC equation.

11.7 Modeling of offsets
11.7.1 Methods for modeling offsets

In the modeling of beams, plates, and shells, the elements are usually defined on the neutral surface (often the geometric middle surface) of the structure, as shown in Figure 11.3. For elements that are not collinear or coplanar, there will be a distance of *offset* between the nodes in the FEM mode, which are connected together in the physical structure. Figure 11.27 shows a typical case of two beams with different thicknesses joined at the corner. In the finite element mesh, the two corner nodes are, however, apart. In this case, there are two offsets, α and β. In such cases, proper techniques may be needed to model the *offset* in order to better represent the connection.

If the offsets are small compared to the length of the beam l, we can often ignore them and the connection is simply modeled by extending the corner nodes to the joint point (see Figure 11.27). If the offsets are too large, they have to be treated using proper modeling techniques. Whether offsets should

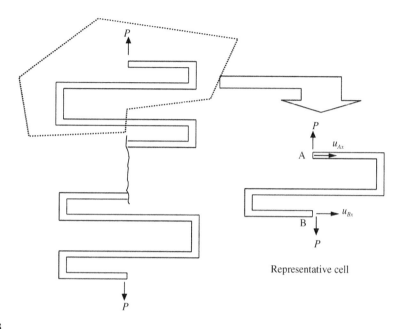

Representative cell

FIGURE 11.26

Representative cell isolated from a repetitive symmetric structure and the repetitive symmetrical conditions on the cell.

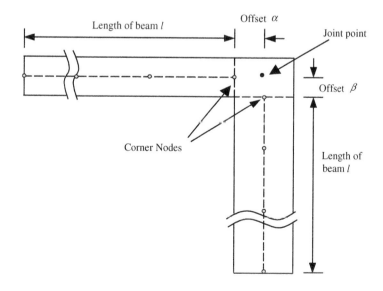

Length of beam l

Offset α

Joint point

Offset β

Corner Nodes

Length of beam l

FIGURE 11.27

Offsets at the joint of two beams with different thickness.

be modeled depends on the engineering judgment of the analyst. A rough guideline shown below can be followed where l is the length of the beam:

$\alpha < \frac{l}{100}$, offset may be ignored

$\frac{l}{100} < \alpha < \frac{l}{5}$, offset may need to be modeled

$\alpha > \frac{l}{5}$, ordinary beam, plate, and shell elements may not be used. Use 2D or 3D solid elements instead.

Modeling of offsets is usually even more crucial in shells as there are both in-plane and out-of-plane forces in the formulation, and a small variation of nodal distance can give rise to a significant difference in results. There are three methods often used to model the offsets:

1. *Very stiff element.* Use an artificial element with very high stiffness (high Young's modulus, large cross-sectional area or second moment of area) to connect the two corner nodes (see Figure 11.28a). This is usually done by increasing the Young's modulus by, say, 10^6 times for the very stiff element. This method is very simple and convenient to use, but is usually not recommended because of the huge difference in stiffness between the stiff element and the normal elements in the same FE model that can lead to ill conditioning of the final global stiffness matrix.
2. *Rigid element.* Use a "rigid element" to connect the corner nodes (see Figure 11.29b). This method can be easily implemented being available in many commercial software packages, and all the analyst needs to do is to create an element in between the two corner nodes and then assign it as a "rigid element." The treatment of the rigid element in the software package is the same as the next method of using MPC equations, but the formulation and implementation is automated in the software package, and therefore the user does not have to formulate these MPC equations.
3. *MPC equations.* If the "rigid element" is not available in the software package, we then need to create MPC equations to establish the relationship between the DOFs at the two corner nodes. This set of MPC equations is then prescribed as an input in the pre-processor. Use of MPC

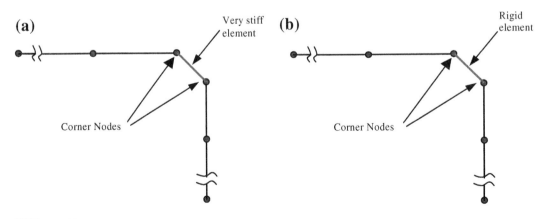

FIGURE 11.28

Modeling of offsets using an artificial element connecting the two corner nodes. (a) Use of very stiff element. (b) Use of rigid element.

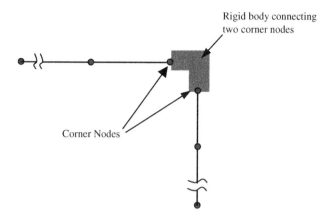

Rigid body connecting
two corner nodes

Corner Nodes

FIGURE 11.29

Modeling of offsets using MPC equations created using a virtual "rigid body" connecting two corner nodes.

equations is supported by most FEM software packages. To establish the MPC equations, we can consider a virtual rigid body that connects these two corner nodes, as shown in Figure 11.29. The detailed process is discussed in the next subsection.

11.7.2 Creation of MPC equations for offsets

The basic idea of using MPC is to create a set of MPC equations that gives the relation between the DOFs of the two separated nodes, i.e., the DOFs of one node depend on the other and vice versa. We first assume that a rigid body connects the two corner nodes. Note that this rigid body is imaginary and is used here to help us establish the MPC equations. The MPC equations are then derived using the simple kinematic relations of the DOFs on the rigid body. The procedure of deriving the MPC equations for this case of Figure 11.30 is described as follows.

First, assume that nodes 1 and 2 are perfectly connected to the rigid body and that the rigid body has a movement of q_1, q_2 (horizontal and vertical translations), and q_3 (rotation) with reference to point 3, as shown in Figure 11.30. Then, enforce points 1 and 2 to follow the rigid body movement, and calculate the resultant displacements at nodes 1 and 2 in terms of q_1, q_2, and q_3 to obtain

$$
\begin{aligned}
d_1 &= q_1 + \beta q_3 \\
d_2 &= q_2 \\
d_3 &= q_3 \\
d_4 &= q_1 \\
d_5 &= q_2 - \alpha q_3 \\
d_6 &= q_3
\end{aligned}
\tag{11.6}
$$

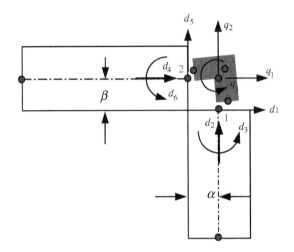

FIGURE 11.30

A rigid body is used for deriving the MPC equations.

Finally, eliminating the DOFs for the rigid body q_1, q_2, and q_3 from the above six equations, we obtain three MPC equations:

$$d_1 - \beta d_3 - d_4 = 0$$
$$d_2 - \alpha d_3 - d_5 = 0 \qquad (11.7)$$
$$d_3 - d_6 = 0$$

which gives the relationship between the six DOFs at nodes 1 and 2. Note that small displacements and rotations are assumed when deriving Eq. (11.6).

The number of the MPC equations one should have can be determined with the following equation:

$$N|_{\text{Equation of MPC}} = N|_{\text{DOFs at all nodes to be connected by the rigid body}} - N|_{\text{DOFs of the rigid body}} \qquad (11.8)$$

Another example of an offset often seen in engineering structures is the stiffened plates or shells, where a beam is fixed to a plate or shell as a stiffener, as shown in Figure 11.31. For the analysis of such a structural component, offset should be modeled if accurate results are required. The offset can be modeled by imagining that the node on the beam is connected to the corresponding node at the plate by a rigid body A-B. For a node in the plate, there are five DOFs (three translational and two rotational about the two axes in the plane of the plate), as shown in Figure 11.32. For a node in the beam, there are four DOFs (three translational and one rotational about the axis perpendicular to the bending plane). Note that in this case, the beam has been defined with its bending plane as the x-z plane. Hence, there are a total of nine DOFs linked by the rigid body between the node on the beam and the node on the plate. To simplify the process of deriving MPC equations, we let the rigid body follow the movement of

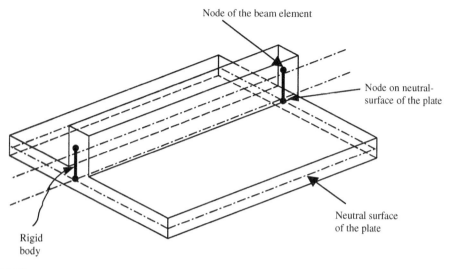

FIGURE 11.31

Stiffened plates with an offset in between the axis of a beam and mid-surface of a plate.

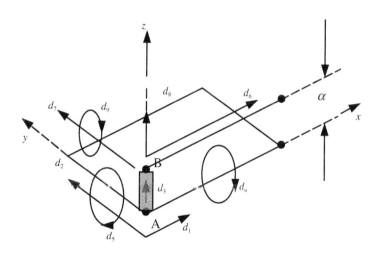

FIGURE 11.32

DOFs at nodes in the plate and beam.

node A in the plate. Therefore, the rigid body should have the same number of DOFs as the node in the plate, which is five: q_1, q_2, q_3, q_4 and q_5. The number of MPC equations should therefore be $9 - 5 = 4$ following Eq. (11.8). As the rigid body follows the movement of A in the plate, what should be done is to force the node belonging to the beam to follow the movement of B on the rigid body. We can write down the 9 DOFs of the nodes involved in terms of q_1 to q_5, then eliminate the five q's from the nine equations. The resulting four constraint equations should therefore be

$$
\begin{aligned}
d_6 &= d_1 + \alpha d_5 \text{ or } d_1 + \alpha d_5 - d_6 = 0 \\
d_7 &= d_2 - \alpha d_4 \text{ or } d_2 - \alpha d_4 - d_7 = 0 \\
d_8 &= d_3 \text{ or } d_3 - d_8 = 0 \\
d_9 &= d_5 \text{ or } d_5 - d_9 = 0
\end{aligned}
\tag{11.9}
$$

The above equations must be specified as constraints for all the pairs of nodes along the length of the beam attached to the plate. This ensures that the beam is perfectly attached to the plate.

11.8 Modeling of supports

Support of a structure is usually very complex and in real physical structures, can be ambiguous. The FEM allows constraints to be imposed differently at different nodes. Different methods can be used to simulate these supports. For example, in the analysis of thick beam structures using 2D plane stress elements, there are three possible ways to model the support conditions at the built-in end, as shown in Figure 11.33.

1. Full constraint is imposed at the boundary nodes only in the horizontal direction.
2. Partial full constraint to the nodes on the boundary.
3. Fully clamped boundary conditions are specified at all the boundary nodes.

It is not possible to say which is the best boundary condition for the support, because the best way is actually the one that best simulates the actual situation. Therefore, good engineering judgment, experience, and the understanding of the actual support play very important roles in creating a suitable boundary condition for the finite element model. Trial and error, parametric studies, and even inverse analysis (Liu and Han, 2003) are often conducted when the physical situation is not clear to the analyst. Similarly for a prop support of a beam, there can be more than one way of modeling the support conditions as shown in Figure 11.34.

For complex support on structures, it might be necessary to use finer meshes near the boundary. Denser nodes provide more flexibility to model the support, as the FEM allows the specification of different constraints at different nodes.

11.9 Modeling of joints

Care should be taken if the joints are between different element types. The complication comes from the difference in DOFs at a node shared by different elements. In many situations, a proper technique or a set of MPC equations are required to model the connection.

Consider a connection between the turbine blade and the turbine disc shown in Figure 11.35a. A turbine blade is usually *perfectly* connected to a disc. The task at hand is to build a FEM model to analyze the blade-disc system. If both the blade and disc are all modeled using 2D solid elements, as shown in

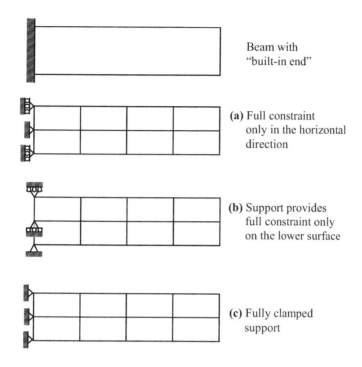

Beam with "built-in end"

(a) Full constraint only in the horizontal direction

(b) Support provides full constraint only on the lower surface

(c) Fully clamped support

FIGURE 11.33

Modeling of built-in end support of beam.

Figure 11.35b, perfect connection is ensured, as long as we ensure that at the interface the blade and the disc share the same nodes, i.e., there is no more than one node at any particular location.

Assume a FEM model where the blade is now modeled using beam elements, and the disc is modeled using 2D elements as shown in Figure 11.36. This model clearly reduces the number of nodes and therefore the DOFs. However, one needs to exercise caution when modeling the connection properly for the following reason.

First, we consider the connection of the beam element and 2D solid elements in the way shown in Figure 11.36a. From Table 11.1, we know that at a node in a 2D solid element, there are 2 DOFs: translational displacement components u and v. At a node in a beam element, there should be 3 DOFs: 2 translational displacement components u and v, and one rotational DOF. Although the beam element and the 2D elements share the common node 1, which ensures that the translational displacements are the same at the connection point for the beam and 2D solid elements, the rotation DOF of the beam element is, however, free. Therefore, the extra DOF can cause the blade to rotate freely since it is not tied (but pinned) to the 2D solid, which is not the perfect connection we wanted to model. The case here is similar to a beam with a pinned support at one end, wherein the rotational DOF is free. The problem, as you may have noticed, is due to the mismatch of the types of DOFs between the beam and 2D solid elements. The rotational DOF cannot be transmitted onto the node on a 2D solid element node, simply

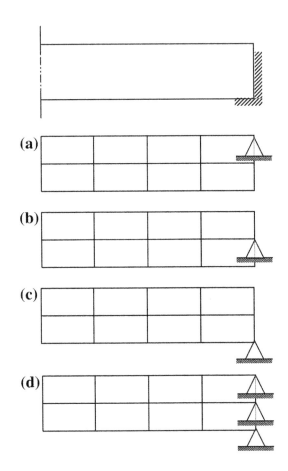

FIGURE 11.34

Modeling of prop support for a beam.

because it does not have a rotational DOF. We therefore require certain modeling techniques to correct this connection between the turbine blade and disc.

A simple method to fix the rotation of the beam on the 2D solid mesh is to extend the beam elements into the disc for at least two nodes, as shown in Figure 11.36b. Burying the nodes of the beam elements into the 2D mesh this way transmits the physical rotation of the beam via the translational DOFs. One may notice that the rotation of the beam at the interface is no longer free like in a pinned connection. The drawback of this simple method, however, is the additional mass introduced near the joint area due to the extension into the 2D mesh, which may affect the results of a dynamic analysis.

Another effective approach is to use MPC equations to create a connection between the two types of elements. The detailed process is given as follows.

First, imagine now that there is a very thin rigid strip connecting the beam and the disc, as shown in Figure 11.37. This strip connects three nodes together, one on the beam and two on the disc. These

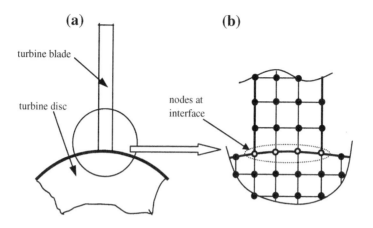

FIGURE 11.35

Modeling of turbine-blade and turbine-disc system. (a) Simplified diagram of a turbine blade connected to a turbine disc. (b) A magnified 2D solid element mesh in at the interface.

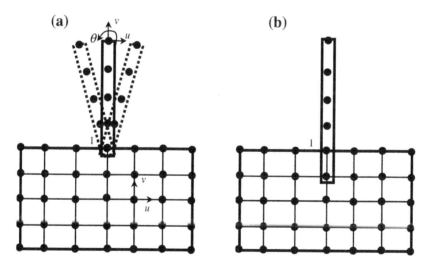

FIGURE 11.36

Joint between beam and 2D elements. (a) Beam is free to rotate in reference to the 2D solid. (b) Perfect connections are modeled by artificially extending beam into 2D element mesh.

three nodes have to move together with the rigid strip. The MPC equations can then be established in a similar manner as that discussed for offsets.

Note that a node on the beam has three DOFs whereas the two nodes on the disc have a total of four DOFs. The total DOFs connected to the rigid strip is thus seven. Since the DOF of the rigid strip is

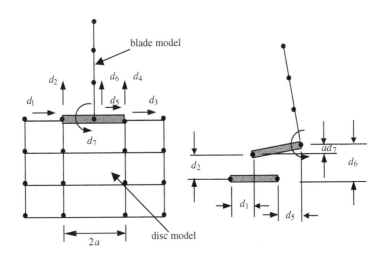

FIGURE 11.37

Modeling of the turbine blade connected to the disc using a rigid strip to create MPC equations.

three, we should have four MPC equations based on Eq. (11.8). The four MPC equations, after eliminating the rigid DOFs, are given as follows:

$$
\begin{aligned}
d_1 &= d_5 \\
d_2 &= d_6 - ad_7 \\
d_3 &= d_5 \\
d_4 &= d_6 + ad_7
\end{aligned}
$$

(11.10)

These equations enforce full compatibility between the beam and the disc.

A similar problem arises when plate or shell elements that possess rotational DOFs are connected to 3D solid elements that do not possess a rotational DOF. Such a connection between a plate and a 3D solid, as shown in Figure 11.38, can be similarly modeled using the above methods.

11.10 Other applications of MPC equations

11.10.1 Modeling of symmetric boundary conditions

When discussing the modeling of symmetric problems, it was mentioned that constraint (boundary) equations are required to ensure that symmetry is properly defined. Note that Eq. (11.2) was obtained when the axis or plane of symmetry is parallel (or perpendicular) to an axis of the Cartesian coordinates. When the axis or plane of symmetry is not parallel (or perpendicular) to x or y axis, as shown in Figure 11.39, the displacement in the normal direction of the axis or plane of symmetry should be zero, i.e.,

$$
d_n = 0
$$

(11.11)

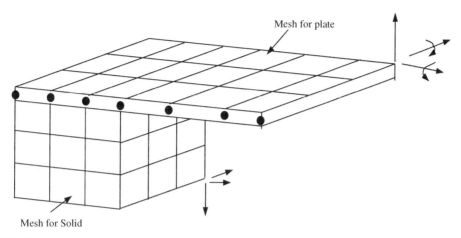

FIGURE 11.38

Joint between plate and 3D solids modeled by extending plate into 3D solid element mesh.

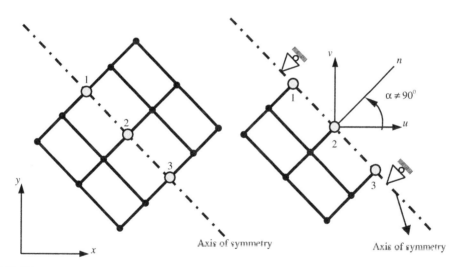

FIGURE 11.39

Imposing mirror symmetry using MPC when the axis of symmetry is not parallel (or perpendicular) to x or y-axis.

which implies

$$u_i \cos\alpha + v_i \sin\alpha = 0 \quad \text{or} \quad u_i + v_i \tan\alpha = 0 \quad \text{for } i = 1, 2, 3. \tag{11.12}$$

where u_i and v_i denote the x and y components of displacement at node i. Equation (11.12) is an MPC equation, as it involves more than one DOF (u_i and v_i). When $\alpha = 0$, the axis of symmetry is parallel to y axis, and Eq. (11.12) will reduce to an SPC as in Eq. (11.2).

11.10.2 **Enforcement of mesh compatibility**

Section 11.5.1 discussed mesh compatibility issues and it was mentioned that to ensure mesh compatibility on the interface of different types of elements, we can make use of MPC equations. We now detail the procedure. With reference to Figure 11.40, the procedure for deriving the MPC equations is as follows:

- Use the lower order shape functions to interpolate the displacements on the edge of the element. In this example, the linear shape functions are used, which gives the displacement at any point on the edge 1-2 as

$$d_x = 0.5(1 - \eta)d_1 + 0.5(1 + \eta)d_3 \tag{11.13}$$

$$d_y = 0.5(1 - \eta)d_4 + 0.5(1 + \eta)d_6 \tag{11.14}$$

- Next substitute the value of η at node 3 into the above two equations to get the constraint equations.

$$0.5\, d_1 - d_2 + 0.5\, d_3 = 0 \tag{11.15}$$

$$0.5\, d_4 - d_5 + 0.5\, d_6 = 0 \tag{11.16}$$

These are the two MPC equations for enforcing the compatibility on the interface between two different types of elements.

When it is not possible to have gradual mesh density transition and a number of elements share a common edge, as shown in Figure 11.41, mesh compatibility can also be enforced by MPC equations. The procedure is given as follows:

- Use the shape functions of the longer element to interpolate the displacements. For the DOF in the x direction, the quadratic shape function gives

$$d_x = -0.5\eta(1 - \eta)\, d_1 + (1 + \eta)(1 - \eta)\, d_3 + 0.5\eta(1 + \eta)\, d_5 \tag{11.17}$$

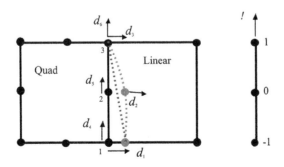

FIGURE 11.40

Enforcing compatibility of two elements of different order using MPC.

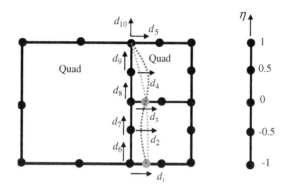

FIGURE 11.41

Enforcing compatibility of two different numbers of elements on a common edge using MPC.

- Substituting the values of η for the two additional nodes of the elements with shorter edges yields

$$d_2 = 0.25 \times 1.5 \, d_1 + 1.5 \times 0.5 \, d_3 - 0.25 \times 0.5 \, d_5 \tag{11.18}$$

$$d_4 = -0.25 \times 0.5 \, d_1 + 0.5 \times 1.5 \, d_3 + 0.25 \times 1.5 \, d_5 \tag{11.19}$$

which can be simplified to the following two constraint equations in the x direction:

$$0.375 \, d_1 - d_2 + 0.75 \, d_3 - 0.125 \, d_5 = 0 \tag{11.20}$$

$$-0.125 \, d_1 + 0.75 \, d_3 - d_4 + 0.375 \, d_5 = 0 \tag{11.21}$$

Similarly, the constraint equations for the displacement in the y direction are obtained as

$$0.375 \, d_6 - d_7 + 0.75 \, d_8 - 0.125 \, d_{10} = 0 \tag{11.22}$$

$$-0.125 \, d_6 + 0.75 \, d_8 - d_9 + 0.375 \, d_{10} = 0 \tag{11.23}$$

Equations (11.20) to (11.23) are a set of MPC equations for enforcing the compatibility of the interface between different numbers of elements.

11.10.3 Modeling of constraints by rigid body attachment

A rigid body attached to an elastic body will impose constraints on it. Figure 11.42 shows a rigid slab sitting on an elastic foundation. Assume that the rigid slab constrains only the vertical movement of the foundation, and it can move freely in the horizontal direction without affecting the foundation.

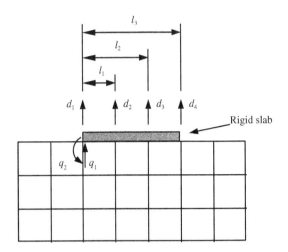

FIGURE 11.42

Modeling of a rigid slab on an elastic foundation using MPC.

To simulate such an effect of the rigid slab sitting on the foundation, MPC equations are required to enforce the four vertical DOFs on the foundation to follow the two DOFs on the slab.

First, the four equations are:

$$\begin{aligned}
d_1 &= q_1 \\
d_2 &= q_1 + q_2\, l_1 \\
d_3 &= q_1 + q_2\, l_2 \\
d_4 &= q_1 + q_2\, l_3
\end{aligned}$$

(11.24)

where q_1 and q_2 are, respectively, the translation and the rotation of the rigid slab. Then, eliminating the DOFs of the rigid body q_1 and q_2 from the above four equations leads to two MPC equations:

$$(l_2/l_1 - 1)\, d_1 - (l_2/l_1)\, d_2 + d_3 = 0$$

(11.25)

$$(l_3/l_1 - 1)\, d_1 - (l_3/l_1)\, d_2 + d_4 = 0$$

(11.26)

In the above example, the DOF in the x direction was not considered because it is assumed that the slab can move freely in the horizontal direction, and hence it will not provide any constraint to the foundation. If the rigid slab is perfectly connected to the foundation, the constraints on the DOFs in the x direction must also be considered.

11.11 Implementation of MPC equations

The previous sections have introduced some procedures for the derivation of MPC equations and some of its applications. However, how are such MPC equations implemented in the process of solving the global FEM system equation? Standard boundary conditions in the form of SPC equations can be easily implemented as each equation concerns only a single DOF. Many examples of implementing such standard boundary conditions of SPC equations can be found in examples and case studies in previous chapters. However, implementation of MPC equations usually requires more complex treatments to the global system equations.

Recall that the global system equation for a static system with DOF of n can be written in the following matrix form.

$$\mathbf{KD} = \mathbf{F} \tag{11.27}$$

where $\mathbf{K}$ is the stiffness matrix of $n \times n$, $\mathbf{D}$ is the displacement vector of $n \times 1$, and $\mathbf{F}$ is the external force vector of $n \times 1$. If the system is constrained by m MPC equations, these equations can always be written in the following general matrix form:

$$\mathbf{CD} - \mathbf{Q} = \mathbf{0} \tag{11.28}$$

where $\mathbf{C}$ is a constant matrix of $m \times n$ ($m < n$), and $\mathbf{Q}$ is a constant matrix of $m \times 1$. For example, let's assume that there is a system of 10 DOFs ($n = 10$), we can then write the displacement vector as

$$\mathbf{D}^T = \left\{ d_1\ d_2\ \cdots\ d_{10} \right\} \tag{11.29}$$

If the two ($m = 2$) MPC equations for this system are given as

$$0.5d_1 - d_2 + 0.5d_3 = 0 \tag{11.30}$$

$$0.5d_4 - d_5 + 0.5d_6 = 0 \tag{11.31}$$

which has the same form as that given by Eqs. (11.15) and (11.16). We then rewrite these MPCs in the following form of

$$0.5d_1 - d_2 + 0.5d_3 + 0 \times d_4 + 0 \times d_5 + 0 \times d_6 + 0 \times d_7 + 0 \times d_8 + 0 \times d_9 + 0 \times d_{10} = 0 \tag{11.32}$$

$$0 \times d_1 + 0 \times d_2 + 0 \times d_3 + 0.5d_4 - d_5 + 0.5d_6 + 0 \times d_7 + 0 \times d_8 + 0 \times d_9 + 0 \times d_{10} - 0 \tag{11.33}$$

From the above two equations, the matrix $\mathbf{C}$ is thus obtained as

$$\mathbf{C} = \begin{bmatrix} 0.5 & -1 & 0.5 & 0 & 0 & 0 & 0 & 0 & 0 & 0 \\ 0 & 0 & 0 & 0.5 & -1 & 0.5 & 0 & 0 & 0 & 0 \end{bmatrix} \tag{11.34}$$

and the vector $\mathbf{Q}$ is a null vector given by

$$\mathbf{Q} = \left\{ \begin{matrix} 0 \\ 0 \end{matrix} \right\} \tag{11.35}$$

The problem at hand is to obtain a solution for Eq. (11.27) subjected to the constraint equations given by Eq. (11.28). The following are two commonly used methods to obtain such a solution.

11.11.1 Lagrange multiplier method

First, the following m additional variables called *Lagrange multipliers* are introduced in this method as

$$\boldsymbol{\lambda} = \left[\lambda_1 \; \lambda_2 \; \cdots \; \lambda_m \right]^T \tag{11.36}$$

corresponding to m MPC equations. Each equation in the matrix equation of Eq. (11.28) is then multiplied by the corresponding λ_i, which yields

$$\boldsymbol{\lambda}^T \{ \mathbf{CD} - \mathbf{Q} \} = 0 \tag{11.37}$$

The left-hand side of this equation is then added to the usual functional (refer to Eq. (3.2) of Chapter 3) to obtain the *modified functional* for the constraint system.

$$\Pi_{\mathrm{p}} = \frac{1}{2}\mathbf{D}^T\mathbf{KD} - \mathbf{D}^T\mathbf{F} + \boldsymbol{\lambda}^T\{\mathbf{CD} - \mathbf{Q}\} \tag{11.38}$$

The solution for the constraint system is found at the stationary point of the modified functional. The stationary condition requires the derivatives of Π_{p} with respect to the $\mathbf{D}_i$ and λ_i to vanish, and together with Eq. (11.37), gives

$$\begin{bmatrix} \mathbf{K} & \mathbf{C}^T \\ \mathbf{C} & \mathbf{0} \end{bmatrix} \begin{Bmatrix} \mathbf{D} \\ \boldsymbol{\lambda} \end{Bmatrix} = \begin{Bmatrix} \mathbf{F} \\ \mathbf{Q} \end{Bmatrix} \tag{11.39}$$

which is the set of algebraic equations for the constraint system. Equation (11.39) is then solved instead of Eq. (11.27) to obtain the solution that satisfies the MPC equations. Note that from Eq. (11.39), $\mathbf{KD} + \mathbf{C}^T\boldsymbol{\lambda} = \mathbf{F}$ is a matrix equation consisting of n equations with $n+m$ unknowns. $\mathbf{CD} = \mathbf{Q}$ therefore provides the additional m equations that allow us to solve for all the unknowns including the m Lagrange multipliers.

The advantage of this method is that the constraint equations are satisfied exactly. However, it can be seen that the total number of unknowns is increased. In addition, the expanded stiffness matrix in Eq. (11.39) is non-positive definitely due to the presence of zero diagonal terms. Therefore, the efficiency of solving the system equations (11.39) is much lower than that of simply solving Eq. (11.27).

11.11.2 Penalty method

For the penalty method, Eq. (11.28) is re-written as

$$\mathbf{t} = \mathbf{CD} - \mathbf{Q} \tag{11.40}$$

so that $\mathbf{t} = \mathbf{0}$ implies that the constraints are fully satisfied. The functional (refer to Eq. (3.2) of Chapter 3) can then be modified as

$$\Pi_p = \frac{1}{2}\mathbf{D}^T\mathbf{KD} - \mathbf{D}^T\mathbf{F} + \frac{1}{2}\mathbf{t}^T\boldsymbol{\alpha}\mathbf{t} \tag{11.41}$$

where $\boldsymbol{\alpha}$ is a diagonal matrix of "penalty numbers" $(\alpha_1, \alpha_2, \ldots, \alpha_m)$ that are constants chosen by the analyst. The stationary condition of the modified functional requires the derivatives of Π_p with respect to the $\mathbf{D}_i$ to vanish, which yields

$$[\mathbf{K} + \mathbf{C}^T\boldsymbol{\alpha}\mathbf{C}]\mathbf{D} = \mathbf{F} + \mathbf{C}^T\boldsymbol{\alpha}\mathbf{Q} \tag{11.42}$$

where $\mathbf{C}^T\boldsymbol{\alpha}\mathbf{C}$ is called a "penalty matrix." From Eq. (11.42), it is seen that if $\alpha=0$, the constraints are ignored and the system equation reduces to the unconstrained equation Eq. (11.27). As α_i becomes large, the penalty of violating the constraints becomes large. The terms with $\boldsymbol{\alpha}$ in Eq. (11.42) overwhelm other terms. Therefore the solution satisfying Eq. (11.42) satisfies the constraints almost completely.

Note that the choice of α_i can be a tricky task, as it depends on many factors in the FEM model. Considering that the discretization errors can be of comparable magnitude to those of not satisfying the constraint, it has been suggested that (Zienkiewicz and Taylor, 2000)

$$\alpha = \text{constant } (1/h)^{p+1} \tag{11.43}$$

where h is the characteristic size of the elements, and p denotes the order of the element used. The following simple method for calculating the penalty factor works well for most problems:

$$\alpha = 1.0 \times 10^{4(\text{to } 6)} \times \max \{\text{diagonal elements in the stiffness matrix}\} \tag{11.44}$$

It is also common to use

$$\alpha = 1.0 \times 10^{5(\text{to } 8)} \times \text{Young's modulus} \tag{11.45}$$

Note that several trials may be required to decide on an appropriate penalty factor.

The advantage of this method is that the total number of unknowns is not changed and the system equations generally behave well. Therefore, the efficiency of solving the system equations Eq. (11.42) is almost the same as that of solving Eq. (11.27). However, the constraint equations can only be satisfied approximately, and the right choice of α can be ambiguous in some cases.

11.12 Review questions

1. What are the conditions of the use of mirror symmetry for reducing the size of the finite element discretization of a problem? Give your answer with reference to geometry, material, boundary conditions, and loading.

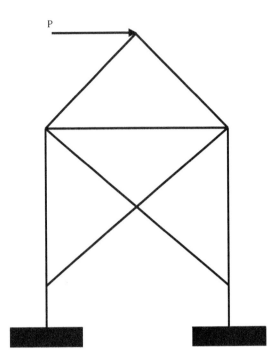

FIGURE 11.43

Frame structure subjected to horizontal load *P*.

2. Indicate how the symmetric and anti-symmetric conditions can be used on a half-model to obtain the natural frequencies of a clamped-clamped beam of length L.

3. Figure 11.43 shows a frame structure subjected to a horizontal force at the top. Explain the use of symmetry to solve the problem with the aid of diagrams.

4. A symmetrical frame structure is under an asymmetric load P as shown in Figure 11.44. Describe with the aid of diagrams how you would use the symmetry to your advantage in modeling. Discuss the merits of using a symmetrical model in terms of the modeling and computational effort.

5. Explain why a transition element or MPCs are needed between the quadratic and linear elements shown in Figure 11.45.

6. Construct the shape functions in natural coordinates for the transition element shown in Figure 11.46. What is another way of solving the problem in Figure 11.45 if a transition element is not used?

7. Figure 11.47 shows a uniform mesh for a plane strain problem. The shaded block is rigid. Derive the MPC equations for nodes 1, 2, 3, and 4.

8. Figure 11.48 shows two planar beams perfectly joined to a rigid square block A (shaded) at right angles. The two beams are modeled using beam elements.
   a. Derive the MPC equations for nodes 1 and 2.
   b. Give an alternative method to model the rigid block, and comment on this method compared with the method using the MPC equations.

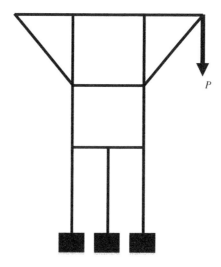

FIGURE 11.44

Frame structure subjected to vertical load *P*.

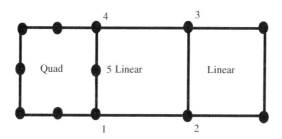

FIGURE 11.45

Mesh consisting of quadrilateral and linear elements.

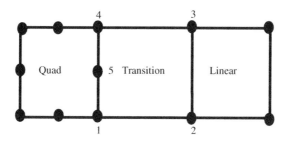

FIGURE 11.46

Mesh consisting of a transition element.

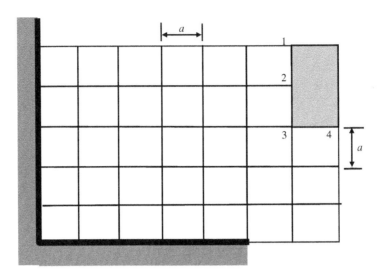

FIGURE 11.47

Mesh consisting of rigid shaded portion.

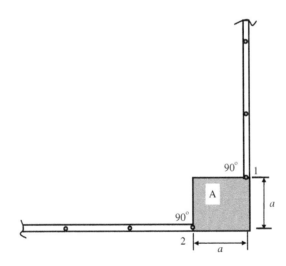

FIGURE 11.48

Two beams perfectly joined to a rigid block A.

9. Consider a 2D plane strain problem, as shown in Figure 11.49, where a long beam is an extension of a main structure. The structure is subjected to dynamic loading.
  a. The global response of the whole structural system is of interest. Design a mesh for the area within the large circle. If MPCs are needed, give the MPC equations.

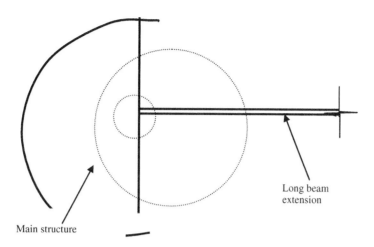

FIGURE 11.49

Long beam extension from a bulky 2D solid.

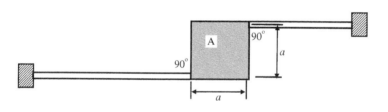

FIGURE 11.50

Two beams perfectly joined to a rigid block A.

   b. The detailed results for the area within the small circle are of interest. Design a mesh for the area within the small circle. If MPCs are needed, give the MPC equations.
10. Figure 11.50 shows a rigid square body A (shaded) connected perfectly by two planar cantilever beams. You are requested to model this problem using a commercial FE package. The two beams should be modeled using beam elements.
   a. Briefly explain three possible methods to model this problem, and the advantages and disadvantages of the methods.
   b. Derive these MPC equations that can be used to model the rigid body A.
11. Figure 11.51 shows a rigid square body A (shaded) connected perfectly to a rectangular 2D solid. You are requested to model this problem using a commercial FE package using 2D solid elements. Derive MPC equations that can be used to properly model the rigid body A.

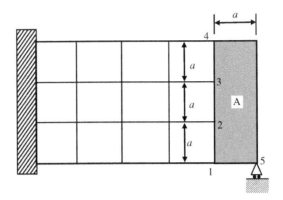

FIGURE 11.51

Rigid square body A (shaded) connected perfectly to a rectangular 2D solid.

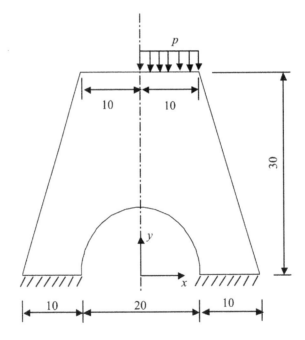

FIGURE 11.52

Structure of a 2D solid.

12. Figure 11.52 shows a bridge pier made of uniform material subjected to a uniformly distributed pressure on the right half of its top surface. The bridge pier can be idealized as a linear elastic 2D plane strain problem (in the x-y plane).

   a. Explain your strategy to model this particular problem.

b. Sketch a mesh for this problem domain of your model using about 10 elements.

c. Provide all the boundary conditions, at relevant nodes, which are needed to solve this problem using your model.

d. How should the pressure be modeled?

13. How are the constraint equations implemented using the penalty method in the system equation $\mathbf{KD} = \mathbf{F}$? Here, $\mathbf{K}$ is the stiffness matrix for the global system, $\mathbf{D}$ is the displacement vector at all the nodes, and $\mathbf{F}$ is the external force vector acting at all the nodes. Assume that the constraint equations are given in the general form $\mathbf{CD} = \mathbf{Q}$, where $\mathbf{C}$ and $\mathbf{Q}$ are given constant matrices.

CHAPTER OUTLINE HEAD

12.1 **Field problems**

This chapter introduces the finite element method to solve steady-state heat transfer problems. Many of the materials in this chapter are from the textbook by Segerlind (1984). Heat transfer problems can be categorized generally under one of the many field problems. A field problem is a boundary value problem with the variables (degrees of freedom (DOFs)) described as a varying field in our simulation domain. The distinction here from earlier chapters is that the field is described in a domain comprising of a static mesh—a Eularian mesh. In the earlier chapters, the equations were formulated in a Lagragian mesh, whereby the nodes are displaced according to the calculated displacement and hence the mesh gets deformed. Other common field problems include torsional deformation of bars, irrotational flow, and acoustic problems. This book will make use of the heat transfer problem to introduce the concepts behind the solving of field problems using FEM. Emphasis will be placed on one-dimensional and two-dimensional heat transfer problems. Three-dimensional problems can be solved in similar ways except for the increase in DOFs due to the dependency of field variables on the third dimension. The approach used to derive FE equations in heat transfer problems is a general approach to solve partial differential equations using FEM. Therefore, the FE equations developed for heat transfer problems are directly applicable to all other types of field problems that are governed by similar types of partial differential equations.

The general form of system equations of 2D linear steady-state field problems can be given by the following general form of the *Helmholtz equation*:

$$D_x \frac{\partial^2 \phi}{\partial x^2} + D_y \frac{\partial^2 \phi}{\partial y^2} - g\phi + Q = 0 \tag{12.1}$$

where ϕ is the field variable, D_x, D_y, g, and Q are given constants whose physical meaning is different for different problems. For one-dimensional field problems, the general form of system equations can be written as

$$D\frac{d^2 \phi}{dx^2} - g\phi + Q = 0 \tag{12.2}$$

The following subsections introduce different physical problems governed by equations in the form given by Eq. (12.1) or (12.2).

12.1.1 **Heat transfer in a two-dimensional fin**

Consider heat transfer in a two-dimensional fin as shown in Figure 12.1. A 2D fin is mounted on a pipe. Heat conduction occurs in the x–y plane, and heat convection occurs on the two surfaces and edges. Assuming that the fin is very thin, so that the temperature does not vary significantly in the thickness (z) direction, therefore the temperature field is only a function of x and y. The governing equation for the temperature field in the fin, denoted by function $\phi(x, y)$, can be given by

$$\underbrace{-\left(k_x t \frac{\partial^2 \phi}{\partial x^2} + k_y t \frac{\partial^2 \phi}{\partial y^2}\right)}_{\text{Heat conduction}} + \underbrace{2h \overbrace{(\phi - \phi_f)}^{\text{Temperature difference}}}_{\text{Heat convectction}} = \underbrace{q}_{\text{Heat supply}} \tag{12.3}$$

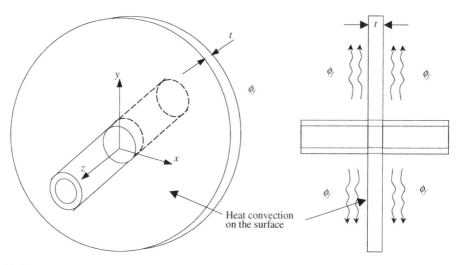

FIGURE 12.1

A 2D fin mounted on a pipe. Heat conduction occurs in the *x–y* plane, and heat convection occurs on the two surfaces and edges.

where k_x and k_y are, respectively, the thermal conductivity coefficients in the x and y directions, h is the convection coefficient, t is the thickness of the fin, and ϕ_f is the ambient temperature of the surrounding fluid. The heat supply that can be a function of x and y is denoted by q. The governing equation (12.3) can be derived simply by using Fourier's laws of heat conduction and heat convection, as well as the conservation law of energy (heat). Equation (12.3) simply states that the heat loss due to heat conduction and heat convection should be equal to the heat supply at any point in the fin.

It can be seen that Eq. (12.3) takes on the general form of a field problem as in Eq. (12.1) with the substitution of,

$$D_x = k_x t, D_y = k_y t, g = 2h, \quad Q = q + 2h\phi_f \tag{12.4}$$

12.1.2 Heat transfer in a long two-dimensional body

If the domain is elongated in the z direction, and both geometry and temperature do not vary in the z direction, as illustrated in Figure 12.2, then a representative 2D "slice" can be used to model the problem. In this case, there is no heat convection occurring on the two surfaces of the 2D slice, and the governing equation becomes

$$\underbrace{k_x \frac{\partial^2 \phi}{\partial x^2} + k_y \frac{\partial^2 \phi}{\partial y^2}}_{\text{Heat conduction}} + \underbrace{q}_{\text{Heat supply}} = 0 \tag{12.5}$$

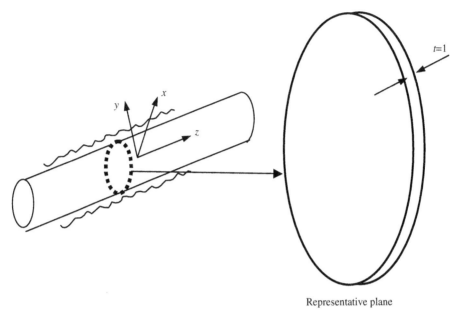

Representative plane

FIGURE 12.2

Heat transfer in a long body.

which relates to the general form of the field equation with the following substitutions:

$$D_x = k_x, D_y = k_y, g = 0 \text{ and } Q = q \tag{12.6}$$

12.1.3 Heat transfer in a one-dimensional fin

Consider a fin of constant cross-section as shown in Figure 12.3. Assuming that the fin is long and slender, the dimension of the cross-section of the fin is much smaller than the length of the fin. The temperature can be assumed to be constant in the cross-section, and hence it is only a function of x. Heat conduction occurs in the x direction, and heat convection occurs on the circumferential surface of the fin. The temperature field in the fin, denoted by function $\phi(x)$, is governed by the following equation

$$\underbrace{kA\frac{d^2\phi}{dx^2}}_{\text{Heat conduction}} \underbrace{- hP\left(\phi - \phi_f\right)}_{\text{Heat convection}} + \underbrace{q}_{\text{Heat supply}} = 0 \tag{12.7}$$

where k is the thermal conductivity, h is the convection coefficient, A is the cross-sectional area of the fin, and P is the perimeter of the fin, as shown in Figure 12.3. Eq. (12.7) can be derived in the same way as its 2D equivalent in Eq. (12.3). Eq. (12.7) can be written in the general form of Eq. (12.2) with the substitutions of

$$D = kA, \quad g = hp, \quad Q = q + hP\phi_f \tag{12.8}$$

FIGURE 12.3

Heat conduction along a thin one-dimensional fin.

12.1.4 **Heat transfer across a composite wall**

Many walls of industrial structures or simple appliances like the thermal flask are composite in nature. They consist of more than one material usually stacked up in layers. By utilizing the thermal conductive properties of the material chosen with a proper stacking sequence, either thermal insulation or effective thermal heat transfer through the walls can be achieved.

Consider the heat transfer across a composite wall as shown in Figure 12.4. The wall is assumed to be infinitely long in the y direction, and hence the heat source and any heat exchanges are also independent of y. In this case, the problem is one-dimensional. Since the wall is infinitely long, it is also not possible to have heat convection along the x axis. Therefore, the system equation in this case is much

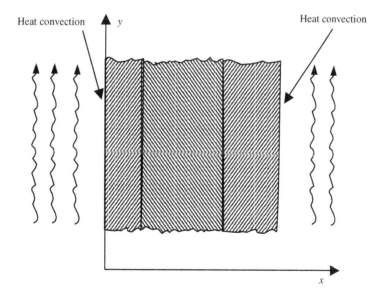

FIGURE 12.4

Heat transfer through a composite wall of three layers assuming that the temperature does not vary in the y direction.

simpler as compared to that of the fin, and it governs only the heat conduction through the composite wall, which can be given by

$$\underbrace{kA\frac{d^2\phi}{dx^2}}_{\text{Heat conduction}} + \underbrace{q}_{\text{Heat supply}} = 0 \qquad (12.9)$$

where ϕ is the known temperature. Eq. (12.9) can be written in a general form of Eq. (12.2) with

$$D = kA, \quad g = 0, \quad Q = q \qquad (12.10)$$

12.1.5 Torsional deformation of a bar

For problems of torsional deformation of a bar with noncircular sections, the field variable will be the stress function ϕ that is governed by a Poisson's equation,

$$\frac{1}{G}\frac{\partial^2\phi}{\partial x^2} + \frac{1}{G}\frac{\partial^2\phi}{\partial y^2} + 2\theta = 0 \qquad (12.11)$$

where G is the shear modulus, and θ is the given angle of twist. The stress function is defined by

$$\sigma_{xz} = \frac{\partial\phi}{\partial y}$$
$$\sigma_{yz} = -\frac{\partial\phi}{\partial x} \qquad (12.12)$$

Equation (12.11) can be easily derived in the following procedure.
    Consider a pure torsional state where

$$\sigma_{xx} = \sigma_{yy} = \sigma_{zz} = \sigma_{xy} = 0 \qquad (12.13)$$

Therefore, there are only two stress components, σ_{xz} and σ_{yz}. Using Hooke's law, we have

$$\gamma_{xz} = \frac{\sigma_{xz}}{G}$$
$$\gamma_{yz} = \frac{\sigma_{yz}}{G} \qquad (12.14)$$

The relationship between displacement and the given angle of twist θ can be given by (Fung, 1965):

$$u = -\theta yz$$
$$v = \theta xz \qquad (12.15)$$

Using Eqs. (2.4), (12.15), (12.14), and (12.12) we obtain

$$\frac{\partial w}{\partial x} = -\frac{\partial u}{\partial z} + \gamma_{xz} = \theta y + \frac{1}{G}\frac{\partial\phi}{\partial y} \qquad (12.16)$$

$$\frac{\partial w}{\partial y} = -\frac{\partial v}{\partial z} + \gamma_{yz} = \theta x - \frac{1}{G}\frac{\partial\phi}{\partial x} \qquad (12.17)$$

Differentiating Eq. (12.16) with respect to y and Eq. (12.17) with respect to x, and equating the two resultant equations, leads to Eq. (12.11).

Once again, it can be seen that Eq. (12.11) takes on the general form of Eq. (12.1) with

$$D_x = 1/G, D_y = 1/G, g = 0, Q = 2\theta \tag{12.18}$$

12.1.6 Ideal irrotational fluid flow

For problems of ideal, irrotational fluid flow, the field variables are the streamline, ψ, and potential, ϕ, functions that are governed by the Laplace's equations,

$$\frac{\partial^2 \psi}{\partial x^2} + \frac{\partial^2 \psi}{\partial y^2} = 0 \tag{12.19}$$

$$\frac{\partial^2 \phi}{\partial x^2} + \frac{\partial^2 \phi}{\partial y^2} = 0 \tag{12.20}$$

The derivation of the above equations can be found in any textbook on fluid flows (see, for example, Daily and Harleman, 1966).

It can be seen that both Eqs. (12.19) and (12.20) take on the form of Eq. (12.1) with

$$D_x = D_y = 1, g = Q = 0 \tag{12.21}$$

12.1.7 Acoustic problems

For problems of a vibrating fluid in a closed volume, as in the case of the vibration of air particles in a room, the field variable, P, is the pressure above the ambient pressure and is governed by the Poisson's equation,

$$\frac{\partial^2 P}{\partial x^2} + \frac{\partial^2 P}{\partial y^2} + \frac{w^2}{c^2} P = 0 \tag{12.22}$$

where w is the wave frequency and c is the wave velocity in the medium. The derivation of the above equations can be found in any book on acoustics (see, for example, Crocker, 1998). It can be seen that Eq. (12.22) takes on the form of Eq. (12.1) with substitutions of

$$g = -\frac{w^2}{c^2}, D_x = D_y = 1, Q = 0 \tag{12.23}$$

The examples above show that the Helmholtz equation (12.1) governs a wide variety of physical phenomena. Therefore, a productive way to solve all these problems using FEM is to derive FE equations for solving the general form of Eq. (12.1) or (12.2). The following sections will thus introduce FEM as a general numerical tool for solving partial differential equations in the form of Eq. (12.1) or (12.2). We will focus on the application of the FEM to solve heat transfer problems for the rest of the chapter.

12.2 **Weighted residual approach for FEM**

In Chapter 3, the Hamilton's principle is used to formulate the FE equations. In that case, one needs to know the functional (the expression of system energy in terms of displacements) for the physical problem. For many engineering problems, one does not know the functional or it is not known as intuitively as in mechanics problems. Instead, the governing equation for the problem would be known. What we want to do now is to establish FEM equations based on the governing equation, but without knowing the functional. For this, it is convenient to use the *weighted residual* approach to formulate the FEM system equations.

The general form of Eq. (12.1) can be rewritten in the form of

$$f(\phi(x, y)) = 0 \tag{12.24}$$

where the function f is given as

$$f(\phi(x, y)) = D_x \frac{\partial^2 \phi}{\partial x^2} + D_y \frac{\partial^2 \phi}{\partial y^2} - g\phi + Q \tag{12.25}$$

In general, it is difficult to obtain the *exact* solution of $\phi(x, y)$ which satisfies Eq. (12.24).

Therefore, an approximated solution of $\phi(x, y)$ is sought for, which satisfies Eq. (12.24) in a weighted integral sense, i.e.,

$$\int_A Wf(\phi(x, y))dxdy = 0 \tag{12.26}$$

where W is the weight function. By solving for $\phi(x, y)$ in Eq. (12.26), we want the solution to be a good approximation to the exact solution of Eq. (12.24). This is the basic idea behind the weighted residual approach. This approach is simple and elegant, and can be used in the finite element method to establish the discretized system of equations as described below.

To ensure a good approximation, the problem domain is divided into smaller sub-domains (elements) as we have done in previous chapters, as shown in Figure 12.5. In each element, it is assumed that

$$\phi^h(x, y) = \mathbf{N}(x, y)\mathbf{\Phi}^{(e)} \tag{12.27}$$

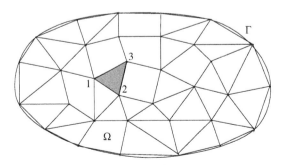

FIGURE 12.5

Division of problem domain Ω bounded by Γ into elements.

where the superscripts indicate that the field variable is approximated, and

$$\mathbf{N} = \begin{bmatrix} N_1 & N_1 \dots N_{n_d} \end{bmatrix} \tag{12.28}$$

in which N_i is the shape function of x and y, and n_d is the number of the nodes of the element. For the triangular element shown in Figure 12.5, $n_d = 3$. In Eq. (12.27), $\Phi^{(e)}$ is the field variable at the nodes of the element. There are a number of ways in which the weight function, W, can be chosen when developing the element equations. When the shape functions are used as the weight function, the method is called the Galerkin method, which is one of the most popular methods for developing FE equations. The shape function N_i in $\mathbf{N}$ is usually developed in exactly the same manner as that discussed in Chapter 3 and other previous chapters. The developed functions should in general satisfy the sufficient conditions discussed in Chapter 3: the delta function property, the partition of unity property, and the linear reproduction property.

Using the Galerkin method, the residual calculated at all the nodes for an element is then evaluated by the following equation.

$$\mathbf{R}^{(e)} = \int_{A_e} \mathbf{N}^T f(\phi(x, y)) dx dy \tag{12.29}$$

Finally, the total residual at each of the nodes in the problem domain is then assembled and enforced to zero to establish the system equation for the whole system. FEM equations will be developed in the following sections for one- and two-dimensional field problems, using a heat transfer problem as an example.

12.3 1D heat transfer problem
12.3.1 One-dimensional fin

Consider a one-dimensional fin as shown in Figure 12.6. The governing differential equation for steady-state heat transfer problems for the fin is given by Eq. (12.7). The boundary conditions associated with Eq. (12.7) usually consist of a specified temperature at $x = 0$

$$\phi(0) = \phi_0 \tag{12.30}$$

and convection heat loss at the free end where $x = H$

$$-kA\frac{d\phi}{dx} = hA\left(\phi_b - \phi_f\right) \text{ at } x = H \tag{12.31}$$

where ϕ_b is the temperature at the end of the fin and is not known prior to obtaining the solution of the problem. Note that the convective heat transfer coefficient in Eq. (12.31) may or may not be the same as the one in Eq. (12.7).

Using the same finite element technique, the fin is divided into elements as shown in Figure 12.6. For each element, the residual equation, $\mathbf{R}^{(e)}$ can be obtained using the Galerkin approach as in Eq. (12.29):

$$
\begin{aligned}
\mathbf{R}^{(e)} &= -\int_{x_i}^{x_j} \mathbf{N}^T \left(D \frac{d^2 \phi^h}{dx^2} - g\phi^h + Q \right) dx \\
&= -\int_{x_i}^{x_j} \mathbf{N}^T \left(D \frac{d^2 \phi^h}{dx^2} + Q \right) dx + \int_{x_i}^{x_j} g\mathbf{N}^T \phi^h dx
\end{aligned}
\tag{12.32}
$$

where $D=kA$, $g=hP$, and $Q=hP\phi_f$ for heat transfer in thin fins. Note that the minus sign is added to the residual mainly for convenience. *Integration by parts* is then performed on the first term of the right-hand side of Eq. (12.32), leading to

$$
\mathbf{R}^{(e)} = -\mathbf{N}^T D \frac{d\phi^h}{dx} \bigg|_{x_i}^{x_j} + \int_{x_i}^{x_j} \frac{d\mathbf{N}^T}{dx} D \frac{d\phi^h}{dx} dx - \int_{x_i}^{x_j} Q\mathbf{N}^T dx + \int_{x_i}^{x_j} g\mathbf{N}^T \phi^h dx
\tag{12.33}
$$

Using the usual interpolation of the field variable, ϕ^h, by the shape functions in the 1D case,

$$
\phi^h(x) = \mathbf{N}(x)\boldsymbol{\Phi}^{(e)}
\tag{12.34}
$$

and substituting Eq. (12.34) into Eq. (12.33), gives

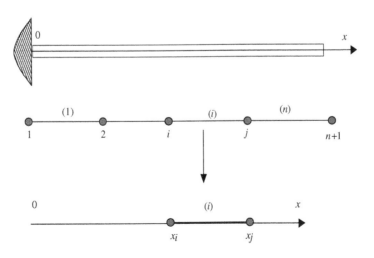

FIGURE 12.6

One-dimensional problem: Heat transfer in a thin fin that is divided into n elements.

$$\mathbf{R}^e = \underbrace{-\mathbf{N}^T D \frac{d\phi^h}{dx}\bigg|_{x_i}^{x_j}}_{\mathbf{b}^{(e)}} + \left(\underbrace{\int_{x_i}^{x_j} \frac{d\mathbf{N}^T}{dx} D \frac{d\mathbf{N}}{dx} dx}_{\mathbf{k}_D^{(e)}}\right) \mathbf{\Phi}^{(e)}$$

$$-\left(\underbrace{\int_{x_i}^{x_j} Q\mathbf{N}^T dx}_{\mathbf{f}_Q^{(e)}}\right) + \left(\underbrace{\int_{x_i}^{x_j} g\mathbf{N}^T N \ dx}_{\mathbf{k}_g^{(e)}}\right) \mathbf{\Phi}^{(e)}$$ (12.35)

or

$$\mathbf{R}^{(e)} = \mathbf{b}^{(e)} + \left[\mathbf{k}_D^{(e)} + \mathbf{k}_g^{(e)}\right]\boldsymbol{\phi}^{(e)} - \mathbf{f}_Q^{(e)}$$ (12.36)

whereby

$$\mathbf{k}_D^{(e)} = \int_{x_i}^{x_j} \frac{d\mathbf{N}^T}{dx} D \frac{d\mathbf{N}}{dx} dx = \int_{x_i}^{x_j} \mathbf{B}^T D\mathbf{B}dx$$ (12.37)

and is the element matrix of thermal conduction. Matrix $\mathbf{K}_g^{(e)}$ in Eq. (12.36) is defined by

$$\mathbf{k}_g^{(e)} = \int_{x_i}^{x_j} g\mathbf{N}^T \mathbf{N}dx$$ (12.38)

which is the matrix of thermal convection on the circumference of the element. Vector $\mathbf{f}_Q^{(e)}$ is associated with the external heat applied on the element defined as

$$\mathbf{f}_Q^{(e)} = \int_{x_i}^{x_j} Q\mathbf{N}^T dx$$ (12.39)

Finally, $\mathbf{b}^{(e)}$ is defined by

$$\mathbf{b}^{(e)} = -\mathbf{N}^T D \frac{d\phi^h}{dx}\bigg|_{x_i}^{x_j}$$ (12.40)

which is associated with the gradient of the temperature (or heat flux) at the two ends of the element.
   In Eq. (12.37), we see the matrix

$$\mathbf{B} = \frac{d\mathbf{N}}{dx}$$ (12.41)

which is the same as the *strain matrix* for structural mechanics problems in earlier chapters. For linear elements, the shape functions are as follows:

$$\mathbf{N}(x) = \begin{bmatrix} N_i & N_j \end{bmatrix} = \begin{bmatrix} \dfrac{x_j - x}{l} & \dfrac{x - x_i}{l} \end{bmatrix}$$ (12.42)

and

$$\mathbf{B} = \frac{d\mathbf{N}}{dx} = \frac{d}{dx}\left[\frac{x_j - x}{l} \quad \frac{x - x_i}{l}\right] = \left[\frac{-1}{l} \quad \frac{1}{l}\right] \tag{12.43}$$

Substituting the above equation into Eq. (12.37), the heat conduction matrix, $\mathbf{k}_D^{(e)}$, is obtained as

$$\mathbf{K}_D^{(e)} = \int_{x_i}^{x_j}\left[\begin{matrix}\frac{-1}{l}\\\frac{1}{l}\end{matrix}\right]D\left[\frac{-1}{l} \quad \frac{1}{l}\right]dx = \frac{kA}{l}\left[\begin{matrix}1 & -1\\-1 & 1\end{matrix}\right] \tag{12.44}$$

The definition of $D=kA$ has been used to derive the above equation. Compared with Eq. (4.15), it is seen that $\mathbf{k}_D^{(e)}$ is in fact an analogy of the stiffness matrix for the 1D truss structure in Chapter 4. The tensile stiffness coefficient $\frac{AE}{l}$ has been replaced by the heat conductivity coefficient of $\frac{kA}{l}$.

Similarly, to obtain the convection matrix $\mathbf{k}_g^{(e)}$ corresponding to the heat convection, substitute Eq. (12.42) in Eq. (12.38),

$$\mathbf{k}_g^{(e)} = \int_{x_i}^{x_j} g\left[\begin{matrix}\frac{x_j-x}{l}\\\frac{x-x_i}{l}\end{matrix}\right]\left[\frac{x_j-x}{l} \quad \frac{x-x_i}{l}\right]dx = \frac{hPl}{6}\left[\begin{matrix}2 & 1\\1 & 2\end{matrix}\right] \tag{12.45}$$

The definition of $g=hP$ has been used to derive the above equation. Compared with Eq. (4.16), it is seen that $\mathbf{k}_g^{(e)}$ is in fact an analogy of the mass matrix of the 1D truss structure. The total mass of the truss element $A\rho l$ corresponds to the total heat convection rate of hPl.

The nodal heat vector $\mathbf{f}_Q^{(e)}$ is obtained when Eq. (12.42) is substituted into Eq. (12.39), giving

$$\mathbf{f}_Q^{(e)} = \int_{x_i}^{x_j} Q\left[\begin{matrix}\frac{x_j-x}{l}\\\frac{x-x_i}{l}\end{matrix}\right]dx = \frac{Ql}{2}\left\{\begin{matrix}1\\1\end{matrix}\right\} = (\underbrace{q}_{\text{Heat supply}} + \underbrace{hP\phi_f}_{\text{Heat convection}})\frac{l}{2}\left\{\begin{matrix}1\\1\end{matrix}\right\} \tag{12.46}$$

The definition of $Q=q+hP\phi_f$ (see, Eq. (12.8)) has been used to derive the above equation. The nodal heat vector consists of the heat supply and the convective heat input to the fin. The nodal heat vector is an analogy of the nodal force vector for the 1D truss element.

Finally, let's analyze the vector $\mathbf{b}^{(e)}$ defined by Eq. (12.40), which is associated with the thermal conditions on the boundaries of the element:

$$\mathbf{b}^{(e)} = -\mathbf{N}^T D\frac{d\phi}{dx}\Big|_{x_i}^{x_j} = \left\{\begin{matrix}kA\frac{d\phi}{dx}\big|_{x=x_i}\\-kA\frac{d\phi}{dx}\big|_{x=x_j}\end{matrix}\right\} = \underbrace{\left\{\begin{matrix}kA\frac{d\phi}{dx}\big|_{x=x_i}\\0\end{matrix}\right\}}_{\mathbf{b}_L^{(e)}} + \underbrace{\left\{\begin{matrix}0\\-kA\frac{d\phi}{dx}\big|_{x=x_j}\end{matrix}\right\}}_{\mathbf{b}_R^{(e)}} \tag{12.47}$$

or

$$\mathbf{b}^{(e)} = \mathbf{b}_L^{(e)} + \mathbf{b}_R^{(e)} \tag{12.48}$$

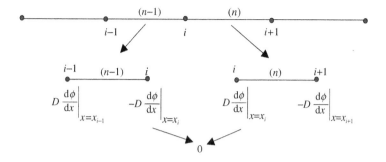

FIGURE 12.7

Vector $\mathbf{b}^{(e)}$ vanishes at internal points after assembly.

where the subscripts "L" and "R" represent the left and right ends of the element. It can be easily proven that at the internal nodes of the fin, $\mathbf{b}_L^{(e)}$ and $\mathbf{b}_R^{(e)}$ vanish when the elements are assembled, as illustrated in Figure 12.7. As a result, $\mathbf{b}$ needs to be determined only for the nodes on the boundary by using the conditions prescribed. At boundaries where the temperature is prescribed, as shown in Eq. (12.30), $\mathbf{b}_L^{(e)}$ or $\mathbf{b}_R^{(e)}$ remain an unknown until the full solution is obtained. In fact, it is not necessary to evaluate $\mathbf{b}_L^{(e)}$ or $\mathbf{b}_R^{(e)}$ while solving the system equation, as the temperature at the node is already known. This situation is very much similar to the prescribed displacement boundary, where the reaction force is usually unknown while solving system equations.

However, when there is heat convection at the ends of the fin as prescribed in Eq. (12.31), $\mathbf{b}_L^{(e)}$ or $\mathbf{b}_R^{(e)}$ is obtained using the boundary conditions, since the heat flux there can be calculated. For example, for boundary conditions defined by Eq. (12.31), we have

$$\mathbf{b}_R^{(e)} = \left\{ \begin{array}{c} 0 \\ hA\left(\phi_b - \phi_f\right) \end{array} \right\} = \left\{ \begin{array}{c} 0 \\ hA\phi_j \end{array} \right\} - \left\{ \begin{array}{c} 0 \\ hA\phi_f \end{array} \right\} \tag{12.49}$$

Since ϕ_b is the temperature of the fin at the boundary point, we have $\phi_b = \phi_j$, which is an unknown. Eq. (12.49) can be rewritten as

$$\mathbf{b}_R^{(e)} = \underbrace{\begin{bmatrix} 0 & 0 \\ 0 & hA \end{bmatrix}}_{\mathbf{k}_M^{(e)}} \left\{ \begin{array}{c} \phi_i \\ \phi_j \end{array} \right\} - \underbrace{\left\{ \begin{array}{c} 0 \\ hA\phi_f \end{array} \right\}}_{\mathbf{f}_S^{(e)}} \tag{12.50}$$

or

$$\mathbf{b}_R^{(e)} = \mathbf{k}_M^{(e)} \boldsymbol{\phi}^{(e)} - \mathbf{f}_S^{(e)} \tag{12.51}$$

where

$$\mathbf{k}_M^{(e)} = \begin{bmatrix} 0 & 0 \\ 0 & hA \end{bmatrix} \qquad (12.52)$$

and

$$\mathbf{f}_s^{(e)} = \left\{ \begin{array}{c} 0 \\ hA\phi_f \end{array} \right\} \qquad (12.53)$$

Note that Eqs. (12.51), (12.52), (12.53) are derived assuming that the convective boundary is on node j, which is the right side of the element. If the convective boundary is on the left side of the element, we then have

$$\mathbf{b}_L^{(e)} = \mathbf{k}_M^{(e)} \boldsymbol{\phi}^{(e)} - \mathbf{f}_S^{(e)} \qquad (12.54)$$

where

$$\mathbf{k}_M^{(e)} = \begin{bmatrix} hA & 0 \\ 0 & 0 \end{bmatrix} \qquad (12.55)$$

and

$$\mathbf{f}_s^{(e)} = \left\{ \begin{array}{c} hA\phi_f \\ 0 \end{array} \right\} \qquad (12.56)$$

Substituting the expressions for $\mathbf{b}^{(e)}$ back to Eq. (12.36), we obtain

$$\mathbf{R}^{(e)} = \underbrace{\left[\mathbf{k}_D^{(e)} + \mathbf{k}_g^{(e)} + \mathbf{k}_M^{(e)} \right]}_{\mathbf{k}^{(e)}} \boldsymbol{\Phi}^{(e)} - \underbrace{\left\{ \mathbf{f}_Q^{(e)} + \mathbf{f}_s^{(e)} \right\}}_{\mathbf{f}^{(e)}} \qquad (12.57)$$

or in a simplified form of

$$\mathbf{R}^{(e)} = \mathbf{k}^{(e)} \boldsymbol{\phi}^{(e)} - \mathbf{f}^{(e)} \qquad (12.58)$$

where

$$\mathbf{k}^{(e)} = \mathbf{k}_D^{(e)} + \mathbf{k}_g^{(e)} + \mathbf{k}_M^{(e)} \qquad (12.59)$$

and

$$\mathbf{f}^{(e)} = \mathbf{f}_Q^{(e)} + \mathbf{f}_S^{(e)} \qquad (12.60)$$

Note that $\mathbf{k}_M^{(e)}$ and $\mathbf{f}_S^{(e)}$ exist only if the node is on the convective boundary, and they are given by Eqs. (12.52) and (12.53) or Eqs. (12.55) and (12.56) depending on the location of the node. If the boundary is insulated, meaning there is no heat exchange occurring there, both $\mathbf{k}_M^{(e)}$ and $\mathbf{f}_S^{(e)}$ vanish, because

$\frac{d\phi}{dx} = 0$ at such a boundary. After obtaining the element matrices, the residual defined by Eq. (12.58) is assembled and enforced to equate to zero, which will lead to the familiar global system of equations:

$$\mathbf{KD} = \mathbf{F} \tag{12.61}$$

The above matrix equation has the same form as that for a static mechanics problem. The detailed assembly process is described in the examples that follow.

12.3.2 Direct assembly procedure

Consider an element equation of residual in the following form of

$$\mathbf{R}^{(e)} = \mathbf{k}^{(e)} \boldsymbol{\phi}^{(e)} - \mathbf{f}^{(e)} \tag{12.62}$$

or in the expanded form of

$$\begin{Bmatrix} R_1^{(e)} \\ R_2^{(e)} \end{Bmatrix} = \begin{bmatrix} K_{11}^{(e)} & K_{12}^{(e)} \\ K_{21}^{(e)} & K_{22}^{(e)} \end{bmatrix} \begin{Bmatrix} \phi_1^{(e)} \\ \phi_2^{(e)} \end{Bmatrix} - \begin{Bmatrix} f_1^{(e)} \\ f_2^{(e)} \end{Bmatrix} \tag{12.63}$$

Consider now two linear, one-dimensional elements as shown in Figure 12.8. For element 1,

$$R_1^{(1)} = k_{11}^{(1)} \phi_1 + k_{12}^{(1)} \phi_2 - f_1^{(1)} \tag{12.64}$$

$$R_2^{(1)} = k_{21}^{(1)} \phi_1 + k_{22}^{(1)} \phi_2 - f_2^{(1)} \tag{12.65}$$

and for element 2,

$$R_1^{(2)} = k_{11}^{(2)} \phi_2 + k_{12}^{(2)} \phi_3 - f_1^{(2)} \tag{12.66}$$

$$R_2^{(2)} = k_{21}^{(2)} \phi_2 + k_{22}^{(2)} \phi_3 - f_2^{(2)} \tag{12.67}$$

The total residual at a node can be obtained by summing all the residuals contributed from all the elements that are connected to the node, and we enforce the total residual to vanish to best satisfy the system equations. We then have at node 1,

$$R_1^{(1)} = 0: \quad k_{11}^{(1)} \phi_1 + k_{12}^{(1)} \phi_2 - f_1^{(1)} = 0 \tag{12.68}$$

at node 2,

$$R_1^{(2)} + R_2^{(1)} = 0: \quad k_{21}^{(1)} \phi_1 + \left(k_{22}^{(1)} + k_{11}^{(2)} \right) \phi_2 + k_{12}^{(2)} \phi_3 - \left(f_2^{(1)} + f_1^{(2)} \right) = 0 \tag{12.69}$$

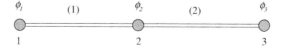

FIGURE 12.8

Assembling of two elements demonstrating the direct assembly principle.

and at node 3,

$$R_2^{(2)} = 0: \quad k_{21}^{(2)}\phi_2 + k_{22}^{(2)}\phi_3 - f_2^{(2)} = 0 \tag{12.70}$$

Writing Eqs. (12.68), (12.69), and (12.70) in matrix form gives

$$\begin{bmatrix} k_{11}^{(1)} & k_{12}^{(1)} & 0 \\ k_{21}^{(1)} & k_{22}^{(1)} + k_{11}^{(2)} & k_{12}^{(2)} \\ 0 & k_{21}^{(2)} & k_{22}^{(2)} \end{bmatrix} \begin{Bmatrix} \phi_1 \\ \phi_2 \\ \phi_3 \end{Bmatrix} \begin{Bmatrix} f_1^{(1)} \\ f_2^{(1)} + f_1^{(2)} \\ f_2^{(2)} \end{Bmatrix} \tag{12.71}$$

which is the same as the assembly procedure we introduced in Section 3.4.7 of Chapter 3 and Example 4.2 of Chapter 4.

12.3.3 **Worked example**

We now present a simple case study of a heat transfer problem. The problem is simple, allowing us to explicitly reveal the detailed FEM procedure. For real-life problems, the FEM model can be much bigger, but the essence and basic procedures will largely stay the same. Therefore, a good understanding of this problem and the solution procedure can be beneficial.

EXAMPLE 12.1

Heat transfer along a 1D fin of rectangular cross-section
The temperature distribution in the fin, shown in Figure 12.9, is to be calculated using the finite element method. The fin is rectangular in shape, 8 cm long, 0.4 cm wide, and 1 cm thick. Assume that convective heat loss occurs from the right end of the fin.

Analysis of the problem: The fin is divided uniformly into 4 elements with a total of 5 nodes. Each element has a length of $l = 2$ cm. The system equation should be 5 by 5. At node 1 the temperature is specified, therefore there is no need to calculate $\mathbf{k}_M^{(e)}$ and $\mathbf{f}_S^{(e)}$ for element 1 as only the temperature is requested. As nodes 2, 3, and 4 are internal, there is also no need to calculate $\mathbf{k}_M^{(e)}$ and $\mathbf{f}_S^{(e)}$ for elements 2 and 3. Since heat convection is occurring on node 5, we have to calculate $\mathbf{k}_M^{(e)}$ and $\mathbf{f}_S^{(e)}$ using Eqs. (12.52) and (12.53) for element 4.

Data preparation:

$$\frac{kA}{l} = \frac{3(0.4)}{2} = 0.6 \frac{W}{^\circ C} \tag{12.72}$$

$$\frac{hPl}{6} = \frac{0.1(2.8)2}{6} = 0.093 \frac{W}{^\circ C} \tag{12.73}$$

$$hA = 0.1(0.4) = 0.04 \frac{W}{^\circ C} \tag{12.74}$$

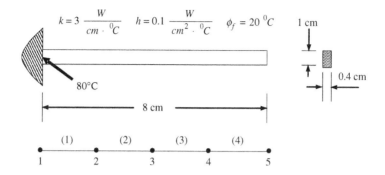

FIGURE 12.9

A one-dimensional fin of rectangular cross-section.

$$\frac{hPl\phi_f}{2} = \frac{0.1(2.8)(20)(2)}{2} = 5.6W \tag{12.75}$$

$$hA\phi_f = 0.1(0.4)(20) = 0.8W \tag{12.76}$$

Solution: The general forms of the element matrices for elements 1, 2, and 3 are

$$\mathbf{k}^{(e)} = \frac{kA}{l}\begin{bmatrix} 1 & -1 \\ -1 & 1 \end{bmatrix} + \frac{hPl}{6}\begin{bmatrix} 2 & 1 \\ 1 & 2 \end{bmatrix} \text{ and} \tag{12.77}$$

$$\mathbf{f}^{(e)} = \frac{hPL\phi_f}{2}\begin{Bmatrix} 1 \\ 1 \end{Bmatrix} \tag{12.78}$$

Substitute the previously prepared data, Eqs. (12.72), (12.73), (12.74), (12.75), (12.76) into the above two equations, and we have

$$\mathbf{k}^{(1,2,3)} = \begin{bmatrix} 0.786 & -0.507 \\ -0.507 & 0.786 \end{bmatrix} \quad \text{and} \quad \mathbf{f}^{(1,2,3)} = \begin{Bmatrix} 5.6 \\ 5.6 \end{Bmatrix} \tag{12.79}$$

The general forms of the element matrices for element 4 are

$$\mathbf{k}^{(e)} = \frac{kA}{l}\begin{bmatrix} 1 & -1 \\ -1 & 1 \end{bmatrix} + \frac{hPL}{6}\begin{bmatrix} 2 & 1 \\ 1 & 2 \end{bmatrix} + \begin{bmatrix} 0 & 0 \\ 0 & hA \end{bmatrix} \text{ and} \tag{12.80}$$

$$\mathbf{f}^{(e)} = \frac{hPL\phi_f}{2}\begin{Bmatrix} 1 \\ 1 \end{Bmatrix} + \begin{Bmatrix} 0 \\ hA\phi_f \end{Bmatrix} \tag{12.81}$$

Substituting the data into the above equations, we have

$$\mathbf{k}^{(4)} = \begin{bmatrix} 0.786 & -0.507 \\ -0.507 & 0.826 \end{bmatrix} \quad \text{and} \quad \mathbf{f}^{(4)} = \begin{Bmatrix} 5.6 \\ 6.4 \end{Bmatrix} \tag{12.82}$$

The next step is to assemble the element matrices together to form the global system of equations. The assembly is carried out using the direct assembly procedure described in the previous section and the resulting global matrix equation is

$$\begin{bmatrix} 0.786 & -0.507 & 0 & 0 & 0 \\ -0.507 & 1.572 & -0.507 & 0 & 0 \\ 0 & -0.507 & 1.572 & -0.507 & 0 \\ 0 & 0 & -0.507 & 1.572 & -0.507 \\ 0 & 0 & 0 & -0.507 & 0.826 \end{bmatrix} \begin{Bmatrix} \phi_1 \\ \phi_2 \\ \phi_3 \\ \phi_4 \\ \phi_5 \end{Bmatrix} = \begin{Bmatrix} 5.6 \\ 11.2 \\ 11.2 \\ 11.2 \\ 6.4 \end{Bmatrix} + \begin{Bmatrix} Q^* \\ 0 \\ 0 \\ 0 \\ 0 \end{Bmatrix} \tag{12.83}$$

Note that at node 1 the temperature is fixed at a particular value. This requires a heat exchanger (or heat source) there for it to happen. The heat source is unknown at this stage, and it is actually not necessary to know it to compute the temperature distribution. However, it is required for balancing the equation. We therefore simply label it as Q^*. Since the temperature at node 1 is given as 80 °C, we actually have four unknown temperatures in a 5 by 5 matrix equation. This implies that we can actually eliminate Q^* in the first system equation as indicated below and still solve the four unknowns with the remaining 4 by 4 matrix equation. Note that the term in the second row, first column (circled term) must also be accounted for in the first equation (it is shown below that $(-0.507 \times \phi_1)$ is brought over to the force vector) of the four remaining system equations.

$$\begin{bmatrix} \overline{=0.786} & \overline{0.507} & 0 & 0 & 0 \\ -0.507 & 1.572 & -0.507 & 0 & 0 \\ 0 & -0.507 & 1.572 & -0.507 & 0 \\ 0 & 0 & -0.507 & 1.572 & -0.507 \\ 0 & 0 & 0 & -0.507 & 0.826 \end{bmatrix} \begin{Bmatrix} \phi_1 = 80 \\ \phi_2 \\ \phi_3 \\ \phi_4 \\ \phi_5 \end{Bmatrix} = \begin{Bmatrix} 5.6 + Q^* \\ 11.2 + 80 \times 0.507 \\ 11.2 \\ 11.2 \\ 6.4 \end{Bmatrix} \tag{12.84}$$

Solving the 4 by 4 equation gives the temperature at the nodes:

$$\Phi^T = \{80.0 \quad 42.0 \quad 28.2 \quad 23.3 \quad 22.1\} \tag{12.85}$$

12.3.4 Remarks

Formulations and examples given in the preceding sections have clearly illustrated that the heat transfer problem is very similar to the structure mechanics problem as far as using the FEM is concerned. The displacement and force correspond to the temperature and heat flux, respectively. We also showed analogies of some of the element matrices and vectors between heat transfer problems and structure mechanics problems. However, we did not mention about the structural mechanics counterparts of the matrix $\mathbf{k}_M^{(e)}$ and vector $\mathbf{f}_S^{(e)}$ that are coming from the heat convection boundary. Since the heat flux on the boundary depends on the unknown field variable (temperature), the resultant term $(\mathbf{k}_M^{(e)})$ needs to be combined with the heat convection matrix. In fact, in structure mechanics problems, we can also have a similar situation when the structure is supported by elastic supports such as springs. The reaction force on the boundary depends on the unknown field variable (displacement) at the boundary. For such a system, we will have an additional matrix corresponding to $\mathbf{k}_M^{(e)}$ and a vector corresponding to $\mathbf{f}_S^{(e)}$. However, in structure mechanics problems, it is often more convenient to formulate such an elastic support using an additional spring element. The stiffness of the support is then assembled to the global stiffness following the same systematic, direct assembly procedure. The reaction force of the elastic support is found after the global equations system is solved for the displacements.

We can therefore conclude that the Galerkin residual formulation gives exactly the same FE equations as those we obtained using the energy principles discussed in previous chapters.

12.3.5 **Composite wall**

Consider heat transfer through a composite wall as shown in Figure 12.10. The governing equation is given by Eq. (12.9). If there is no heat source or sink in a layer ($q=0$ within the layer), one linear element is enough for modeling the entire layer. (Why?)[1] For the case shown in Figure 12.10, a total of three elements, one for each layer, should be used if there is a heat source within these layers. This argument holds even if there are heat sources or sinks in between the layers.

As for the boundary conditions, at any one or at both outer surfaces, the temperature or heat convection, or heat insulation can be specified. The convective boundary conditions are given as

$$kA\frac{d\phi}{dx} = hA\left(\phi_b - \phi_f\right) \text{ at } x = 0 \tag{12.86}$$

and

$$-kA\frac{d\phi}{dx} = hA\left(\phi_b - \phi_f\right) \text{ at } x = H \tag{12.87}$$

Therefore, all the elements developed in the section for 1D fin are also valid for the case of the composite wall except that $\mathbf{k}_G^{(e)}$ and $\mathbf{f}_Q^{(e)}$ do not exist because the g and Q vanish in the case of the composite wall (see, Eq. (12.10)).

The general form of the element stiffness matrix is given by

$$\mathbf{k}^{(e)} = \frac{kA}{l}\begin{bmatrix} 1 & -1 \\ -1 & 1 \end{bmatrix} + \mathbf{k}_M^{(e)} \tag{12.88}$$

where $\mathbf{k}_M^{(e)}$ is obtained either by Eq. (12.52) or (12.55), if there is heat convection occurring on the surface, or else $\mathbf{k}_M^{(e)} = 0$ if the surface is insulated. As for a surface with a prescribed temperature, it is not necessary to calculate $\mathbf{k}_M^{(e)}$. For the force vector, $\mathbf{f}_S^{(e)}$, it is obtained either by Eq. (12.53) or (12.56),

---

[1] Hint: see Example 4.1.

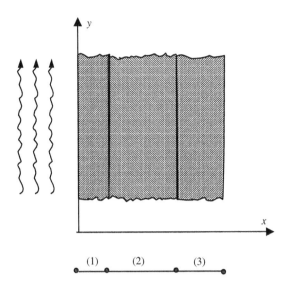

FIGURE 12.10

Heat transfer through a composite wall of three layers. One element is sufficient to model a layer if there is no heat supply/sink in the layers.

if there is heat convection occurring on the surface, or zero if the surface is insulated. In the case where the surface has a prescribed temperature, again it is not necessary to have $\mathbf{f}_S^{(e)}$ for calculating the temperature. It can always be calculated after the temperature field is found.

12.3.6 Worked example

EXAMPLE 12.2

Heat transfer through a composite wall of two layers
Figure 12.11 shows a composite wall consisting of two layers of different materials. Various heat transfer parameters are shown in the figure. The temperature distribution through the composite wall is to be calculated by the finite element method.

Analysis of problem: The wall is divided into 2 elements with a total of 3 nodes. Hence, the global system of equations should be a 3 by 3 matrix equation. At node 3, the temperature is specified, therefore there is no need to calculate $\mathbf{k}_M^{(e)}$ and $\mathbf{f}_S^{(e)}$ for element 2. Since the heat convection is occurring on node 1, $\mathbf{k}_M^{(e)}$ and $\mathbf{f}_S^{(e)}$ have to be calculated for element 1 using Eqs. (12.55) and (12.56), respectively.

Data preparation:
For element (1),

$$\frac{kA}{L} = \frac{0.2(1)}{2} = 0.1 \frac{W}{^\circ C} \tag{12.89}$$

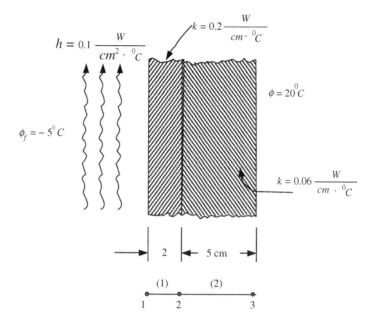

FIGURE 12.11

Heat transfer through a composite wall of two layers.

$$hA = 0.1(1) = 0.1\frac{W}{^\circ C} \tag{12.90}$$

$$hA\phi_f = 0.1(1)(-5) = -0.5W \tag{12.91}$$

For element (2),

$$\frac{kA}{L} = \frac{0.06(1)}{5} = 0.012\frac{W}{^\circ C} \tag{12.92}$$

Solution: The element matrices for element (1) are

$$\mathbf{K}^{(1)} = \begin{bmatrix} 0.2 & -0.1 \\ -0.1 & 0.1 \end{bmatrix} \quad \text{and} \quad \mathbf{f}_S^{(1)} = \begin{Bmatrix} -0.5 \\ 0 \end{Bmatrix} \tag{12.93}$$

The element matrix for element (2) is

$$\mathbf{K}^{(2)} = \begin{bmatrix} 0.012 & -0.012 \\ -0.012 & 0.012 \end{bmatrix} \tag{12.94}$$

The next step is to assemble the element matrices of the two elements together to form the global system of equations, which leads to the matrix equation:

$$
\begin{bmatrix}
0.20 & -0.10 & 0 \\
-0.10 & 0.112 & -0.012 \\
0 & -0.012 & 0.012
\end{bmatrix}
\begin{Bmatrix}
\phi_1 \\
\phi_2 \\
\phi_3(=20)
\end{Bmatrix}
=
\begin{Bmatrix}
-0.5 \\
0 \\
0
\end{Bmatrix}
+
\begin{Bmatrix}
0 \\
0 \\
Q^*
\end{Bmatrix}
\tag{12.95}
$$

Note once again that Q^* is still an unknown but is required to balance the equation, as the temperature at node 3 is fixed. Having only two unknown temperatures, ϕ_1 and ϕ_2, the first two equations in the above give

$$
\begin{bmatrix}
0.20 & -0.10 \\
-0.10 & 0.112
\end{bmatrix}
\begin{Bmatrix}
\phi_1 \\
\phi_2
\end{Bmatrix}
=
\begin{Bmatrix}
-0.5 \\
0.24(=20\times0.012)
\end{Bmatrix}
\tag{12.96}
$$

The above matrix equation (two simultaneous equations with two unknowns) is solved and the solution is given as

$$
\mathbf{\Phi}^T \{-2.5806 \quad -0.1613 \quad 20\}
\tag{12.97}
$$

EXAMPLE 12.3

Heat transfer through layers of thin films

Figure 12.12 shows schematically the process of producing layers of thin film of different materials on a substrate using physical deposition techniques. A heat supply is provided on the upper surface of the glass substrate. At the stage shown in Figure 12.12, a layer of iron and a layer of platinum have been formed. The thickness of these layers and the thermal conductivities for the materials of these layers are also shown in Figure 12.12. Convective heat loss occurs on the lower platinum surface, and the ambient temperature is 150 °C. A heat supply is provided to maintain the temperature on the upper surface of the glass substrate at 300 °C. The temperature distribution through the thickness of the thin film layers is to be calculated by the finite element method.

Analysis of problem: This problem is actually similar to the previous example on the composite wall. Hence, 1D elements can be used for this purpose. The layers are divided into 3 elements with a total of 4 nodes as shown in Figure 12.12. Since the temperature is specified at node 1, there is no need to calculate $\mathbf{k}_M^{(e)}$ and $\mathbf{f}_S^{(e)}$ for element 1. $\mathbf{k}_M^{(e)}$ and $\mathbf{f}_S^{(e)}$ for element 2 are zero since there is no heat convection occurring at either node 2 or 3. Since there is heat convection occurring at node 4, $\mathbf{k}_M^{(e)}$ and $\mathbf{f}_S^{(e)}$ have to be calculated for element 3 using Eqs. (12.55) and (12.56) respectively.

Data preparation:
For element (1),

$$
\frac{kA}{L} = \frac{(0.1)(1)}{0.2} = 0.5\frac{W}{{}^\circ C}
\tag{12.98}
$$

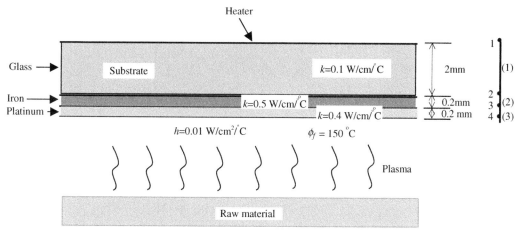

FIGURE 12.12

Heat transfer during a thin film deposition process.

For element (2),

$$\frac{kA}{L} = \frac{(0.5)(1)}{0.02} = 25\frac{W}{°C} \tag{12.99}$$

For element (3),

$$\frac{kA}{L} = \frac{0.4(1)}{0.02} = 20\frac{W}{°C} \tag{12.100}$$

$$hA = 0.01(1) = 0.01\frac{W}{°C} \tag{12.101}$$

$$hA\phi_f = 0.01(1)(150) = 1.5W \tag{12.102}$$

Solution: The element matrix for element (1) is

$$\mathbf{k}^{(1)} = \begin{bmatrix} 0.5 & -0.5 \\ -0.5 & 0.5 \end{bmatrix} \tag{12.103}$$

The element matrix for element (2) is

$$\mathbf{k}^{(2)} = \begin{bmatrix} 25 & -25 \\ -25 & 25 \end{bmatrix} \tag{12.104}$$

And finally the element matrices for element (3) are

$$\mathbf{k}^{(3)} = \begin{bmatrix} 20 & -20 \\ -20 & 20.01 \end{bmatrix} \text{ and } \mathbf{f}_S^{(3)} = \begin{Bmatrix} 0 \\ 1.5 \end{Bmatrix} \tag{12.105}$$

The next step is to assemble the element matrices of the three elements together to form the global system equations, which leads to

$$\begin{bmatrix} 0.5 & -0.5 & 0 & 0 \\ -0.5 & 25.5 & -25 & 0 \\ 0 & -25 & 45 & -20 \\ 0 & 0 & -20 & 20.01 \end{bmatrix} \begin{Bmatrix} \phi_1 \\ \phi_2 \\ \phi_3 \\ \phi_4 \end{Bmatrix} = \begin{Bmatrix} Q^* \\ 0 \\ 0 \\ 1.5 \end{Bmatrix} \tag{12.106}$$

Since ϕ_1 is given to be $300\,^\circ\text{C}$, we can therefore reduce the above equation to a 3 x 3 matrix to solve for the remaining 3 unknown temperatures

$$\begin{bmatrix} 25.5 & -25 & 0 \\ -25 & 45 & -20 \\ 0 & -20 & 20.01 \end{bmatrix} \begin{Bmatrix} \phi_2 \\ \phi_3 \\ \phi_4 \end{Bmatrix} = \begin{Bmatrix} 150 \\ 0 \\ 1.5 \end{Bmatrix} \tag{12.107}$$

The above matrix equation is solved and the solution can be obtained as:

$$\mathbf{\Phi}^T = [300.0 \quad 297.1 \quad 297.0 \quad 296.9] \tag{12.108}$$

To calculate the heat flux, Q^* on the top of the glass substrate, we can now substitute the temperatures into the first equation of the matrix equation in Eq. (12.106) to obtain

$$Q^* = 0.5 \times 300 - 0.5 \times 297.1 = 1.45 \text{ W/cm}^2 \tag{12.109}$$

12.4 2D heat transfer problem

12.4.1 Element equations

This section deals with heat transfer problems in 2D that is governed by Eq. (12.1). The procedure for obtaining the FEM equations for 2D heat transfer problems is the same as that for the 1D problems described in the previous sections, except that the mathematical manipulation becomes slightly more involuted due to the additional dimension.

Let us assume that the problem domain is divided into elements as shown in Figure 12.5. For one element in general, the residual can be obtained by the Galerkin method as

$$\mathbf{R}^{(e)} = -\int_{A_e} \mathbf{N}^T \left(D_x \frac{\partial^2 \phi^h}{\partial x^2} + D_y \frac{\partial^2 \phi^h}{\partial y^2} - g\phi^h + Q \right) dA \tag{12.110}$$

Note that the minus sign is added to the residual mainly for convenience and has no consequence for the end result since we eventually enforce the residual to zero. The integration in Eq. (12.110) for the residual must be evaluated so as to obtain the element matrices, but in this case, the integration is more complicated than the 1D case because the integration is performed over the area of the element. Recall that in the 1D case, the integral is evaluated by parts. In the 2D case, we need to use the Gauss's divergence theorem instead.

First, using the product rule for differentiation, the following expression can be obtained:

$$\frac{\partial}{\partial x}\left(\mathbf{N}^T \frac{\partial \phi}{\partial x}\right) = \mathbf{N}^T \frac{\partial^2 \phi}{\partial x^2} + \frac{\partial \mathbf{N}^T}{\partial x}\frac{\partial \phi}{\partial x} \tag{12.111}$$

The first integral in Eq. (12.110) can then be obtained by

$$-\int_{A_e} \mathbf{N}^T D_x \frac{\partial^2 \phi}{\partial x^2} dA = -\int_{A_e} D_x \frac{\partial}{\partial x}\left(\mathbf{N}^T \frac{\partial \phi}{\partial x}\right) dA + \int_{A_e} D_x \frac{\partial \mathbf{N}^T}{\partial x}\frac{\partial \phi}{\partial x} dA \tag{12.112}$$

where A_e is the area of the element. Gauss's divergence theorem can be stated mathematically for this case as

$$\int_{A_e} \frac{\partial}{\partial x}\left(\mathbf{N}^T \frac{\partial \phi}{\partial x}\right) dA = \int_{\Gamma_e} \mathbf{N}^T \frac{\partial \phi}{\partial x} \cos\theta \, d\Gamma \tag{12.113}$$

where θ is the angle of the outwards normal on the boundary Γ_e of the element with respect to the x axis. Eq. (12.113) is thus substituted into Eq. (12.112) to obtain

$$-\int_A \mathbf{N}^T D_x \frac{\partial^2 \phi}{\partial x^2} dA = -\int_{\Gamma_e} D_x \mathbf{N}^T \frac{\partial \phi}{\partial x} \cos\theta \, d\Gamma + \int_A D_x \frac{\partial \mathbf{N}^T}{\partial x}\frac{\partial \phi}{\partial x} dA \tag{12.114}$$

In a similar way, the second integral in Eq. (12.110) can be evaluated to obtain

$$-\int_A \mathbf{N}^T D_y \frac{\partial^2 \phi}{\partial y^2} dA - \int_{\Gamma_e} D_y \mathbf{N}^T \frac{\partial \phi}{\partial y} \sin\theta \, d\Gamma + \int_A D_y \frac{\partial \mathbf{N}^T}{\partial y}\frac{\partial \phi}{\partial y} dA \tag{12.115}$$

The two integrals in Eqs. (12.114) and (12.115) are substituted back into the residual in Eq. (12.110) to give

$$\mathbf{R}^{(e)} = -\int_{\Gamma_e} \mathbf{N}^T \left(D_x \frac{\partial \phi^h}{\partial x}\cos\theta + D_y \frac{\partial \phi^h}{\partial y}\sin\theta\right) d\Gamma$$

$$+ \int_{A_e} \left(D_x \frac{\partial \mathbf{N}^T}{\partial x}\frac{\partial \phi^h}{\partial x} + D_y \frac{\partial \mathbf{N}^T}{\partial y}\frac{\partial \phi^h}{\partial y}\right) dA \tag{12.116}$$

$$+ \int_{A_e} g\mathbf{N}^T \phi^h dA - \int_{A_e} Q\mathbf{N}^T dA$$

The field variable ϕ is now interpolated from the nodal variables by shape functions as in Eq. (12.34), which is then substituted into Eq. (12.116) above to give

$$\mathbf{R}^{(e)} = \underbrace{- \int_{\Gamma_e} \mathbf{N}^T \left(D_x \frac{\partial \phi^h}{\partial x} \cos\theta + D_y \frac{\partial \phi^h}{\partial y} \sin\theta \right) d\Gamma}_{\mathbf{b}^{(e)}}$$

$$+ \underbrace{\left[\int_{A_e} \left(D_x \frac{\partial \mathbf{N}^T}{\partial x} \frac{\partial \mathbf{N}}{\partial x} + D_y \frac{\partial \mathbf{N}^T}{\partial y} \frac{\partial \mathbf{N}}{\partial y} \right) dA \right]}_{\mathbf{k}_D^{(e)}} \Phi^{(e)} \qquad (12.117)$$

$$+ \underbrace{\left(\int_{A_e} g\mathbf{N}^T \mathbf{N} dA \right)}_{\mathbf{k}_g^{(e)}} \phi^{(e)} - \underbrace{\int_{A_e} Q\mathbf{N}^T dA}_{\mathbf{f}_Q^{(e)}}$$

or in matrix form of

$$\mathbf{R}^{(e)} = \mathbf{b}^{(e)} + \left[\mathbf{k}_D^{(e)} + \mathbf{k}_g^{(e)} \right] \Phi^{(e)} - \mathbf{f}_Q^{(e)} \qquad (12.118)$$

where

$$\mathbf{b}^{(e)} = - \int_{\Gamma_e} \mathbf{N}^T \left(D_x \frac{\partial \phi^h}{\partial x} \cos\theta + D_y \frac{\partial \phi^h}{\partial y} \sin\theta \right) d\Gamma \qquad (12.119)$$

$$\mathbf{k}_D^{(e)} = \int_{A_e} \left(\frac{\partial \mathbf{N}^T}{\partial x} D_x \frac{\partial \mathbf{N}}{\partial x} + \frac{\partial \mathbf{N}^T}{\partial y} D_y \frac{\partial \mathbf{N}}{\partial y} \right) dA \qquad (12.120)$$

$$\mathbf{k}_g^{(e)} = \int_{A_e} g\mathbf{N}^T \mathbf{N} dA \qquad (12.121)$$

$$\mathbf{f}_Q^{(e)} = \int_{A_e} Q\mathbf{N}^T dA \qquad (12.122)$$

Like in the 1D case, the vector $\mathbf{b}^{(e)}$ is related to the variation or the derivatives of temperature (heat flux) on the boundaries of the element. It will be evaluated in detail in the next section. For now, let us evaluate and analyze Eqs. (12.120), (12.121), (12.122).

The integral in Eq. (12.120) can be rewritten in the matrix form by defining

$$\mathbf{D} = \begin{bmatrix} D_x & 0 \\ 0 & D_y \end{bmatrix} \qquad (12.123)$$

and the gradient vector as

$$\nabla \phi = \begin{Bmatrix} \dfrac{\partial \phi}{\partial x} \\ \dfrac{\partial \phi}{\partial y} \end{Bmatrix} = \begin{bmatrix} \dfrac{\partial \mathbf{N}}{\partial x} \\ \dfrac{\partial \mathbf{N}}{\partial y} \end{bmatrix} \mathbf{\Phi}^{(e)} = \mathbf{B} \mathbf{\Phi}^{(e)} \qquad (12.124)$$

where $\mathbf{B}$ is the strain matrix given by

$$\mathbf{B} = \begin{bmatrix} \dfrac{\partial \mathbf{N}}{\partial x} \\ \dfrac{\partial \mathbf{N}}{\partial y} \end{bmatrix} = \begin{bmatrix} \dfrac{\partial N_1}{\partial x} & \dfrac{\partial N_2}{\partial x} & \cdots & \dfrac{\partial N_{n_d}}{\partial x} \\ \dfrac{\partial N_1}{\partial y} & \dfrac{\partial N_2}{\partial y} & \cdots & \dfrac{\partial N_{n_d}}{\partial y} \end{bmatrix} \qquad (12.125)$$

Note that we use the usual shape function given by Eq. (12.28) to obtain the above equations. Using Eqs. (12.123), (12.124), (12.125), it can be easily verified that

$$\mathbf{B}^T \mathbf{D} \mathbf{B} = D_x \frac{\partial \mathbf{N}^T}{\partial x} \frac{\partial \mathbf{N}}{\partial x} + D_y \frac{\partial \mathbf{N}^T}{\partial y} \frac{\partial \mathbf{N}}{\partial y} \qquad (12.126)$$

Therefore, the general element "stiffness" matrix for 2D elements given by Eq. (12.120) becomes

$$\mathbf{k}_D^{(e)} = \int_{A_e} \mathbf{B}^T \mathbf{D} \mathbf{B} \, \mathrm{d}A \qquad (12.127)$$

which is exactly the same form as Eq. (3.71) that is obtained using the Hamilton's principle, except that the matrix of material elasticity is replaced by the matrix of heat conductivity. Note also that in Eq. (12.121), $\mathbf{k}_g^{(e)}$ is similar to the matrix given by Eq. (3.75) for mechanics problems. We observe once again that the Galerkin weighted residual formulation produces the same set of FE equations as those produced by the energy principle.

12.4.2 Triangular elements

Using shape functions of the 2D triangular element, the field function of temperature, ϕ, can be interpolated as follows:

$$\phi^{(e)} = \mathbf{N} \mathbf{\Phi}^{(e)} = \begin{bmatrix} N_1 & N_2 & N_3 \end{bmatrix} \begin{Bmatrix} \phi_1 \\ \phi_2 \\ \phi_3 \end{Bmatrix} \qquad (12.128)$$

where N_i ($i = 1, 2, 3$) are the three shape functions defined by Eq. (7.22) and (7.23), and ϕ_i ($i = 1, 2, 3$) are the nodal values of temperature at the 3-nodes of the triangular element shown in Figure 12.13.

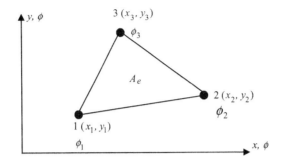

FIGURE 12.13

Linear triangular element.

Note that the strain matrix **B** is constant for triangular elements, and can be evaluated similar to Eq. (7.38). Using Eq. (12.127), $\mathbf{k}_D^{(e)}$ can be easily evaluated as the integrand and is a constant matrix if the material constants D_x and D_y do not change within the element.

$$\mathbf{k}_D^{(e)} = \int_{A_e} \mathbf{B}^T \mathbf{DB}\, dA = \mathbf{B}^T \mathbf{DB} \int_{A_e} dA = \mathbf{B}^T \mathbf{DB} A_e \tag{12.129}$$

Expanding the matrix product yields

$$\mathbf{K}_D^{(e)} = D_x A_e \begin{bmatrix} b_1^2 & b_1 b_2 & b_1 b_3 \\ b_1 b_2 & b_2^2 & b_2 b_3 \\ b_1 b_3 & b_2 b_3 & b_3^2 \end{bmatrix} + D_y A_e \begin{bmatrix} c_1^2 & c_1 c_2 & c_1 c_3 \\ c_1 c_2 & c_2^2 & c_2 c_3 \\ c_1 c_3 & c_2 c_3 & c_3^2 \end{bmatrix} \tag{12.130}$$

It is noted that the stiffness matrix is symmetrical.

The matrix, $\mathbf{k}_g^{(e)}$ defined by Eq. (12.121) can be evaluated as

$$\mathbf{k}_g^{(e)} = \int_{A_e} g\mathbf{N}^T \mathbf{N} dA = g \int_{A_e} \begin{Bmatrix} N_1 \\ N_2 \\ N_3 \end{Bmatrix} \begin{bmatrix} N_1 & N_2 & N_3 \end{bmatrix} dA$$

$$\tag{12.131}$$

$$= g \int_{A_e} \begin{bmatrix} N_1^2 & N_1 N_2 & N_1 N_3 \\ N_1 N_2 & N_2^2 & N_2 N_3 \\ N_1 N_3 & N_2 N_3 & N_3^2 \end{bmatrix} dA$$

The above integral can be carried out using the factorial formula given in Eq. (7.43), since the shape functions are equal to the area coordinates—just as we did for the mass matrix, Eq. (7.44), in Chapter 7. For example,

$$
\begin{aligned}
\int_{A_e} N_1 N_2 dA &= \int_{A_e} L_1^1 L_2^1 L_3^0 \, dA \\
&= \frac{1!1!0!}{(1+1+0+2)!} 2A_e \\
&= \frac{1}{4 \times 3 \times 2 \times 1} 2A_e \\
&= \frac{A_e}{12}
\end{aligned}
\tag{12.132}
$$

Using the area coordinates and the factorial formula in Eq. (7.43), the matrix $\mathbf{k}_g^{(e)}$ is found as

$$
\mathbf{k}_g^{(e)} = \frac{gA_e}{12}
\begin{bmatrix}
2 & 1 & 1 \\
1 & 2 & 1 \\
1 & 1 & 2
\end{bmatrix}
\tag{12.133}
$$

The element force vector $\mathbf{f}_Q^{(e)}$ defined in Eq. (12.122) also involves the integration of shape functions and can also be obtained using the factorial formula in Eq. (7.43):

$$
\mathbf{f}_Q^{(e)} = \int_{A_e} Q\mathbf{N}^T dA = Q \int_{A_e}
\begin{Bmatrix}
N_i \\
N_j \\
N_k
\end{Bmatrix}
dA = Q \int_{A_e}
\begin{Bmatrix}
L_1 \\
L_2 \\
L_3
\end{Bmatrix}
dA = \frac{QA_e}{3}
\begin{Bmatrix}
1 \\
1 \\
1
\end{Bmatrix}
\tag{12.134}
$$

It is assumed that Q is a constant within the element. Note that the heating rate, QA, is equally shared by the three nodes of the triangular element.

12.4.3 Rectangular elements

Consider now a 4-nodal, rectangular element as shown in Figure 7.8. The field function ϕ is interpolated over the element as follows.

$$
\phi^{(e)} = \mathbf{N}\boldsymbol{\Phi}^{(e)} = \begin{bmatrix} N_1 & N_2 & N_3 & N_4 \end{bmatrix}
\begin{Bmatrix}
\phi_1 \\
\phi_2 \\
\phi_3 \\
\phi_4
\end{Bmatrix}
\tag{12.135}
$$

Note that for rectangular elements, the natural coordinate system is again adopted as for the case in structure mechanics problem, as shown in Figure 7.8. The shape functions are given by Eq. (7.51) and are known as bilinear shape functions. The strain matrix for the rectangular element can be evaluated in a form similar to Eq. (7.55). Note that for bilinear elements, the strain matrix is no longer constant. Using Eqs. (12.127) and (7.51) $\mathbf{k}_D^{(e)}$ can be evaluated as

$$
\begin{aligned}
\mathbf{k}_D^{(e)} &= \int_{A_e} \mathbf{B}^T \mathbf{D} \mathbf{B} \, dA = \int_{-1}^{+1} \int_{-1}^{+1} ab \mathbf{B}^T \mathbf{D} \mathbf{B} d\xi \, d\eta \\
&= \frac{D_x b}{6a} \begin{bmatrix} 2 & -2 & -1 & 1 \\ -2 & 2 & 1 & -1 \\ -1 & 1 & 2 & -2 \\ 1 & -1 & -2 & 2 \end{bmatrix} + \frac{D_y a}{6b} \begin{bmatrix} 2 & 1 & -1 & -2 \\ 1 & 2 & -2 & -1 \\ -1 & -2 & 2 & 1 \\ 1 & -1 & 1 & 2 \end{bmatrix}
\end{aligned}
\tag{12.136}
$$

The matrix $\mathbf{k}_g^{(e)}$ defined by Eq. (12.121) can be evaluated as

$$
\begin{aligned}
\mathbf{k}_g^{(e)} &= \int_{A_e} g \mathbf{N}^T \mathbf{N} dA = \int_{-1}^{+1} \int_{-1}^{+1} ab g \mathbf{N}^T \mathbf{N} d\xi \, d\eta \\
&= abg \int_{A_e} \begin{bmatrix} N_1^2 & N_1 N_2 & N_1 N_3 & N_1 N_4 \\ N_1 N_2 & N_2^2 & N_2 N_3 & N_2 N_3 \\ N_1 N_3 & N_2 N_3 & N_3^2 & N_3 N_4 \\ N_1 N_4 & N_2 N_4 & N_3 N_4 & N_4^2 \end{bmatrix} d\xi \, d\eta
\end{aligned}
\tag{12.137}
$$

which results in

$$
\mathbf{k}_g^{(e)} = \frac{g A_e}{36} \begin{bmatrix} 4 & 2 & 1 & 2 \\ 2 & 4 & 2 & 1 \\ 1 & 2 & 4 & 2 \\ 2 & 1 & 2 & 4 \end{bmatrix}
\tag{12.138}
$$

The element force vector, $\mathbf{f}_Q^{(e)}$, defined in Eq. (12.122) also involves the integration of the shape functions and with the substitution of the shape functions in Eq. (7.51), it can be obtained as

$$
\mathbf{f}_Q^{(e)} = \int_{A_e} Q \mathbf{N}^T dA = Q \int_{A_e} \begin{Bmatrix} N_1 \\ N_2 \\ N_3 \\ N_4 \end{Bmatrix} dA = \frac{QA}{4} \begin{Bmatrix} 1 \\ 1 \\ 1 \\ 1 \end{Bmatrix}
\tag{12.139}
$$

Note that in the above, Q is assumed to be constant within the element. The heating rate, QA, is equally shared by the four nodes of the rectangular element.

In this subsection, matrices $\mathbf{k}_D^{(e)}, \mathbf{k}_g^{(e)}, \mathbf{f}_Q^{(e)}$ have been evaluated exactly and explicitly for rectangular elements. In engineering practice, however, it is very rare to use rectangular elements unless the geometry of the problem domain is also a rectangular one. Very often, the more general, quadrilateral elements with four nodes and four non-parallel sides are used to mesh the problem domain defined by a complex geometry. Formulating FEM equations for quadrilateral elements has been detailed in Section 7.4. Note that with quadrilateral elements, it is difficult to work out the exact explicit form of the element matrices. Therefore, the integrals are carried out in most of the commercial software packages using numerical integral schemes such as the Gauss integration scheme discussed in Chapter 7 for 2D solid elements.

12.4.4 Boundary conditions and vector b$^{(e)}$

Previously, it is mentioned that the vector, $\mathbf{b}^{(e)}$, for the 2D element as defined by Eq. (12.119) is associated with the variation or derivatives of temperature (or heat flux) on the boundaries of the element. In this section, the relationship of the vector, $\mathbf{b}^{(e)}$, with the boundaries of the element, and hence the boundaries of the problem domain, will be studied in detail.

The vector $\mathbf{b}^{(e)}$ defined in Eq. (12.119) is first split into two parts:

$$\mathbf{b}^{(e)} = \mathbf{b}_I^{(e)} + \mathbf{b}_B^{(e)} \tag{12.140}$$

where $\mathbf{b}_I^{(e)}$ comes from integration of the element boundaries lying inside the problem domain, and $\mathbf{b}_B^{(e)}$ is that which lies on the boundary of the problem domain. It can then be proven that $\mathbf{b}_I^{(e)}$ should vanish, which we have seen for the one-dimensional case.

Figure 12.14 shows two adjacent elements numbered, 1 and 2. In evaluating the vector, $\mathbf{b}^{(e)}$, as defined in Eq. (12.119), the integration needs to be done on all the edges of these elements. As Eq. (12.119) involves a line integral, the results will be direction-dependent. The direction of integration has to be consistent for all the elements in the system, either clockwise or counter-clockwise. For elements 1 and 2, their directions of integration are assumed counter-clockwise as shown by arrows in Figure 12.14. Note that on their common edge j-k the value of $\mathbf{b}_I^{(e)}$ obtained for element 2 is the same as that obtained for element 1, except that their signs are opposite because the directions of integration on this common edge for both elements are opposite. Therefore, when these elements are assembled together, values of $\mathbf{b}_I^{(e)}$ will cancel each other out and vanish. This happens for all other element edges that fall in the interior of the problem domain. Obviously, when the edge lies on the boundary of the problem domain (where it is not shared), $\mathbf{b}_B^{(e)}$ has to be evaluated.

The boundary of the problem domain can be divided broadly into two categories. One is the boundary where the field variable temperature ϕ is specified, as noted by Γ_1 in Figure 12.15, which is known as the *essential boundary condition*. The other is the boundary where the derivatives of the field variable of temperature (heat flux) are specified, as shown in Figure 12.15. This second type of boundary condition is known as the *natural boundary condition*. For the essential boundary condition, we do not need to evaluate $\mathbf{b}_B^{(e)}$ at the stage of formulating and solving the FEM equations, as the temperature is already known, and the corresponding columns and rows will be removed from the global FEM equations in order to solve for the other unknowns. We have seen such a treatment in examples such as Example 12.1. Because $\mathbf{b}^{(e)}$ is derived *naturally* from the weighted residual weak form, it is termed the *natural boundary condition*. Therefore, our concern is only for elements that are on the natural boundaries, where the derivatives of the field variable are specified, and special methods of evaluating the integral are required just as in the 1D case (Example 12.2).

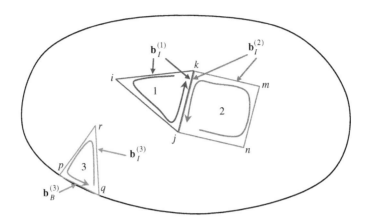

FIGURE 12.14

Direction of integration path for evaluating $\mathbf{b}^{(e)}$. For element edges that are located in the interior of the problem domain, $\mathbf{b}^{(e)}$ vanishes after assembly of the elements, because the values of $\mathbf{b}^{(e)}$ obtained for the same edge of the two adjacent elements possess opposite signs.

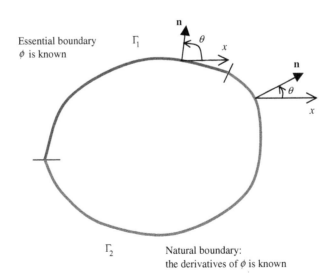

FIGURE 12.15

Types of boundary conditions. Γ_1: Essential boundary where the temperature is known; Γ_2: Natural boundary where the heat flux (derivative of temperature) is known.

In problems of heat transfer, natural boundary often refers to a boundary where heat convection occurs. The integrand in Eq. (12.119) can be generally rewritten in the following form:

$$D_x \frac{\partial \phi^h}{\partial x} \cos \theta + D_y \frac{\partial \phi^h}{\partial y} \sin \theta = -M\phi_b + S \text{ on natural boundary } \Gamma_2 \tag{12.141}$$

where θ is the angle of the outwards normal on the boundary with respect to the x axis, M and S are given constants depending on the type of the natural boundaries, and ϕ_b is the unknown temperature on the boundary. Note that the left-hand side of Eq. (12.141) is in fact the heat flux across the boundary, and can therefore be rewritten as

$$k \frac{\partial \phi^h}{\partial n} = -M\phi_b + S \text{ on } \Gamma_2 \tag{12.142}$$

where k is the heat conductivity at the boundary point in the direction of the boundary normal. For heat transfer problems, there are the following types of boundary conditions:

Heat Insulation boundary: On the boundary where the heat is insulated from heat exchange, there will be no heat flux across the boundary and the derivatives of temperature there will be zero. In such cases, we have $M = S = 0$, and the value of $\mathbf{b}_B^{(e)}$ is simply zero.

Convective boundary condition: Figure 12.16 shows the situations whereby there are exchanges of heat via convection. Following the Fourier's heat convection law, the heat flux across the boundary due to the heat conduction can be given by

$$q_k = -k \frac{\partial \phi}{\partial n} \tag{12.143}$$

where k is the heat conductivity at the boundary point in the direction of the boundary normal. On the other hand, following Fourier's heat convection law, the heat flux across the boundary due to the heat convection can be given by

$$q_h = h \left(\phi_b - \phi_f \right) \tag{12.144}$$

where h is the heat convection coefficient at the boundary point in the direction of the boundary normal. At the same boundary point the heat flux by conduction should be the same as that by convection, i.e, $q_k = q_h$, which leads to

$$k \frac{\partial \phi}{\partial n} = - \underbrace{h}_{M} \phi_b + \underbrace{h\phi_f}_{S} \tag{12.145}$$

The values of M and S for heat convection boundary are then found to be

$$M = h, \quad S = h\phi_f \tag{12.146}$$

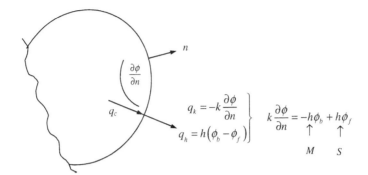

FIGURE 12.16

Heat convection on the boundary.

Specified heat flux on boundary: When there is a heat flux specified on the boundary as shown in Figure 12.17. The heat flux across the boundary due to the heat conduction can be given by Eq. (12.143). The heat flux by conduction should be the same as the specified heat flux, i.e, $q_k = q_s$, which leads to

$$k\frac{\partial\phi}{\partial n} = \underbrace{0}_{M} \times \phi_b \underbrace{-q_s}_{S} \tag{12.147}$$

The values of M and S for heat convection boundary are then found to be

$$M = 0, \quad S = -q_s \tag{12.148}$$

From Figure 12.17, it can be seen that

$$S = \begin{cases} \text{Positive} & \text{if heat flows into the boundary} \\ \text{Negative} & \text{if heat flows out of the boundary} \\ 0 & \text{insulated} \end{cases} \tag{12.149}$$

For other cases whereby M and/or S is not zero, $\mathbf{b}_B^{(e)}$ can be given by

$$\mathbf{b}_B^{(e)} = -\int_{\Gamma_2} \mathbf{N}^T \left(D_x \frac{\partial\phi^h}{\partial x}\cos\theta + D_y \frac{\partial\phi^h}{\partial y}\sin\theta \right) d\Gamma$$

$$= -\int_{\Gamma_2} \mathbf{N}^T (-M\phi_b + S)d\Gamma \tag{12.150}$$

where ϕ_b can be expressed using shape function as follows

$$\phi_b^{(e)} = \mathbf{N}\boldsymbol{\Phi} \tag{12.151}$$

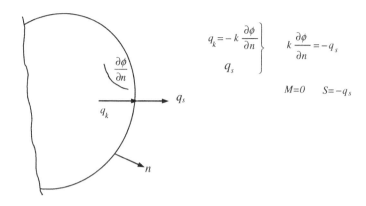

FIGURE 12.17

Specified heat flux applied on the boundary.

Substituting Eq. (12.151) back into Eq. (12.150) leads to

$$
\begin{aligned}
\mathbf{b}_B^{(e)} &= -\int_{\Gamma_2} \mathbf{N}^T \left(-M\mathbf{N}\boldsymbol{\Phi}^{(e)} + S \right) d\Gamma \\
&= \underbrace{\left(\int_{\Gamma_2} \mathbf{N}^T M \mathbf{N}\, d\Gamma \right)}_{\mathbf{k}_M^{(e)}} \boldsymbol{\Phi}^{(e)} - \underbrace{\int_{\Gamma_2} \mathbf{N}^T S\, d\Gamma}_{\mathbf{f}_S^{(e)}}
\end{aligned}
\tag{12.152}
$$

or

$$
\mathbf{b}_B^{(e)} = \mathbf{k}_M^{(e)} \boldsymbol{\Phi}^{(e)} - \mathbf{f}_S^{(e)}
\tag{12.153}
$$

in which,

$$
\mathbf{k}_M^{(e)} = \int_{\Gamma_2} \mathbf{N}^T M \mathbf{N}\, d\Gamma
\tag{12.154}
$$

is the contribution by the natural boundaries to the "stiffness" matrix, and

$$
\mathbf{f}_S^{(e)} = \int_{\Gamma_2} \mathbf{N}^T S\, d\Gamma
\tag{12.155}
$$

is the force vector contribution from the natural boundaries.

Let us now calculate the force vector $\mathbf{f}_S^{(e)}$ for a rectangular element shown in Figure 7.8. Assuming that S is specified over side 1–2,

$$
\mathbf{f}_S^{(e)} = \int_{\Gamma_{1-2}} S\mathbf{N}^T\, d\Gamma = \int_{-1}^{1} S \begin{Bmatrix} N_1 \\ N_2 \\ N_3 \\ N_4 \end{Bmatrix} a\, d\xi
\tag{12.156}
$$

where the shape functions are given by Eq. (7.51) in the natural coordinate system. Note, however, that $N_3 = N_4 = 0$ along edge 1–2. Substituting the nonzero shape functions into the above equation, we obtain

$$\mathbf{f}_S^{(e)} = \int_{-1}^{1} \frac{Sa}{2} \begin{Bmatrix} (1 - \xi) \\ (1 + \xi) \\ 0 \\ 0 \end{Bmatrix} d\xi = Sa \begin{Bmatrix} 1 \\ 1 \\ 0 \\ 0 \end{Bmatrix} \tag{12.157}$$

The above equation implies that the quantity of $(2aS)$ is shared equally between the two nodes 1 and 2 on the edge. This even distribution among the nodes on the edge is valid for all the elements with linear shape function. Therefore, if the natural boundary is on the other three edges of the rectangular element, the force vector can be simply written as follows:

$$\mathbf{f}_{S,2-3}^{(e)} = Sb \begin{Bmatrix} 0 \\ 1 \\ 1 \\ 0 \end{Bmatrix}, \mathbf{f}_{S,3-4}^{(e)} = Sb \begin{Bmatrix} 0 \\ 0 \\ 1 \\ 1 \end{Bmatrix}, \mathbf{f}_{S,1-4}^{(e)} = Sb \begin{Bmatrix} 1 \\ 0 \\ 0 \\ 1 \end{Bmatrix} \tag{12.158}$$

Note that if S is specified on more than one side of an element, the values for $\left\{f_S^{(e)}\right\}$ for the appropriate sides are added together.

The same principle of equal sharing can be applied to the linear triangular element shown in Figure 12.13. The expression for the force vectors on the three edges can be simply written as

$$\mathbf{f}_{S,1-2}^{(e)} = \frac{SL_{12}}{2} \begin{Bmatrix} 1 \\ 1 \\ 0 \end{Bmatrix}, \quad \mathbf{f}_{S,2-3}^{(e)} = \frac{SL_{23}}{2} \begin{Bmatrix} 0 \\ 1 \\ 1 \end{Bmatrix}, \quad \mathbf{f}_{S,1-3}^{(e)} = \frac{SL_{13}}{2} \begin{Bmatrix} 1 \\ 0 \\ 1 \end{Bmatrix} \tag{12.159}$$

The quantities L_{12}, L_{23}, and L_{13} are the lengths of the respective edges of the triangular element. Using Eq. (12.154) to derive $\mathbf{k}_M^{(e)}$ for the rectangular element shown in Figure 7.8, we obtain

$$\mathbf{k}_M^{(e)} = \int_{\Gamma_2} M \begin{bmatrix} N_1^2 & N_1N_2 & N_1N_3 & N_1N_4 \\ N_1N_2 & N_2^2 & N_2N_3 & N_2N_4 \\ N_1N_3 & N_2N_3 & N_3^2 & N_3N_4 \\ N_1N_4 & N_2N_4 & N_3N_4 & N_4^2 \end{bmatrix} d\Gamma \tag{12.160}$$

Note that the line integration is performed round the edge of the rectangular element. If we assume that M is specified over edge 1–2, then $N_3 = N_4 = 0$ and the above equation becomes

$$\mathbf{k}_{M,1-2}^{(e)} = aM \int_{-1}^{1} \begin{bmatrix} N_1^2 & N_1 N_2 & 0 & 0 \\ N_2 N_1 & N_2^2 & 0 & 0 \\ 0 & 0 & 0 & 0 \\ 0 & 0 & 0 & 0 \end{bmatrix} d\xi \tag{12.161}$$

Evaluation of the individual coefficients after noting that $\eta = -1$ for edge 1–2 gives

$$\int_{-1}^{1} N_1^2 d\xi = \int_{-1}^{1} \frac{(1-\xi)^2}{4} d\xi = \frac{2}{3}$$

$$\int_{-1}^{1} N_1 N_2 d\xi = \int_{-1}^{1} \frac{(1-\xi)(1+\xi)}{4} d\xi = \frac{2}{6} \tag{12.162}$$

$$\int_{-1}^{1} N_2^2 d\xi = \int_{-1}^{1} \frac{(1-\xi)^2}{4} d\xi = \frac{2}{3}$$

Eq. (12.161) thus becomes

$$\mathbf{k}_{M,1-2}^{(e)} = \frac{2aM}{6} \begin{bmatrix} 2 & 1 & 0 & 0 \\ 1 & 2 & 0 & 0 \\ 0 & 0 & 0 & 0 \\ 0 & 0 & 0 & 0 \end{bmatrix} = (2aM) \begin{bmatrix} 2/6 & 1/6 & 0 & 0 \\ 1/6 & 2/6 & 0 & 0 \\ 0 & 0 & 0 & 0 \\ 0 & 0 & 0 & 0 \end{bmatrix} \tag{12.163}$$

It is observed that the amount of $(2aM)$ is shared by four components k_{11}, k_{12}, k_{21}, and k_{22} in ratios of $\frac{2}{6}, \frac{1}{6}, \frac{1}{6}$, and $\frac{2}{6}$. This sharing principle can be used to obtain the matrices $\mathbf{k}_M^{(e)}$ directly for situations where M is specified on the other three edges. They are

$$\mathbf{k}_{M,2-3}^{(e)} = \frac{M2b}{6} \begin{bmatrix} 0 & 0 & 0 & 0 \\ 0 & 2 & 1 & 0 \\ 0 & 1 & 2 & 0 \\ 0 & 0 & 0 & 0 \end{bmatrix}, \mathbf{k}_{M,3-4}^{(e)} = \frac{M2a}{6} \begin{bmatrix} 0 & 0 & 0 & 0 \\ 0 & 0 & 0 & 0 \\ 0 & 0 & 2 & 1 \\ 0 & 0 & 1 & 2 \end{bmatrix},$$

$$\mathbf{k}_{M,1-4}^{(e)} = \frac{M2b}{6} \begin{bmatrix} 2 & 0 & 0 & 1 \\ 0 & 0 & 0 & 0 \\ 0 & 0 & 0 & 0 \\ 1 & 0 & 0 & 2 \end{bmatrix} \tag{12.164}$$

This sharing principle can also be applied to linear triangular elements, since the shape functions are also linear, and the corresponding matrices are:

$$
\mathbf{k}_{M,i-j}^{(e)} = \frac{ML_{ij}}{6}
\begin{bmatrix}
2 & 1 & 0 \\
1 & 2 & 0 \\
0 & 0 & 0
\end{bmatrix}, \mathbf{k}_{M,j-k}^{(e)} = \frac{ML_{jk}}{6}
\begin{bmatrix}
0 & 0 & 0 \\
0 & 2 & 0 \\
0 & 1 & 2
\end{bmatrix},
$$

$$
\mathbf{k}_{M,i-k}^{(e)} = \frac{ML_{ik}}{6}
\begin{bmatrix}
2 & 0 & 1 \\
0 & 0 & 0 \\
1 & 0 & 2
\end{bmatrix}
$$

(12.165)

12.4.5 **Point heat source or sink**

If there is a heat source or sink in the domain of the problem, it is best recommended in the modeling that a node is placed at the point where the source or sink is located, so that the source or sink can be directly added into the force vector, as shown in Figure 12.18. If, for some reason, this cannot be done, then we have to distribute the source or sink to the nodes of the element in which the source or sink is located. To do this, we have to go back to Eq. (12.122), which is once again rewritten below

$$
\mathbf{f}_Q^{(e)} = \int_A Q\mathbf{N}^T \, dA
$$

(12.166)

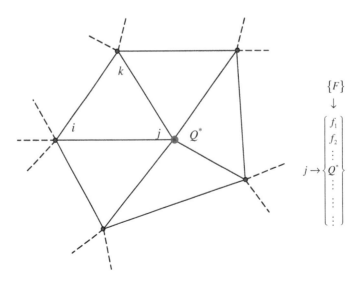

FIGURE 12.18

A heat source or sink at a node of the FE model.

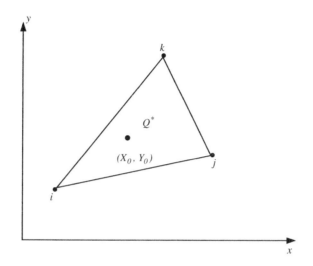

FIGURE 12.19

A heat source or sink in a triangular element.

Consider a point source or sink in a triangular element as shown in Figure 12.19. The source or sink can be mathematically expressed using delta functions

$$Q = Q^* \delta \left(x - X_0 \right) \delta \left(y - Y_0 \right) \tag{12.167}$$

where Q^* represents the strength of the source or sink, and (X_0, Y_0) is the location of the source or sink. Substitute Eq. (12.167) into Eq. (12.166) and we obtain

$$\mathbf{f}_Q^{(e)} = Q^* \int_{A_e} \begin{Bmatrix} N_i \\ N_j \\ N_k \end{Bmatrix} \delta(x - X_0)\delta(y - Y_0)\mathrm{d}x\mathrm{d}y \tag{12.168}$$

which becomes

$$\mathbf{f}_Q^{(e)} = Q^* \begin{Bmatrix} N_i(X_0, Y_0) \\ N_j(X_0, Y_0) \\ N_k(X_0, Y_0) \end{Bmatrix} \tag{12.169}$$

This implies that the source or sink is shared by the nodes of the elements in the ratios of shape functions evaluated at the location of the source or sink. This sharing principle can be applied to any type of elements, and also other types of physical problems, for example, a concentrated force applied in the middle of a 2D element.

12.5 Summary

Finite element formulation for field problems that are governed by the general form of *Helmholtz equation* can be summarized as follows.

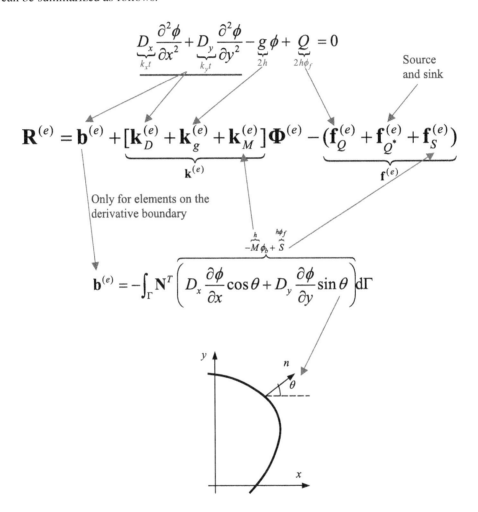

12.6 Case study: Temperature distribution of heated road surface

Figure 12.20 shows the cross-section of a road with heating cables to prevent the surface of the road from freezing. The cables are 4 cm apart and 2 cm below the surface of the road. The slab rests on a thick layer of insulation and the heat loss from the bottom can be neglected. The conductivity coefficients are $k_x = k_y = 0.018$ W/cm°C and the surface convection coefficient is $h = 0.0034$ W/cm°C. The latter corresponds to about a 30–35 km/hr of wind velocity. The surface temperature of the road is to be determined when the cable produces 0.080 W/cm of heat and the air temperature is -6°C.

12.6.1 **Modeling**

Since the road is very long in the horizontal direction, a representative section shown in Figure 12.20 can be used to model the entire problem domain. The FE mesh is shown in Figure 12.21 together with boundary conditions specified.

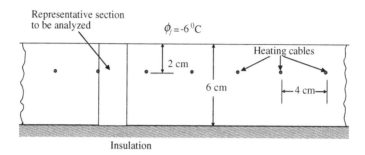

FIGURE 12.20

Cross-section of road with heating cables.

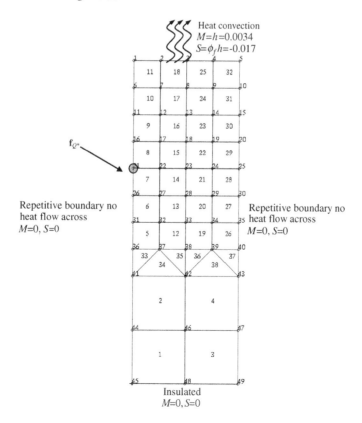

FIGURE 12.21

2D finite element mesh with boundary conditions.

The mesh shown in Figure 12.21 demonstrates mesh transition from an area consisting of a sparse mesh to an area of denser mesh. The analyst has chosen to mesh it this way since the temperature distribution at the bottom of the model is not critical. Hence, computational time is reduced as a result. The transition is done with the use of triangular elements in between larger rectangular elements and smaller rectangular elements. Note that all the elements used are linear elements and hence the mixture of elements here is compatible.

12.6.2 ABAQUS input file

Part of the ABAQUS input file is shown here:

```
*HEADING
Calculation of 2D heat transfer
**
*NODE
1, 0., 6.
2, 0.5, 6.
3, 1., 6.
4, 1.5, 6.
5, 2., 6.

46, 1., 1.
47, 2., 1.
48, 1., 0.
49, 2., 0.
**
*ELEMENT, TYPE=DC2D4,
ELSET=QUAD
1,45,48,46,44
2,48,49,47,46
31,14,15,10,9
32,9,10,5,4
**
*ELEMENT, TYPE=DC2D3,
ELSET=TRI
33,41,37,36
34,41,42,37
37,39,43,40
38,42,43,39
**
**
*SOLID SECTION, ELSET=QUAD,
MATERIAL=ROAD
1.,
**
*SOLID SECTION, ELSET=TRI,
MATERIAL=ROAD
```

Nodal cards

Node I.D., x-coordinate, y-coordinate

Element (connectivity) cards

Element type here is DC2D4 which represents a 2D, 4-node quadrilateral, heat transfer element.
(Element I.D., node 1, node 2, node 3, node4)

Element (connectivity) cards

Element type here is DC2D3 which represents a 2D, 3-nodal triangular, heat transfer element.
(Element I.D., node 1, node 2, node 3)

Property cards

Define properties to the elements of sets "QUAD" and "TRI." They will have the material properties defined under "ROAD."

```
1.,
**
**
*MATERIAL, NAME=ROAD
**
*CONDUCTIVITY, TYPE=ISO
0.018,
**
**
*STEP
**
*HEAT TRANSFER, STEADY STATE
0.1, 1.
**
*ELSET, ELSET=SURFACE
11, 18, 25, 32
*ELSET, ELSET=LEFT_QUAD, GENERATE
1, 2, 1
5, 11, 1
*ELSET, ELSET=RIGHT_QUAD, GENERATE
3, 4, 1
26, 32, 1
*ELSET, ELSET=BASE
1, 3
*ELSET, ELSET=LEFT_TRI
33
*ELSET, ELSET=RIGHT_TRI
37
*NSET, NSET=SOURCE
21
**
*FILM, OP=NEW
SURFACE, F3, -6., 0.0034
**
** insulated edges
**
*DFLUX, OP=NEW
LEFT_QUAD, S4, 0.
RIGHT_QUAD, S2, 0.
BASE, S1, 0.
LEFT_TRI, S3, 0.
RIGHT_TRI, S2, 0.
**
```

Material cards

Define material properties under the name "ROAD." Thermal conductivity coefficient is being defined. TYPE=ISO represents isotropic properties.

Control cards

Indicate the steady state, heat transfer analysis procedure.

Element sets

Group elements into sets to be referenced when defining boundary conditions.

BC cards

The keyword, *FILM is used to define the heat convection properties. In the data line, the first input refers to the element set SURFACE, the second refers to the surface or edge where the convection is occurring, the fourth is the sink temperature and lastly, the convection coefficient.

*DFLUX is for specifying distributed heat flux. In this case, the left, right and bottom edges are all insulated (= 0)

```
** heat source
**
*CFLUX, OP=NEW
SOURCE, 11, 0.08
**
*NODE PRINT, FREQ=1
NT,
*NODE FILE, FREQ=1
NT,
**
*END STEP
```

Load cards
The load here is a concentrated heat flux or source defined by *CFLUX and applied on node 21 or node set SOURCE. Note that in this case the DOF for the temperature is defined by the number 11.

Output control cards
Define the requested output. In this case, NT is the nodal temperature.

The information provided in the above input file is being used by the software in a similar manner as that discussed in the previous case studies in previous chapters.

12.6.3 Results and discussion

Running the above problem in ABAQUS, the nodal temperatures can be calculated. Figure 12.22 shows a fringe plot of the distribution of the temperatures in the model. It can be seen clearly how the temperature varies from a maximum at the heat source (the heating cables) to other parts of the road cross-section.

In the analysis, the temperatures at all the nodes are calculated. For this problem, we would be interested in only the temperature of the road surface. Table 12.1 shows the nodal temperature on the surface of the road. It can be seen here how the presence of the heating cables under the road is able to keep the road surface at a temperature above the freezing point of water $(0\,^\circ C)$. This would prevent the build-up of ice on the road surface during winter, which makes it safer for drivers on the road. The usefulness of the finite element method is demonstrated here as there are actually many parameters involved when it comes to designing such a system. For example, how deep should the cables be buried underground, what should be the distance between cables, what is the amount of heat generated by the heating cables that is sufficient for the purpose, and so on. The finite element method used here can effectively aid the engineer in deciding all these parameters.

12.7 Review questions

1. **a.** A fin with a length L has a uniform cross-sectional area A and thermal conductivity k as shown in Figure 12.23. A linearly distributed heat supply is applied on the fin. The temperature at the left end is fixed at T_0, and the heat flux at the right end is q_0. The governing equation for the fin is given by

$$Ak\frac{\partial^2 \phi}{\partial x^2} + Q = 0, \quad 0 \le x \le L$$

Develop the finite element equation for a two-node element.

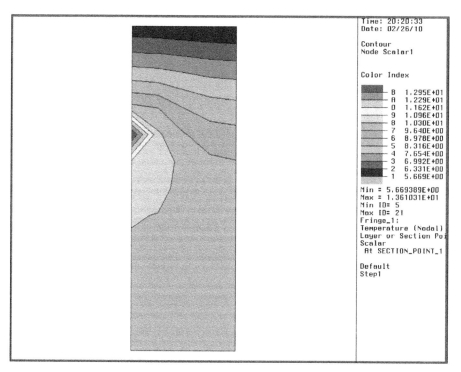

FIGURE. 12.22

Temperature distribution of cross-section of road.

Table 12.1 Nodal temperatures of road surface.	
Node	**Temperature (°C)**
1	5.861
2	5.832
3	5.764
4	5.697
5	5.669

b. If $L = 8$ m, $A = 1$ m^2, $k = 5$ J/°Cms, $Q = 100$ J/sm, $T_0 = 0$, and $q_0 = 15$ J/m^{2}s, determine the temperatures at the nodes by using two linear elements.

2. Figure 12.24 shows a one-dimensional fin of length L and a varying cross-sectional area $A(x)$. The thermal conductivity of the material that is constant is denoted by k. Heat convection occurs on

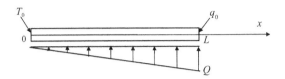

FIGURE 12.23

1D fin with linearly distributed heat supply.

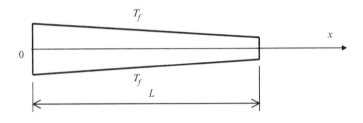

FIGURE 12.24

1D fin with a varying cross-sectional area.

the surfaces of the fin, and the ambient temperature is denoted by T_f. The governing equation for the temperature T in the fin is given by

$$k\frac{\partial}{\partial x}\left(A(x)\frac{\partial T}{\partial x}\right) + hP(x)(T_f - T) = 0, \quad 0 \le x \le L$$

where h is a given convection coefficient. The area of the fin is given by

$$A(x) = A_0 + A_d\frac{x}{L}, \quad 0 \le x \le L$$

where A_0 and A_d are constants. The circumference of the cross-section of the fin is given by

$$P(x) = P_0 + P_d\frac{x}{L}, \quad 0 \le x \le L$$

where P_0 and P_d are constants.

Develop the finite element equations for a two-node linear element of length L.

3. Figure 12.25 shows a sandwiched composite wall. Convection heat loss occurs on the left surface, and the temperature on the right surface is constant. Considering a unit area, and with the parameters given in Figure 12.25, use three linear elements (one for each layer) to:

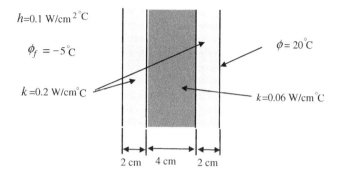

$h=0.1$ W/cm^2 $^\circ$C

$\phi_f = -5^\circ$C

$k =0.2$ W/cm$^\circ$C

$\phi= 20^\circ$C

$k=0.06$ W/cm$^\circ$C

2 cm   4 cm   2 cm

FIGURE 12.25

Sandwiched composite wall.

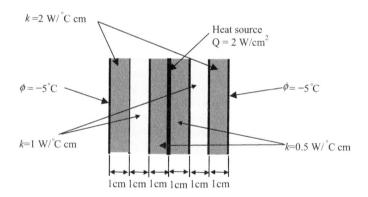

$k =2$ W/$^\circ$C cm

Heat source
$Q = 2$ W/cm^2

$\phi= -5^\circ$C

$\phi= -5^\circ$C

$k=1$ W/$^\circ$C cm

$k=0.5$ W/$^\circ$C cm

1cm 1cm 1cm 1cm 1cm 1cm

FIGURE 12.26

Layered composite wall.

a. determine the temperature distribution through the composite wall, and
b. calculate the flux on the right surface of the wall.
4. Figure 12.26 shows a system of a composite wall. The temperature of the two outer surfaces of the wall is kept constant. A heat source is located in the middle surface of the wall. Using linear elements:
a. determine the temperature distribution across the composite wall,
b. calculate the flux on the surfaces of the wall,

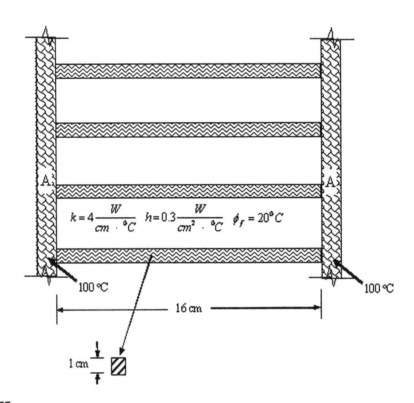

$$k = 4\frac{W}{cm \cdot {}^{\circ}C} \quad h = 0.3\frac{W}{cm^2 \cdot {}^{\circ}C} \quad \phi_f = 20^{\circ}C$$

100 °C   100 °C

16 cm

1 cm

FIGURE 12.27

An array of identical fins with square cross-section used to dissipate the heat of two vertical walls connected to these fins.

    c. calculate the temperature distribution across the wall, when $\phi = -2.5\,^{\circ}C$ while all other parameters remain the same, and

    d. explain whether or not higher order elements lead to more accurate results.

5. Figure 12.27 shows an array of identical fins with square cross-section used to dissipate the heat of two vertical walls connected to these fins. At the steady-state, the temperature on these walls is kept at 100 °C. The dimensions and the thermal conductivity of the fins, the convection coefficient on the surfaces of the fine, and the ambient temperature are given in the figure.

    a. Using four linear elements for each fin, determine the temperature distribution along the fins.

    b. Calculate all the heat dissipated by each of the fins.

    c. Is the FEM capable of producing the exact solution? Justify your answer.

    d. What is the possible ways to improve the accuracy of the solution? Which way is the most preferred?

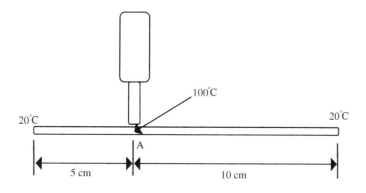

FIGURE 12.28

Soldering of copper wire.

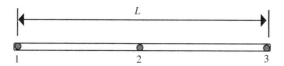

FIGURE 12.29

Quadratic 1D element.

6. Consider a soldering situation where the tip raises the temperature at point A of a copper wire to 100 °C, shown in Figure 12.28. The wire is 15 cm long and 0.02 cm in diameter. The temperature at both ends is 20 °C. The thermal conductivity k of the copper wire is 26 J/°Cms. Assume the circumferential surface of the wire is adiabatic. Using three linear elements of equal length,
   a. determine the heat flux into the wire at point A,
   b. determine the heat flux at both ends of the wire,
   c. explain whether three elements are really needed for this problem, and
   d. repeat (b) for a given heat flux of 4×10^{-3} J/s instead of the temperature.
7. Consider a quadratic heat-conduction line element with 3 equally spaced nodes as shown in Figure 12.29.
   a. Using the quadratic element, determine the heat conduction matrix.
   b. Using one linear element for the left portion, and one quadratic element for the right portion of the wire shown in Figure 12.28 of question 5 above, determine the heat flux at point A and both ends of the wire.
   c. Comment on the results by comparing them with the results of question 3.

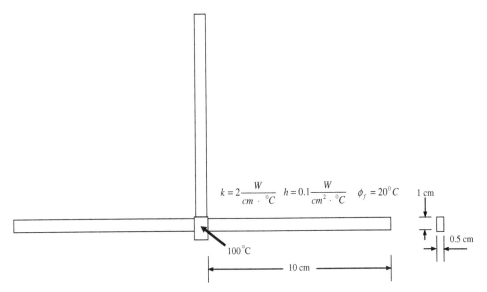

FIGURE 12.30

Three identical fins used to dissipate the heat.

8. Figure 12.30 shows three identical fins used to dissipate the heat. At the steady state, the temperature at the roots of these three fins is kept at $100\,^\circ C$.

a. Using two linear elements for each fin, determine the temperature distribution along all three fins.

b. Calculate all the heat dissipated by all three fins.

c. Explain how many elements in a fin are required to obtain an accurate solution. What is the other way to improve the accuracy of the solution without increasing the number of elements? Which way is more efficient?

Using FEM Software Packages

13

13.1 Introduction

Realistic finite element problems might consist of hundreds of thousands and even several millions of elements and nodes and, therefore, they are usually solved in practice using commercially available software packages. There are currently a large number of commercial software packages available for solving a wide range of problems: Solid and structural mechanics, heat and mass transfer, fluid mechanics, acoustics and multi-physics, which might be static, dynamic, linear or non-linear. Most of these software packages use the FEM as a stand-alone approach or combine it with other numerical methods to solve physical problems. All FEM software packages are developed based on the same basic methodology described in this book, but usually extend their capabilities for more complex problems with the development of detailed and fine-tuned algorithms and numerical schemes. Table 13.1 lists some of the commercially available software packages using FEM, the finite volume method (FVM), and the boundary element method (BEM). This chapter introduces the use of an FEM package primarily through ABAQUS FEA®, currently owned by Dassault Systèmes, due to its strong capabilities, in

Table 13.1   Some of the commercially available software packages.

Software Package	Methods Used	Application Problems
ABAQUS	FEM (implicit, explicit)	Structural analysis, acoustics, thermal analysis
ANSYS	FEM (implicit, explicit)	Structural analysis, acoustics, thermal analysis, multi-physics
I-deas	FEM (implicit)	Structural analysis, acoustics, thermal analysis
LS-DYNA	FEM (explicit)	Structural dynamics, computational fluid dynamics, fluid-structural interaction
Sysnoise	FEM/BEM	Acoustics (frequency domain)
NASTRAN	FEM (implicit)	Structural analysis, acoustics, thermal analysis
MARC	FEM (implicit)	Structural analysis, acoustics, thermal analysis
MSC-DYTRAN	FEM+FVM (explicit)	Structural dynamics, computational fluid dynamics, fluid-structural interaction
ADINA	FEM (implicit)	Structural analysis, computational fluid dynamics, Fluid-structural interaction

general, in dealing with non-linear problems. At the end of the chapter, we also remark on the modeling similarities across FEM packages by highlighting basic steps using a graphical user interface (GUI) via another popular FEM package, ANSYS.

With the development of convenient user interfaces, most finite element software can be used as a "black box" by users without extensive knowledge of FEM. The authors have witnessed many cases where the FEM packages were misused, resulting in what is termed "garbage in, garbage out" simulations. The danger here is that the "garbage" output is often obscured by beautiful pictures and animations that can lead to erroneous decisions in designing an engineering system.

An understanding of the materials covered in the previous chapters in this book should shed some light into this black box, leading to the proper use of most software packages. As mentioned, this chapter will primarily use ABAQUS FEA as an example to highlight the general steps involved in using commercial software packages from a user point of view. Chapters 1 to 12 have highlighted various basic concepts in the finite element method and the case studies actually relate these concepts to examples solved using typical FEM packages (primarily ABAQUS/Standard). There are other modules in the ABAQUS finite element package including ABAQUS/Explicit, ABAQUS/CFD, and ABAQUS/CAE. ABAQUS/Explicit is used for mainly explicit dynamic analysis. ABAQUS/CFD provides computational fluid dynamics capabilities. ABAQUS/CAE is an interactive pre- and post-processor that can be used to create finite element models and the associated input file for ABAQUS. In this chapter, however, the focus will be on the writing of a basic ABAQUS/Standard input file and ABAQUS/Standard will from now on just be called ABAQUS. This book cannot and will not try to in any way replace the extensive and excellent manuals provided together with the ABAQUS software. This chapter will just serve as a general guide especially suited for beginner users to have a quick understanding of using an FEM package, without going through the often daunting details covered in comprehensive user manuals. It is hoped that, after reading this chapter, readers will have an even better understanding of the basic finite element concepts being implemented in finite element software packages.

13.2 Basic building block: keywords and data lines

The first step to writing an ABAQUS input file is to know the way data is included in the input file. In ABAQUS, the data definitions are expressed in what are termed option blocks or groups of cards. Basically, it is thus named because the user has the option to choose particular data blocks that are relevant for the model to be defined. Each option block can be considered to be a basic building block that builds up the entire input file. The option block is introduced by a *keyword* line and if the option block requires *data lines*, these will follow directly below the keyword line. The general layout of a particular option block is shown below with the definition of beam elements as an example:

```
Example 1
*ELEMENT, TYPE=B23, ELSET=BEAM  ⎤
   1, 1, 2                      ⎬  Keyword line  begins with *
   2, 2, 3                      ⎦
                                   Data lines
```

Keyword lines begin with an asterisk, *, followed by the name of the block. In this case, we define elements in the element block identified with the keyword, ELEMENT. Subsequent information on the keyword line provides additional parameters associated with the block being used. In this case, it is necessary to tell ABAQUS what type of element is being used (B23 – 2 node, 1D Euler-Bernoulli beam element) and the elements are grouped into a particular set with the arbitrary given name, "BEAM." This grouping of entities into sets is a very convenient tool, which the analyst will use very often. It enables the analyst to make references to the set when defining certain option blocks.

The data lines basically provide data, if required, that is associated with the option block used. In the above example, element identification (i.d.), and the nodes that make up the element are the necessary data required. Note that the information provided in the data lines would vary with some of the parameters defined in the keyword line. For example, if the element type being used is a 2D plane stress element, then the data lines would require different data as shown below.

```
Example 2
*ELEMENT, TYPE=CPS4, ELSET=PLSTRESS
  1, 1, 3, 4, 2  ⎤
  2, 3, 5, 6, 4  ⎦  Element i.d., 1ˢᵗnode, 2ⁿᵈ node, 3ʳᵈ node, 4ᵗʰ node
```

Note that in this case, the element type CPS4 which represents 4-nodal, 2D solid elements is being used and, correspondingly, the data lines must include the element i.d. (as before), and four nodes that make up each element instead of two for the case of the beam element previously.

13.3 Using sets

In the last section, it can be seen in Eg. 1 that elements can be grouped into a *set* for future reference by other option blocks. A set can be a grouping of nodes or a grouping of elements. The analyst will usually provide a name for the set that contains between 1 to 80 characters. For example, in Eg. 1 above, "BEAM" is the name of the set containing elements 1 and 2; and in Eg. 2, "PLSTRESS" is the name of the set containing elements 1 and 2. In both examples, the sets are defined together with the definition of the elements themselves in the element block. However, sets can also be defined as a separate block on their own. In Eg. 3 below, the pinned support of the 1D beam is to be defined. Nodes 1 and 11 (provided in the data line for the "NSET" block) are first grouped in the node set called "SUPPORT." Then using the "BOUNDARY" option block, the node set, "SUPPORT" is referenced to constrain the DOFs 1 and 2 (*x* translation and *y* translation) to zero. In other words, rather than having four data lines for the two nodes 1 and 11 (each node having to constrain two DOFs), we now have only two lines with the reference to the node set support. Another thing to note in this example is the usage of comment lines. Comment lines exist in ABAQUS too just like in most programming languages. Comment lines begin with two asterisks, **, and whatever follows in that line after the asterisks will not be read by ABAQUS as input information defining the model.

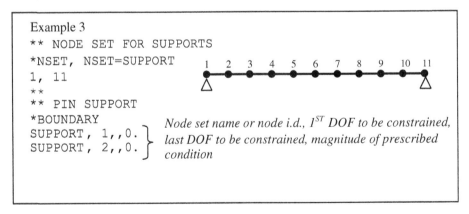

This is of course a very simple case and the reduction of the number of data lines from four to two may not seem very significant. However, imagine if it were a huge 3D model and one whole surface containing about 100 nodes were to be prescribed a boundary condition. If the nodes on this surface were not grouped up into node sets, then the user would end up with $100 \times$ (*No. of DOFs to be constrained*) data lines just to prescribe a boundary condition. A more efficient way of course is to group the nodes in this particular surface in a node set and then, as in the above example, write down the data lines referencing the node set to be constrained. In this way, the number of data lines for the "BOUNDARY" block would be equal to the number of DOFs to be constrained. Similar usage of sets can be applied to elements as well. One common usage of element sets (ELSET) is the referencing of element properties to the elements in a particular set (see example in Section 13.5.3). Sets are thus the basic referencing tool in ABAQUS.

13.4 **ABAQUS input syntax rules**

The previous section has introduced the way data are organized in the ABAQUS input file. Like most programming language, there are certain rules which the entries into the input file must follow. A violation of these rules would generally result in a syntax error when ABAQUS is run, and most of the time, the analysis will not be carried out. So far, it has been briefed that ABAQUS generally has three types of entries in the input file: the keyword lines, the data lines, and the comment lines. Comment lines generally do not need many rules except that they must begin with two asterisks (**) in columns 1 and 2. This section therefore describes the rules that apply to all keyword and data lines.

13.4.1 **Keyword lines**

The following rules apply when entering a keyword line:

- The first non-blank character of each keyword line must be an asterisk (*).
- The keyword must be followed by a comma (,), if any further parameters are to be included in the line.
- Parameters are separated by commas.
- Blanks on a keyword line are ignored.
- A line can include no more than 256 characters including blanks.
- Keywords and parameters are not case sensitive.
- Parameter values are usually not case sensitive. The only exceptions to this rule are those imposed externally to ABAQUS, such as file names on case-sensitive operating systems.
- If a parameter has a value, the equal sign (=) is used. The value can be an integer, a floating point number, or a character string, depending on the context. For example,

```
*EL FILE, POS=INTEG, FREQ=1
```

- Continuation of a keyword line is sometimes necessary especially when there are a large number of parameters. If the last character of a keyword line is a comma, the next line is treated as a continuation line. For example, the example stated in the previous rule can also be written as

```
*EL FILE, POS=INTEG,
FREQ=1
```

13.4.2 **Data lines**

The data lines must immediately follow a keyword line if they are required. The following rules apply when entering a data line:

1. A data line can include no more than 256 characters including blanks.
2. All data items are separated by commas. An empty data field is specified by omitting data between commas. ABAQUS will use values of zeros for any required numeric data that are omitted unless

there is a default value allocated. If a data line contains only a single data item, the data item should be followed by a comma.

3. A line must contain only the number of items specified.
4. Empty data fields at the end of a line can be ignored.
5. Floating point numbers can occupy a maximum of 20 spaces including the sign, decimal point, and any exponential notation.
6. Floating point numbers can be given with or without an exponent. Any exponent, if input, must be preceded by E or D and an optional (–) or (+) to indicate the sign of the exponent.
7. Integer data items can occupy a maximum of 10 digits.
8. Character strings can be up to 80 characters long and are not case sensitive.
9. Continuation lines are allowed in specific instances such as when defining elements with a large numbers of nodes. If allowed, such lines are indicated by a comma as the last character of the preceding line.

13.4.3 Labels

Examples of labels are set names, surface names and material names and they are case sensitive unlike the other entries in the keyword line. Labels can be up to 80 characters long. All spaces within a label are ignored unless the label is enclosed in quotation marks, in which case all spaces within the label are maintained. A label that is not enclosed within quotation marks may not include a period (.) and should not contain characters such as commas and equal signs. If a label is defined using quotation marks, the quotation marks are stored as part of the label. Any subsequent reference or use of the label should also include the quotation marks. Labels cannot begin and end with a double underscore (e.g., __ALU__). This label format is reserved for use internally within ABAQUS.

13.5 Defining a finite element model in ABAQUS

Though very often, the use of a GUI such as in ABAQUS/CAE or PATRAN can be helpful in creating the finite element model and generating the input file for complex models, the analyst may still find certain limitations in pre-processors whereby certain functions, element types, material definitions, and other modeling requirements cannot be automatically defined. Many times, a specific analysis will require more than just the usual analysis steps and this is when an analyst will find that knowing the basic concepts of writing an input file will enable him or her to either write a whole input file for the specific problem or to modify the existing input file that is generated by the pre-processor.

An ABAQUS input file is an ASCII data file and can be created or edited by using any text editor. The input file contains two main sets of data: *model data* and *history data*. The model data consists of data defining the finite element model. This part of the input file defines the elements, the nodes, the element properties, the material properties, and any other data relating to the model itself. Looking at the input files provided for the case studies included in previous chapters, it should be noted that all the data provided before the "*STEP" line is considered as the model data. The history data, on the other hand, defines what happens to the finite element model. It tells ABAQUS the events the model has gone through, the loadings the model have, the type of response that should be sought for, and so on. In ABAQUS, the history data is made up of one or more *steps*. Each step defines the analysis procedures by providing the required parameters. It is possible and in fact quite common to have multiple steps to define a whole analysis procedure. For example, to obtain a steady-state dynamic response due to a harmonic excitation at a given frequency by modal analysis, one must first obtain the eigenvalues and eigenvectors. This can be defined in a step defined by *FREQUENCY, which calls for the eigenvector extraction analysis procedure. Following that, another step defined by *STEADY STATE DYNAMICS is necessary that calls for the modal analysis procedure to solve for the response under a harmonic excitation. As such, the history data can be said to be made up of a series of steps, which in a way tells the history of the analysis procedure. This section will describe in more detail how a basic model can be defined in ABAQUS. Input files defining most problems have the same basic structure:

1. An input file must begin with the *HEADING option block, which is used to define a title for the analysis. Any number of data lines can be used to give the title.
2. After the heading lines, the input file would usually consist of the model data, which consists of the node definition, element definition, material definition, initial conditions, and so on.
3. Finally the input file would consist of the history data, in which is defined the analysis type, loading, output requests, and so on. Usually, the *STEP option is the dividing point in the input file between the model and history data. Everything appearing before the *STEP option will be considered model data and everything after will be considered the history data.

The following will list some of the options and data that can be included in the model and history data. This book will not elaborate on all the available options and, if required, the user is recommended to refer to the software's user manual (ABAQUS keywords manual, 2000; ABAQUS user's manual, 2000; ABAQUS theory manual, 2000). Elaboration will be done on some of the necessary options for a basic finite element model later.

13.5.1 **Model data**

Some of the items that must be included in the model data are as follows:

- Geometry of the model: The geometry of the model is described by its elements and nodes.
- Material definitions, which are usually associated with parts of the geometry.

Other optional data in the model data section are:

- Parts and an assembly: The geometry can be divided into parts, which are positioned relative to one another in an assembly.
- Initial conditions: Non-zero initial conditions such as initial stresses, temperatures, or velocities can be specified.
- Boundary conditions: Zero-valued boundary conditions (including symmetry conditions) can be imposed on the model.
- Constraints: Linear constraint equations or multipoint constraints can be defined.
- Contact interactions: Contact conditions between surfaces or parts can be defined.
- Amplitude definitions: Amplitude curves which the loads or boundary conditions are to follow can be defined.
- Output control: Options for controlling model definition output to the data file can be included.
- Environment properties: Environment properties such as the attributes of a fluid surrounding the model can be defined.

- User subroutines: User-defined subroutines, which allow the user to customize ABAQUS can be defined.
- Analysis continuation: It is possible to write restart data or to use the results from a previous analysis and continue the analysis with new a model or history data.

13.5.2 History data

As mentioned, in the history data, the entries are divided into steps. Each step will begin with the *STEP option and ends with the *END STEP option. There are generally two kinds of steps that can be defined in ABAQUS—the general response analysis steps (can be linear or non-linear); and the linear perturbation steps. A general analysis step is one in which the effects of any nonlinearities present in the model can be included. The starting condition for each general analysis step is the ending condition from the last general analysis step. The response of each general analysis step contributes to the overall history of the response of the model. A linear perturbation analysis step, on the other hand, is used to calculate the linear perturbation response from the *base state*. The base state is the present state of the model at the end of the last general analysis response. For the perturbation step, the response does not contribute to the history of the overall response and hence can be called for at any time in between general steps. For cases where the general step or the linear perturbation step is the first step, then the initial conditions defined will define the starting condition or the base state, respectively. Following is a list of the analysis types that use linear perturbation procedures:

- *BUCKLE (Eigenvalue buckling prediction)
- *FREQUENCY (Natural frequency extraction)
- *MODAL DYNAMIC (Transient modal dynamic analysis)
- *STEADY STATE DYNAMICS (Modal steady-state dynamic analysis)
- *STEADY STATE DYNAMICS, DIRECT (Direct steady-state analysis)
- *RESPONSE SPECTRUM (Response spectrum analysis)
- *RANDOM RESPONSE (Random response analysis)

Except for the above analysis types and for the *STATIC (where both general and perturbation can be used), all other analysis types are general analysis steps.

Some of the data that must be included in the history data or within a step are:

- Analysis type: An option to define the analysis procedure type which ABAQUS will perform. This must appear immediately after the *STEP option.

Other optional data include:

- Loading: Some form of external loading can be defined. Loadings can be in the form of concentrated loads, distributed loads, thermal loads, and so on. Loadings can also be prescribed as a function of time following the amplitude curve defined in the model data. If an amplitude curve is not defined, ABAQUS will assume that the loading varies linearly over the step (ramp loading) or that the loading is applied instantaneously at the beginning of the step (step loading).
- Boundary conditions: Zero-valued or non-zero boundary conditions can be added, modified or removed. Note that if defined in the model data, only zero-valued and symmetrical boundary conditions can be included.
- Output control: Controls the requested output from the analysis. Output variables depend on the type of analysis and the type of elements used.

- Auxiliary controls: Options are provided to allow the user to overwrite the solution controls that are built into ABAQUS.
- Element and surface removal/reactivation: Portions of the model can be removed or reactivated from step to step.

13.5.3 An example of a cantilever beam problem

One of the best ways of learning about and understanding the ABAQUS input file is to follow an example. We will use a simple example of modeling a cantilever beam subjected to a downward force as shown in Figure 13.1. This problem can be modeled using 1D beam elements and the finite element model using an ABAQUS input file.

As mentioned, the first thing to include in the input file is the *HEADING option. The data line after the *HEADING keyword line briefly describes the problem.

```
*HEADING
Model of a cantilever beam with a downward force
```

Next we will write the model data. First, the nodes of the problem must be defined since elements must be made up of nodes and both nodes and elements make up the geometry of the problem.

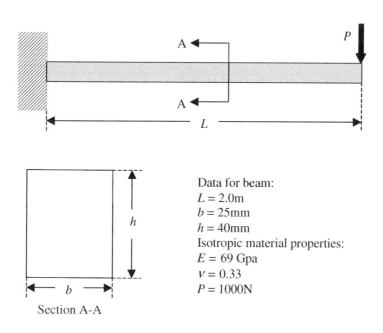

Data for beam:
$L = 2.0$m
$b = 25$mm
$h = 40$mm
Isotropic material properties:
$E = 69$ Gpa
$v = 0.33$
$P = 1000$N

Section A-A

FIGURE 13.1

Cantilever beam subject to a downward force.

```
*NODE
1, 0.
11, 2.0
*NGEN
1, 11
```

Using the *NODE option, the nodes at the end are first defined. We then use the option *NGEN to generate evenly distributed nodes between the first and the last nodes. *NGEN is one of the several mesh generation capabilities provided by ABAQUS. We could also define all 11 nodes individually by specifying their coordinates, but using *NGEN would be more efficient for large problems. So we have now defined 11 nodes uniformly along the length of the beam. Next, the elements will be defined:

```
*ELEMENT, TYPE=B23
1, 1, 2
*ELGEN, ELSET=RECT_BEAM
1, 10
```

Here, the *ELEMENT option is used to define the first element that consists of nodes 1 and 2. The TYPE parameter is included to specify what type of element is being defined. In this case, B23 refers to a 1D beam element in a plane with cubic interpolation. Users can refer to the ABAQUS manual for the element library to check the codes for other element types. Similar to the definition of nodes, *ELGEN is used to generate 1 to 10 elements. The elements are then grouped into a node set called RECT_BEAM. This will make the referencing of element properties much easier later on. So, we have now defined 11 nodes and 10 elements (see Figure 13.2).

The next step will be to define the element properties:

```
*BEAM SECTION, ELSET=RECT_BEAM, SECTION=RECT, MATERIAL=ALU
0.025, 0.040
0., 0., -1.0
```

In the *BEAM SECTION keyword line, the element set RECT_BEAM defined earlier is now referenced, meaning that the elements grouped under RECT_BEAM will all have the properties defined in this option block. We also provide the information that the beam has a rectangular (RECT) cross-section. There are other cross-sections available in ABAQUS such as circular cross-sections (CIRC), trapezoidal

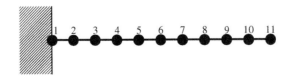

FIGURE 13.2

Cantilever beam meshed with 1-D, 2-nodal beam elements.

cross-sections (TRAPEZOID), closed thin-walled sections (BOX, HEX, and PIPE), and open thin-walled sections (I-section, and L-section). ABAQUS also provides for a "general" cross-section by specifying geometrical quantities necessary to define the section. The material associated with the elements is also defined as aluminum (ALU), the properties of which will be defined later. It is a good time to note that, unlike most programming languages, the ABAQUS input file need not follow a top-down approach when ABAQUS is assessing the file. For example, the material ALU is already referenced at this point under the *BEAM SECTION option block though its material properties are actually defined further down the input file. There will not be any error stating that the material ALU is invalid regardless of where the material is defined, unless it is not defined at all throughout the input file. This is true for all other entries into the input file. Let us now look at the data lines. The first data line in the *BEAM SECTION basically defines the dimensions of the cross-section (0.025 m × 0.04 m). Note that the dimensions here are converted to meters to be consistent with the coordinates of the nodes. The second data line basically defines the direction cosines indicating the local beam axis. What are given are the default values and this line can actually be omitted in this case. The next entry in the model data would be the material properties definition:

```
*MATERIAL, NAME=ALU
*ELASTIC, TYPE=ISOTROPIC
69.E9, 0.33
```

The material for our example is aluminum and we will name it ALU for short. All properties option blocks will follow after the *MATERIAL option block which does not require any data lines by itself. The *ELASTIC option defines elastic properties and TYPE=ISOTROPIC defines the material as an isotropic material, that is, the material properties are the same in all directions. The data line for the *ELASTIC option includes the values for the Young's modulus and the Poisson's ratio. Depending on the type of analysis carried out or the type of material being defined, other properties may need to be defined. For example, if a dynamic analysis is required, then the *DENSITY option would also need to be included; or when the material exhibits viscoelastic behavior, then the *VISCOELASTIC option would be required.

At this point, we have almost completed describing the model in the model data. What are left are the boundary conditions. Note that the boundary conditions can also be defined in the history data. What can be defined in the model data are only the zero-valued conditions.

```
*BOUNDARY
1, 1, 6, 0.
```

There is actually more than one way of defining a *BOUNDARY. What is shown is the most direct way. The first entry into the data line is the node i.d., or the name of the node set, if one is defined. In this case, since it is only one single node, there is no need for a node set. But many times, a problem might involve a whole set of nodes whereby the same boundary conditions are applied. It would, thus, be more convenient to group these nodes into a set and have it referenced in the data line. The second entry is the first DOF of the node to be constrained, while the third entry is the last DOF to be constrained. In ABAQUS, for displacement DOFs, the numbers 1,2, and 3 would represent the translational displacements in the x, y, and z directions, respectively, while the numbers 4, 5, and 6 would represent the rotations about the x, y, and z axis, respectively. Of course, depending on the type of element used and the type of analysis carried out, there may be other DOFs represented by other numbers (for more information, refer to the ABAQUS manual). For example, if a piezoelastic analysis is carried out using piezoelastic elements, there is an additional DOF (other than the displacement DOFs) number 9 representing the electric potential of the node. In this case, all the DOFs from 1 to 6 will be constrained to zero (fourth entry in data line). Strictly speaking, the DOFs available for the 1D planar beam element in ABAQUS are only 1, 2, and 6, since the others are considered out of plane displacements. Since we constrained all 6 DOFs, ABAQUS will just give a warning during analysis that the constraints on DOFs 3, 4, and 5 will be ignored since they do not exist in this context.

There are numerous parameters that can actually be included in the keyword line of the *BOUNDARY option if they are required (refer to the ABAQUS manual for further details). For example, the boundary condition can be made to follow an amplitude curve by including AMPLITUDE=*Name of amplitude curve definition* in the keyword line. ABAQUS also provides for certain standard types of zero-valued boundary conditions. For example, the above boundary condition can also be written as

```
*BOUNDARY
1, ENCASTRE
```

The word ENCASTRE used in the data line represents a fully built-in condition, which also means that DOFs 1 to 6 are constrained to zero. Other standard boundary conditions are listed in Table 13.2. After defining the boundary condition, we have now completed what is required for the model data of the input file.

We would now need to define the history data. As mentioned, the history data would begin with the *STEP option. In this example, we would be required to obtain the displacements of the beam as well as the stress along the beam due to the downward force. One step would be sufficient here and the loading will be static.

```
*STEP, PERTURBATION
*STATIC
```

Table 13.2 Standard boundary condition types in ABAQUS.

Boundary Condition Type	Description
XSYMM	Symmetry about a plane x=constant (DOFs 1, 5, 6=0)
YSYMM	Symmetry about a plane y=constant (DOFs 2, 4, 6=0)
ZSYMM	Symmetry about a plane z=constant (DOFs 3, 4, 5=0)
ENCASTRE	Fully clamped (DOFs 1 to 6=0)
PINNED	Pinned joint (DOFs 1, 2, 3=0)
XASYMM	Antisymmetry about a plane x=constant (DOFs 2, 3, 4=0)
YASYMM	Antisymmetry about a plane y=constant (DOFs 1, 3, 5=0)
ZASYMM	Antisymmetry about a plane z=constant (DOFs 1, 2, 6=0)

The perturbation parameter following the *STEP option basically tells ABAQUS that only a linear response should be considered. The *STATIC option specifies that a static analysis is required. The next thing to include will be the loading conditions:

```
*CLOAD
11, 2, -1000.
```

ABAQUS offers many types of loading. *CLOAD represents concentrated loading which is the case for our problem. Other types of loading include *DLOAD for distributed loading, *DFLUX for distributed thermal flux in thermal-stress analysis, *CECHARGE for concentrated electric charge for nodes of piezoelectric elements. The first entry in the data line is the node i.d. or the name of the node set, if defined, the second is the DOF the load is applied to, and the third is the value of the load. In our case, since the force is acting downward, it is acting on DOF 2 with a negative sign following the convention in ABAQUS. Most loadings can also follow an amplitude curve varying with time by including AMPLITUDE = *Name of amplitude curve definition* in the keyword line. This is especially so if transient, dynamic analysis is required.

For this problem, there is not much more data to include in the history data other than the output requests. The user can request the type of outputs he or she wants by indicating these as follows:

```
*NODE PRINT, FREQ=1
U,
*NODE FILE, FREQ=1
U,
*ELEMENT PRINT, FREQ=1
S,
E
*ELEMENT FILE, FREQ=1
S,
E
```

From what we learned from the finite element method, we can actually deduce that certain output variables are direct nodal variables like displacements, while others such as stress and strain, are actually determined as a distribution in the element using the shape functions. In ABAQUS, this difference is categorized into nodal output variables and element output variables. *NODE PRINT outputs the results of the required nodal variables in an ASCII text file (.dat file) while the *NODE FILE outputs the results in a binary format (.fil file). The binary format can be read by post-processors in which the results can be displayed. Similarly, *ELEMENT PRINT outputs element variables in ASCII format, while *ELEMENT FILE outputs them in binary format. A list of the different output variables can be obtained from the ABAQUS manuals. For our case, "U" in the data lines for *NODE PRINT and *NODE FILE will output all the components of the nodal displacements. "S" and "E" represent all components of stress and strain, respectively. So if the analysis is run, there will be in total three tables: One showing the nodal displacements, one showing the stresses in the elements, and the last one showing the strains in the elements. The last thing to do now is to end the step by including *ENDSTEP. If multiple steps are present, this would separate the different steps in the history data.

Running the analysis

The whole input file defining the problem of the cantilever beam is shown below:

```
*HEADING
Model of a cantilever beam with a downward force
**
*NODE
1, 0.
11, 2.0
*NGEN
1, 11
**
*ELEMENT, TYPE=B21
1, 1, 2
*ELGEN, ELSET=RECT_BEAM
1, 10
**
*BEAM SECTION, ELSET=RECT_BEAM, SECTION=RECT, MATERIAL=ALU
0.025, 0.040
0., 0., -1.0
**
*MATERIAL, NAME=ALU
*ELASTIC, TYPE=ISOTROPIC
69.E9, 0.33
**
*BOUNDARY
1, 1, 6, 0.
**
*STEP, PERTURBATION

*STATIC
**
*CLOAD
11, 2, -1000.
**
*NODE PRINT, FREQ=1
U,
*NODE FILE, FREQ=1
U,
*ELEMENT PRINT, FREQ=1
S,
E
*ELEMENT FILE, FREQ=1
S,
E
**
*ENDSTEP
```

ABAQUS input files end with the extension .inp. So if we name this file as beam.inp, we can run this example in ABAQUS using the following command at the command prompt of a Linux/Unix machine:

$$\text{ABAQUS job} = \text{beam}$$

Users can check up the full syntax of the ABAQUS execution command in the manuals (note that the execution command may vary depending on how the system administrator sets it up).

Results

After executing the analysis, there could be several results files being generated. The beam.dat file would be in ASCII text format. ABAQUS outputs its results in ASCII format in the file ending with the extension .dat. As mentioned, the binary format would be in the file with the .fil extension and is generally used as inputs for post-processors. The .dat file can of course be viewed by any text editor and it will show lots of numbers associated with the input processing, the analysis steps and lastly the requested outputs (*NODE PRINT and *EL PRINT). These output data can of course be used for plotting graphs or as inputs to other programming codes depending on the user. Many users would view the results using post-processors like ABAQUS/CAE, PATRAN and so on. The choice is entirely up to the preference of the user and of course the availability of these post-processors. In this book, the results are viewed using PATRAN and the results are shown below.

Figure 13.3 shows the deformation plot of the cantilever beam as obtained in PATRAN. This plot shows how the cantilever beam deforms under the applied loads. The actual displacements of the nodes can also be included in the deformation plot, but if there are many nodes, it makes viewing them on screen difficult. Figure 13.4 shows an XY-plot obtained in PATRAN of the stress, σ_{xx}, on the top and bottom surface of the beam. The plot shows clearly a tensile stress on the top and a compressive stress at the bottom. XY-plots of strain and displacements can be similarly obtained in PATRAN.

13.6 General procedures

The way to write the ABAQUS input file of a simple problem of a cantilever beam has been shown in the previous section. This chapter will not go through the many keywords that are available in ABAQUS. Readers and users should consult the manuals for more information regarding the keywords. This section thus aims to provide a general guide, not just for using ABAQUS, but generally for using most finite element software.

From the previous example, it is seen that certain information must be provided for the software to carry out the analysis. This information is required to solve the finite element problem, and it has been highlighted throughout this book that the information is mainly used to formulate the necessary matrices. Of course, there are certain parameters that govern the algorithm and the way the equations are solved in the program as well. So, in this sense, there should not be much difference between different software other than the format of the way information is supplied and the way results are presented. Figure 13.5 shows a summary of the general information that finite element software requires to solve most problems. The keywords provided on the left are some of the keywords used in ABAQUS to provide that particular information. To summarize, we would first need to define the geometry by

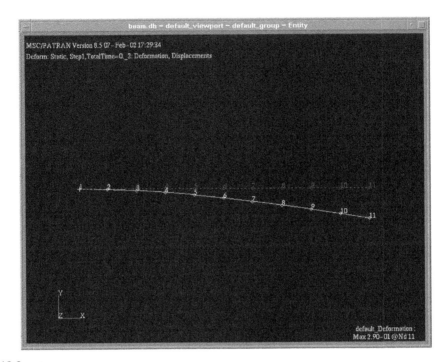

FIGURE 13.3

Deformation plot from PATRAN.

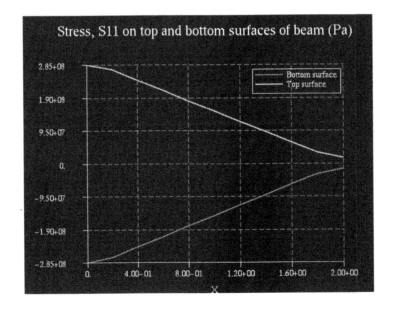

FIGURE 13.4

XY-plot of stress, σ_{xx} along the beam.

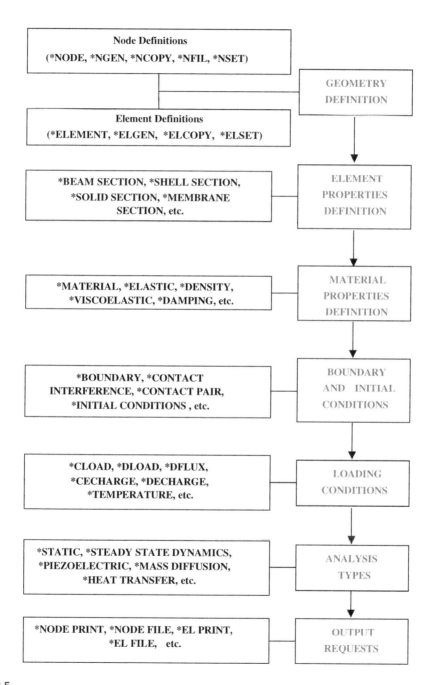

FIGURE 13.5

General information required by finite element software.

defining the nodes and the elements. Remember that in the finite element method, the whole domain is discretized into small elements. This is generally called meshing. Next, we would need to provide some properties for the elements used. For example, using 1D beam elements would require one to provide the type of cross-section and the cross-section dimension; or when using 2D plate elements the thickness of the plate would be required, and so on. After that, we would need to define the properties of the material or materials being used and associated with the elements. We would then need to provide information regarding the boundary and initial conditions the model is under. This is necessary for the solver to evaluate the equations. Similarly so for the loading conditions, which must also be provided, unless there are no loads on the model, as is the case in many analyses involving natural frequencies extraction. After all this, the model is more or less defined. The next step would be to tell the software what type of problem or analysis this is. For example, is the problem a static analysis, or a transient dynamic analysis, or a heat transfer analysis? The software would require this information and the user must provide it with the analysis type. Lastly, the user can also tell the software what results he or she is seeking. For example, for most applied mechanics problems, the displacements are the true nodal variables that the solver will compute. The software, however, can also compute the stress and strain from interpolation of these nodal displacements automatically and this can be done by specifying them in the input file.

13.7 Remarks (example using a *GUI*: ANSYS)

The information provided in the previous sections serves to highlight the typical steps in using a finite element software package through the description of the ABAQUS input file. In practice, most FEM packages including ABAQUS (ABAQUS/CAE) provide a well-integrated user interface that allows modelers to create or input geometry; perform meshing over the geometry; prescribe the necessary material properties; impose the appropriate boundary and loading conditions; select the appropriate analysis type for the problem; perform the execution of the FEM job; and even perform post-processing and visualization of the results.

Another popular FEM software package that encompasses this idea of integrating all the procedures in a nice GUI is ANSYS. Originally developed essentially for structural mechanics and heat transfer problems, ANSYS has since been developed to offer solutions for a wide range of physical problems ranging from fluid dynamics, electromagnetics, hydrodynamics, multi-physics, and so on. This section now highlights the use of ANSYS, as an example of a typical process of creating an FEM model using a user-friendly GUI. We will repeat the earlier example of the cantilever beam using ANSYS in this section. Similar procedures can also be performed in ABAQUS/CAE.

In ANSYS, one typically starts the modeling by specifying the type of elements to be used in the current model that is going to be built. Figure 13.6 shows the ANSYS environment and the dialog windows that allow one to select the appropriate element type for the problem. The figure also shows that a beam element named in ANSYS as "BEAM188" is selected. "BEAM188" is a 2-node beam in 3D space based on the Timoshenko beam theory, i.e., shear deformation is included. To model the Euler-Bernoulli beam introduced in this book (Chapter 5), one may use the "BEAM3" element, implemented via the command line. The next step after selecting the element type is to define the section properties for the beam element type. Figure 13.7 shows the pop-up window that allows one to define the section

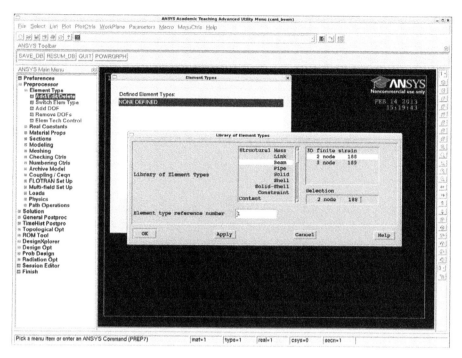

FIGURE 13.6

Selecting element types in ANSYS.

properties in ANSYS, and in it, different cross-section profiles (for the beam elements in this case) can be selected which require input of the corresponding dimensions and properties.

Typically, one then needs to provide information on the material(s) involved, and these can be prescribed in ANSYS easily via the user-friendly GUI as shown in Figure 13.8. The relevant material information is prescribed accordingly to the problem to be solved, and in this particular case, isotropic, elastic properties (Young's modulus and Poisson ratio) are keyed in.

In the ANSYS environment, there are basic capabilities of geometry creation that allow modelers to create simple geometry from points, lines, and surfaces, to solids, as shown in the screen shot in Figure 13.9. In the figure, we are creating a straight line for modeling the 1D cantilever beam. For more complex geometry, ANSYS also allows the import of compatible CAD files that can be used for meshing within the software.

Recall that in the ABAQUS input file, we perform meshing of the cantilever beam structure by using the *NGEN and *ELGEN. In a GUI, the geometry is selected (in this example, a line) and appropriate mesh properties (the number of elements/nodes and/or the size of the elements) can be prescribed via the GUI as shown in Figure 13.10a. The "MeshTool" window shown in Figure 13.10b then allows one to select and mesh the geometry (in this case, the line we created).

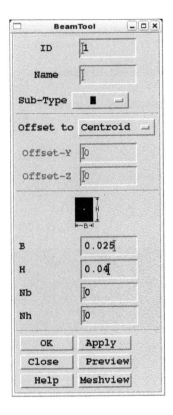

FIGURE 13.7

Defining beam section properties in ANSYS.

To prescribe the boundary and loading conditions, the geometry (e.g., surface, edge, etc.) is selected and the nodes associated with the geometry will be assigned with the boundary and loading conditions prescribed. Alternatively, the boundary and loading conditions can be applied directly to the nodes or elements by selecting them directly. Figure 13.11a shows the pop-up window for prescribing the displacement boundary conditions. In this case, "All DOF" is selected and the value of the displacement is set to "0" for the node at the clamped end of the cantilever beam. Figure 13.11b shows the window for prescribing the force. Note that the direction of the force is selected as "Fy" and "−1000" is entered as the constant value. Finally, we need to select the type of analysis to carry out. As shown in Figure 13.12, there are a variety of analyses available for selection and, in this case, "Static" analysis is chosen. Most integrated FEM packages can start the computation for the solution directly within the software GUI as shown in Figure 13.13a. Note that a summary of the model input is also shown in a separate window in the background. Once the computation is done, the results are automatically imported into the same ANSYS software environment for viewing as can be seen in Figure 13.13b.

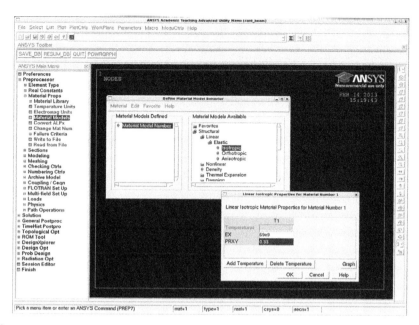

FIGURE 13.8

Prescribing material properties in ANSYS.

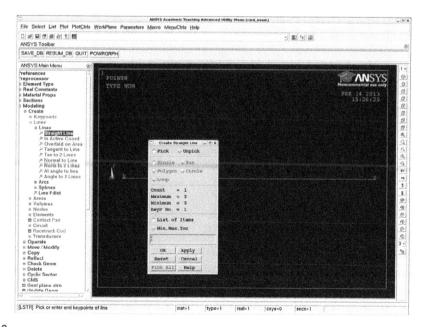

FIGURE 13.9

Basic geometry creation environment in ANSYS.

(a)

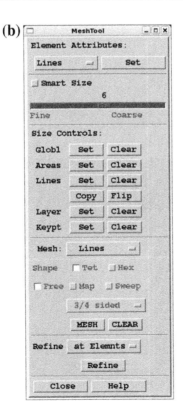

(b)

FIGURE 13.10

Meshing in ANSYS: (a) specifying element size and/or number of elements to be used and (b) the mesh tool window for meshing the geometry.

(a)
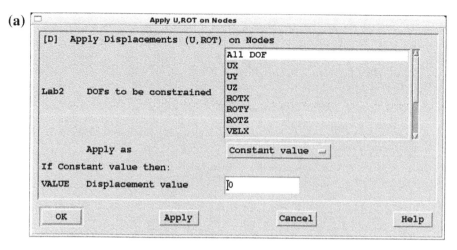

(b)

FIGURE 13.11

Prescribing (a) displacement boundary and (b) force loading conditions.

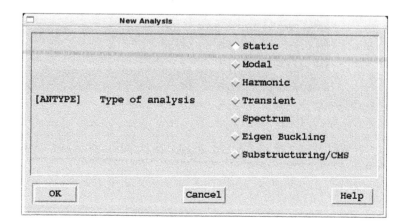

FIGURE 13.12

Selecting the analysis type in ANSYS.

(a)

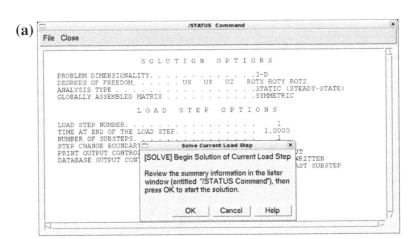

(b)

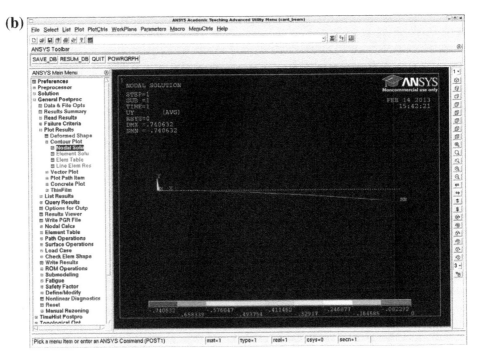

FIGURE 13.13

(a) The computation is started directly in the ANSYS GUI; and (b) the results can be viewed once the computation is completed.

Note that in this section, the general steps for creating and solving a simple finite element model using the GUI are highlighted via the introduction of the ANSYS software package. Our discussion is brief and is meant to provide only a rough idea of how an FEM model can be created in a commercial software package. It is of course much more sophisticated in practice, and for more details, readers are referred to the ANSYS user manual.

References

ABAQUS keywords manual, version 6.1, 2000. Hibbitt, Karlsson & Sorensen, Inc.

ABAQUS theory manual, version 6.1, 2000. Hibbitt, Karlsson & Sorenson, Inc.

ABAQUS user's manual, volumes I, II, and III, version 6.1, 2000. Hibbitt, Karlsson & Sorensen, Inc.

Anderson, T.L., 1995. Fracture Mechanics: Fundamentals and Applications. CRC Press, Boston.

Argyris, J.H., Fried, I., Scharpf, D.W., 1968. The TET 20 and TEA 8 elements for the matrix displacement method. Aeronautical Journal 72, 618–625.

Barsoum, R.S., 1976. On the use of isoparametric finite elements in linear fracture mechanics. International Journal for Numerical Methods in Engineering 10, 25–38.

Barsoum, R.S., 1977. Triangular quarter point elements as elastic and perfectly elastic crack tip elements. International Journal for Numerical Methods in Engineering 11, 85–98.

Bathe, K.J., 1996. Finite Element Procedures. Prentice Hall, Englewood Cliffs.

Belytschko, T., Liu, K.L., Moran, B., 2000. Nonlinear Finite Elements for Continua and Structures. John Wiley & Sons, Ltd.

Bettess, P., 1992. Infinite Elements. Penshaw Press.

Cheung, Y.K., 1976. Finite Strip Method in Structured Analysis. Pergamon Press.

Clough, R.W., Penzien, J., 1975. Dynamics of Structures. McGraw-Hill, New York.

Cook, R.D., 1981. Concepts and Applications of Finite Element Analysis, second ed. John Wiley & Sons.

Cook, R.D., 1995. Finite Element Modeling for Stress Analysis. John Wiley & Sons, Inc.

Crandall, S.H., 1956. Engineering Analysis: A Survey of Numerical Procedures. McGraw-Hill, New York.

Crocker, M.J. (Ed.), 1998. Handbook of Acoustics. John Wiley & Sons (Chapter 1).

Daily, J.W., Harleman, D.R.F., 1966. Fluid Dynamics. Addison-Wesley, Reading, Mass.

Eisenberg, M.A., Malvern, L.E., 1973. On finite element integration in natural coordinates. International Journal for Numerical Methods in Engineering 7, 574–575.

Finlayson, B.A., 1972. The Method of Weighted Residuals and Variational Principles. Academic Press, New York.

Finlayson, B.A., Scriven, L.E., 1966. The method of weighted residuals—a review. Applied Mechanics Review 19, 735–748.

Fung, Y.C., 1965. Foundations of Solid Mechanics. Prentice-Hall, Englewood Cliffs.

Henshell, R.D., Shaw, K.G., 1975. Crack tip elements are unnecessary. International Journal for Numerical Methods in Engineering 9, 495–509.

Hilber, H.M., Hughes, T.J.R., Taylor, R.L., 1978. Collocation, dissipation and "overshoot" for time integration schemes in structural dynamics. Earthquake Engineering and Structural Dynamics 6, 99–117.

Hu, H.C., 1982. Variational Principles in Elasticity and Applications. Scientific Publication (in Chinese).

Hughes, J.R.T., 1987. The Finite Element Method. Prentice-Hall International, Inc.

Kardestuncer, H. (Ed.), 1987. Finite Element Handbook. McGraw-Hill.

Kausel, E., Roësset, J.M., 1977. Semianalytic hyperelement for layered strata. Journal of the Engineering Mechanics Division 103 (4), 569.

Liu, G.R., 2002. A combined finite element/strip element method for analyzing elastic wave scattering by cracks and inclusions in laminates. Computational Mechanics 28, 76–81.

Liu, G.R., 2009. Meshfree Methods: Moving Beyond the Finite Element Method, second ed. CRC Press, Boca Raton.

Liu, G.R., Achenbach, J.D., 1994. A strip element method for stress analysis of anisotropic linearly elastic solids. ASME Journal of Applied Mechanics 61, 270–277.

Liu, G.R., Achenbach, J.D., 1995. A strip element method to analyze wave scattering by cracks in anisotropic laminated plates. ASME Journal of Applied Mechanics 62, 607–613.

Liu, G.R., Han, X., 2003. Computational Inverse Techniques in Nondestructive Evaluation. CRC Press.

Liu, G.R., Liu, M.B., 2003. Smoothed Particle Hydrodynamics: A Meshfree Particle Method. World Scientific Publishing Company.

Liu, G.R., Nguyen, Thoi Trung, 2010. Smoothed Finite Element Methods. CRC Press, Boca Raton.

Liu, G.R., Quek, S.S., 2002. A finite element study of the stress and strain fields of InAs quantum dots embedded in GaAs. Semiconductor Science and Technology 17, 630–643.

Liu, G.R., Quek, S.S., 2003. A non-reflecting boundary for analyzing wave propagation using the finite element method. Finite Elements in Analysis and Design 39 (5), 403–417.

Liu, G.R., Xi, Z.C., 2001. Elastic Waves in Anisotropic Laminates. CRC Press.

Liu, G.R., Zhang, G.Y., 2013. Smoothed Point Interpolation Methods – G Space Theory and Weakened Weak Forms. World Scientific, New York.

Liu, G.R., Achenbach, J.D., Kim, J.O., Li, Z.L., 1992. A combined finite element method/boundary element method for V(z) curves of anisotropic-layer/substrate configurations. Journal of the Acoustical Society of America 92 (5), 2734–2740.

Liu, G.R., Xu, Y.G., Wu, Z.P., 2001. Total solution for structural mechanics problems. Computer Methods in Applied Mechanics and Engineering 191, 989–1012.

Logan, D.L., 1992. A first course in the finite element method, second ed. PWS Publishing, Boston.

Mindlin, R.D., 1951. Influence of rotary inertia and shear on flexural motion of isotropic elastic plates. Journal of Applied Mechanics 18, 31–38.

MSC/Dytran user's manual, version 4: The MacNeal-Schwendler Corporation, USA, 1997.

Murnaghan, F.D., 1951. Finite Deformation of an Elastic Solid. John Wiley & Sons.

NAFEMS, 1986. A Finite Element Primer, Department of Trade and Industry, National Engineering Laboratory, Glasgow, UK.

Newmark, N.M., 1959. A method of computation for structural dynamics. ASCE Journal of Engineering Mechanics Division 85, 67–94.

Ottosen, N.S., Petersson, H., 1992. Introduction to the Finite Element Method. Prentice Hall, New York.

Peterson, L.A., Londry, K.J., 1986. Finite-element structural analysis: a new tool for bicycle frame design. Bike Tech 5 (2).

Petyt, M., 1990. Introduction to Finite Element Vibration Analysis. Cambridge University Press, Cambridge.

Quek, S.S. NUS industrial attachment report for session 1997/98: DSO National Laboratories.

Rao, S.S., 1999. The Finite Element in Engineering, third ed. Butterworth-Heinemann.

Reddy, J.N., 1984. Energy and Variational Methods in Engineering. John Wiley, New York.

Reddy, J.N., 1993. Finite Element Method. John Wiley & Sons Inc., New York.

Reissner, E., 1945. The effect of transverse shear deformation on the bending of elastic plates. Journal of Applied Mechanics 67, A67–A77.

Segerlind, L.J., 1984. Applied Finite Element Analysis, second ed. John Wiley & Sons, Inc.

Tassoulas, J.L., Kausel, E., 1983. Elements for the numerical analysis of wave motion in layered strata. International Journal for Numerical Methods in Engineering 19, 1005–1032.

Timoshenko, S., 1940. Theory of Plates and Shells. McGraw-Hill, London.

Timoshenko, S.P., Gere, J., 1972. Mechanics of Materials. Van Nostrand Reinhold Company.

Timoshenko, S.P., Goodier, J.N., 1970. Theory of Elasticity, third ed. McGraw-Hill, New York.

Washizu, K., 1974. Variational Methods in Elasticity and Plasticity, second ed. Pergamon Press, Oxford.

Washizu, K. et al., 1981. Finite Elements Handbook, vols. 1–2. Baitukan, Japan (in Japanese).

Xu, Z.L., 1979. Elasticity, vols. 1–2. People's Education Press (in Chinese).

Zienkiewicz, O.C., 1989. The Finite Element Method, fourth ed. McGraw-Hill, London.

Zienkiewicz, O.C., Taylor, R.L., 2000. The Finite Element Method, fifth ed. Butterworth-Heinemann.

Index

Printed and bound by CPI Group (UK) Ltd, Croydon, CR0 4YY

03/10/2024

01040315-0004